Encyclopaedia of Mathematical Sciences

Volume 90

Editor-in-Chief: R.V. Gamkrelidze

Springer
Berlin
Heidelberg
New York
Barcelona
Budapest
Hong Kong
London
Milan
Paris
Santa Clara
Singapore
Tokyo

R. Osserman (Ed.)

Geometry V

Minimal Surfaces

With 30 Figures

Springer

Title of the Russian edition:
Itogi nauki i tekhniki, Sovremennye problemy matematiki,
Fundamental'nye napravleniya, Vol. 90, Geometriya 5
Publisher VINITI, Moscow, in preparation

Mathematics Subject Classification (1991):
53A10, 49Q05, 35J60, 30D35

ISSN 0938-0396

ISBN 3-540-60523-1 Springer-Verlag Berlin Heidelberg New York

Typesetting: Editing and reformatting of the authors' input files by Adam Leinz, Karlsruhe using a Springer TEX macro package.

SPIN 10499544 41/3143 - 5 4 3 2 1 0 – Printed on acid-free paper

List of Editors and Authors

Editor-in-Chief

R. V. Gamkrelidze, Russian Academy of Sciences, Steklov Mathematical Institute, ul. Gubkina 8, 117966 Moscow; Institute for Scientific Information (VINITI), ul. Usievicha 20a, 125219 Moscow, Russia; e-mail: gam@ipsun.ras.ru

Consulting Editor

R. Osserman, Deputy Director, Mathematical Sciences Research Institute, 1000 Centennial Drive, Berkeley, CA 94720, USA; e-mail: osserman@msri.org

Authors

H. Fujimoto, Department of Mathematics, Faculty of Sciences, Kanazawa University, Kakuma-machi, 920-11 Kanazawa, Japan; e-mail: fujimoto@kappa.s.kanazawa-u.ac.jp

S. Hildebrandt, Mathematisches Institut, Rheinische Friedrich-Wilhelms Universität Bonn, Beringstr.1, D-53115 Bonn, Germany

D. Hoffman, Mathematical Sciences Research Institute, 1000 Centennial Drive, Berkeley, CA 94720, USA; e-mail: david@msri.org

H. Karcher, Mathematisches Institut, Rheinische Friedrich-Wilhelms Universität Bonn, Beringstr.1, D-53115 Bonn, Germany; e-mail: unm416@ibm.rhrz.uni-bonn.de

L. Simon, Department of Mathematics, Stanford University, Stanford, CA 94305, USA; e-mail: lms@math.stanford.edu

This volume is dedicated to

Lars V. Ahlfors
1907–1996

twentieth century giant of geometric function theory,
whose influence is clearly visible here, as in many other parts
of contemporary mathematics.

Contents

Introduction

Few people outside of mathematics are aware of the varieties of mathematical experience – the degree to which different mathematical subjects have different and distinctive flavors, often attractive to some mathematicians and repellant to others.

The particular flavor of the subject of minimal surfaces seems to lie in a combination of the concreteness of the objects being studied, their origin and relation to the physical world, and the way they lie at the intersection of so many different parts of mathematics. In the past fifteen years a new component has been added: the availability of computer graphics to provide illustrations that are both mathematically instructive and esthetically pleasing.

During the course of the twentieth century, two major thrusts have played a seminal role in the evolution of minimal surface theory. The first is the work on the Plateau Problem, whose initial phase culminated in the solution for which Jesse Douglas was awarded one of the first two Fields Medals in 1936. (The other Fields Medal that year went to Lars V. Ahlfors for his contributions to complex analysis, including his important new insights in Nevanlinna Theory.) The second was the innovative approach to partial differential equations by Serge Bernstein, which led to the celebrated Bernstein's Theorem, stating that the only solution to the minimal surface equation over the whole plane is the trivial solution: a linear function.

The subsequent history of these two problems provides a fascinating glimpse into the way a single result can provide the seed for a profusion of growth in many directions. In the case of Bernstein's Theorem, the two obvious directions for generalization were toward a wider class of equations, and to higher dimensions. Both of those took a while to be realized, since Bernstein's original proof did not lend itself easily to generalizations of any sort. The first step was the discovery of alternative proofs, particularly by Bers, Heinz, and Nitsche. They made possible the introduction of equations of "minimal surface type" with many of the properties of the minimal surface equation, including the Bernstein Theorem. These results are described below in the article by Leon Simon, himself one of the leading contributors to those developments. Heinz's proof involved a generalization in a different direction: bounds on the second derivatives at a point for any solution defined over a disk centered at the point. Those bounds tend to zero as the radius of the disk tends to infinity, showing that a solution over the whole plane must have all second derivatives identically zero and therefore be a linear function. The second-derivative bounds can also be reformulated as a bound on the Gauss curvature, and that approach proved useful in extensions to higher dimensions. Yet another approach to Bernstein's Theorem, due to Fleming, provided the key ideas that yielded the first results for higher dimensions. The big surprise was the discovery that Bernstein's Theorem turned out to be true for solutions in up to seven variables, and then false. That again led to a whole series of new questions about global behavior of the minimal surface equation in higher dimensions, all of which are explored in the article by Simon.

A generalization of Bernstein's Theorem in quite another direction was suggested by Nirenberg. His idea was to think of the theorem in more geometric terms, concerning surfaces not so much as solutions of a differential equation, but as surfaces with zero mean curvature, and replacing the assumption that the surface projects onto the whole plane by the pair of assumptions that the surface be complete and the Gauss map omit a neighborhood of some point. Formulated in those terms, the result was reminiscent of the Weierstrass Theorem stating that the values of a nonconstant entire function must be everywhere dense. That led Nirenberg to a second conjecture that there should also be a Picard Theorem for minimal surfaces, saying that for a complete minimal surface, if the Gauss map omits more than two values, then the Gauss map must be constant, and the surface must be a plane. It turned out that his first conjecture was correct, and the second one false. But it was Nirenberg's formulation that was most crucial to further developments, since it led to a whole new theory, that of complete minimal surfaces, first in three dimensions, and then of arbitrary codimension. Also, it was natural to think of replacing Weierstrass and Picard-type results by a Nevanlinna Theory for the Gauss map. An early result in that direction was the theorem of Ahlfors-Osserman that the Gauss map of a complete nonplanar minimal surface not only could not be bounded, but could not belong to Nevanlinna's class of "bounded type". That provided a considerable strengthening of the Weierstrass result that the Gauss map must be everywhere dense to the fact that the set of omitted points had to have logarithmic capacity zero. But the full flowering of the Nevanlinna theory for complete minimal surfaces came with the work of Fujimoto, and is explained in detail in his paper in this volume. Fujimoto was also the one to provide the definitive answer to the second Nirenberg conjecture. After Xavier showed that the image under the Gauss map could omit at most a finite number of points, Fujimoto proved that the precise maximum for the number of omitted points possible was four.

The theory of complete minimal surfaces blossomed in quite another direction following the work of Jorge and Meeks devoted to analyzing the possibilities for a complete minimal surface to be embedded in 3-space, with no self intersections. Before 1980, the only known examples of complete embedded minimal surfaces were the plane, the catenoid, and various periodic surfaces, such as the helicoid. The latter necessarily had infinite total curvature, and it seemed not unlikely that there were no others of finite total curvature. Then Costa came up with an example of a surface which was of genus one, with three embedded ends, and opened the floodgate for further examples, after Hoffman and Meeks showed that Costa's surface was in fact embedded. In the decade since the publication of Costa's example and Hoffman and Meeks' first paper, the subject has seen phenomenal growth, all of which is described in the paper by Hoffman and Karcher in this volume.

These are by no means the only developments that can be traced directly back to Bernstein's seminal theorem, but they illustrate the manner in which a single result can have a powerful influence on the future course of mathematics.

The Plateau problem has been at least equally influential. The general existence theorems of Douglas left some important questions unanswered. First, Douglas was unable to rule out the possibility that the surfaces obtained had certain kinds of singularities, called branch points. Second, he left open the question of regularity of the surface at the boundary. And third, there was the question of how many different solution surfaces might be bounded by a single curve. Much progress has been made on these questions, but much also remains to be done. A full account of the results to date is given in the paper by Hildebrandt in this volume.

One of the most important roles of Plateau's problem was as a spark to the development of the powerful new tool known as geometric measure theory. There were two rather different approaches, one due basically to de Giorgi, and the other to Federer and Fleming. After leading separate lives for a while, both from each other and from the "classical" approach to minimal surfaces, all three methods gradually intermingled, so that methods and results from geometric measure theory find their way into a number of topics discussed in this book. Perhaps surprisingly, it has proved especially critical in the PDE aspect of the subject, where it has played a major role in connection with Bernstein's Theorem.

Taken together, the articles in this volume provide a fairly broad spectrum of recent activity in the field of minimal surfaces. Needless to say, it is far from comprehensive. The lists of references at the end of the different articles point in a number of further directions. A combined index to all four articles at the end of this volume will allow a reader to track down topics of particular interest, sometimes seen from different perspectives in different articles. That should help to provide at least some of the flavor that has given the subject of minimal surfaces such an enduring appeal over the years.

January 1997 *Robert Osserman*

I. Complete Embedded Minimal Surfaces of Finite Total Curvature

David Hoffman[1] and Hermann Karcher[2]

Contents

---

[1] Supported by research grant DE-FG03-95ER25250 of the Applied Mathematical Science subprogram of the Office of Energy Research, U.S. Department of Energy, and National Science Foundation, Division of Mathematical Sciences research grant DMS-9596201. Research at MSRI is supported in part by NSF grant DMS-90-22140.

[2] Partially supported by Sonderforschungsbereich SFB256 at Bonn.

1. Introduction

We will survey what is known about minimal surfaces $S \subset \mathbf{R}^3$, which are complete, embedded, and have finite total curvature: $\int_S |K| dA < \infty$. The only classically known examples of such surfaces were the plane and the catenoid. The discovery by Costa [14, 15] early in the last decade, of a new example that proved to be embedded sparked a great deal of research in this area. Many new examples have been found, even families of them, as will be described below. The central question has been transformed from whether or not there are any examples except surfaces of rotation to one of understanding the structure of the space of examples.

Up to this point, *every new example* of a complete, embedded minimal surface of finite total curvature has been discovered first by using the global version of the Enneper-Riemann-Weierstrass representation, which is essentially due to Osserman [58, 59].[3] This involves knowledge of the compact Riemann surface structure of the minimal surface, as well as its Gauss map and other geometric-analytic data. One of our goals is to show how this is done in the case that has been most completely analyzed, namely surfaces with genus ≥ 1 and three topological ends. An important quality of this construction is that the Riemann surface and the meromorphic data are constructed simultaneously under the assumption of symmetry. Moreover, once this is done, parameters must be found in order to have a well-defined finite-total-curvature surface. This parameter search is typically done by computer using a combination of relatively simple numerical routines and relatively complex graphics tools ([8, 41]). In many cases a full theoretical analysis, as is done here in Section 4 for the three-ended surfaces of Theorem 3.3 has yet to be carried out. Moreover, solution of the period problem does not at all guarantee that the surfaces are embedded. In Section 5 we present examples of Callahan, Hoffman, Karcher, Meeks and Wohlgemuth that lie in one parameter families containing both embedded and immersed surfaces. In fact the period problem and embeddedness are totally independent issues. There are examples of Weierstrass data meeting all necessary conditions (Proposition 2.4) for embeddedness, for which the period problem is not solvable (a genus-one example with two catenoid ends does not exist but Weierstrass data for such a surface – even a very symmetric one – does) and others for which the period

[3] *Added in Proof* (November, 1996). This was written in early 1995. In the Fall of 1995, Nikolaos Kapouleas constructed new examples using methods similar to those he used to construct higher genus surfaces of constant mean curvature. Briefly, he proves that the construction imagined and described at the beginning of Section 5.2 can be carried out in detail (Kapouleas, N.: Complete embedded minimal surfaces of finite total curvature, preprint). Martin Traizet has carried out an analogous construction for periodic minimal surfaces (Traizet, M.: Construction de surfaces minimales en recollant des surfaces de Scherk, Annales de l'Institut Fourier (to appear)).

problem is solvable for a family of surfaces that are embedded outside of a compact set of $\mathbf{R}^3$, but are not embedded.

The survey is organized as follows. In Section 2 we present the basic tools of the subject, the most important of which is the Weierstrass-Enneper representation. In Section 2.2 we describe the construction of Chen and Gackstatter [10], which produces a genus-one surface with the symmetries and end behavior of Enneper's surface. To our knowledge this complete minimal surface of finite total curvature was the first one explicitly constructed by first specifying a geometric property – in this case, end behavior – and then deriving the necessary Weierstrass data. It is not, of course, embedded but its construction, as presented here, has most of the features of the construction of the three-ended examples in Section 4. In Section 2.3, the hypothesis of embeddedness is used to derive relationships between the geometry of the surface and its analytic representation. Propositions 2.3, 2.4 and 2.5 gather together all necessary conditions including the relationship between flux, logarithmic growth rates of the ends and residues of the complex differential of the height function.

In Section 3, we present the few global rigidity theorems that are known. (Theorems 3.1 and 3.4, due to Lopez-Ros [51], Schoen [67] and Costa [16].) We present a proof (in Section 3.1) of the Lopez-Ros theorem, which states that a complete minimal surface of genus zero and finite total curvature must be the plane or the catenoid. Our proof follows that of Perez-Ros [60]. We also state the existence result, Theorem 3.3, for the three-ended, complete minimal surfaces with genus $k-1$ and k vertical planes of symmetry ([30]). The details of the construction of these surfaces are presented in Section 4. We include here the estimation of the parameters that solve the period problem when $k > 2$. The values of the parameters that close the periods determine the logarithmic growth rates of the ends of these surfaces. For the surface to be embedded, they must lie in a certain range, which they (happily) do. This is done in Sections 4.5–4.9.

In Section 5, we survey other known examples and discuss what little is known about the structure of the space of complete embedded minimal surfaces of finite total curvature. Section 5.2 presents some conjectures about this space.

Finite total curvature implies finite topology, even without the additional assumption of embeddedness. The converse is not true; the helicoid is simply connected, nonflat and periodic, so its curvature is infinite, while its topology is finite. Up until recently, the helicoid was the only known embedded minimal surface with finite topology and infinite total curvature. In 1992, we discovered, with Fusheng Wei, a complete embedded minimal surface of genus one with one end – asymptotic to the helicoid – that has infinite total curvature [32, 33]. The details of this construction are outside the scope of this work.

However, the extent to which finite topology implies finite total curvature is discussed in Section 6.[4]

Section 7 discusses the index of stability of a complete minimal surface. The basic results [22, 25, 26] the equivalence of finite index and finite total curvature, and the fact that the index is completely defined by the Gauss map, are discussed. There is, as yet, no known relationship between embeddedness and properties of the index. This final section is therefore, strictly speaking, misplaced in this survey. However the ideas and techniques may, in the long run, prove useful in the study of embeddedness of minimal surfaces.

Acknowledgements. The authors would like to thank Jim Hoffman of MSRI and Ed Thayer, now in the Department of Molecular Biotechnology at the University of Washington, for the preparation of the computer graphics illustrations in this article. Both were formerly at the Center for Geometry, Analysis, Numerics and Graphics (GANG) at the University of Massachusetts, Amherst where one of us was the Director. Peter Norman and Pascal Romon read earlier versions and gave us useful critical comments. Harold Rosenberg helped clarify some important points in Section 2.3. We are grateful to M. Kotani and S. Nayatani for useful comments on Section 7. Robert Bryant and Michael Wolf gave us many useful comments. The authors enjoyed the hospitality of MSRI where most of the final work on this paper was done.

2. Basic Theory and the Global Weierstrass Representation

Let $\mathcal{S}, \langle\rangle$ be an oriented surface with a Riemannian metric, and let ∇ denote its Riemannian connection. For a smooth function $f: \mathcal{S} \to \mathbf{R}$, we will denote its differential by df and its gradient by grad f. They are related by $Vf = df(V) = \langle \operatorname{grad} f, V\rangle, V \in T\mathcal{S}$, which can be thought of as the defining equation for grad f. If w is a one form, its covariant derivative ∇w is defined by the relation

$$(\nabla_U w)V = Uw(V) - w(\nabla_U V)\,. \tag{2.1}$$

The divergence operator is div $= \operatorname{tr} \nabla$ and the Laplacian of a smooth function f is given by $\Delta f := \operatorname{div} \operatorname{grad} f$. Note that

$$\langle \nabla_U \operatorname{grad} f, V\rangle = U\langle \operatorname{grad} f, V\rangle - (\nabla_U V)f = (\nabla_U df)(V)\,. \tag{2.2}$$

[4] *Added in Proof* (November, 1996). Pascal Collin recently showed that a complete embedded minimal surface with finite topology and *more than one end* has to have finite total curvature. He did this by proving the Nitsche Conjecture. See Section 6.2. (Collin, P.: Topologie et courbure des surfaces minimales proprement plongées de R^3, Ann. Math. **145** (1997), 1–31.

Hence

$$\Delta f = \Sigma(\nabla_{e_i} df)(e_i)\,. \tag{2.3}$$

The operator $\mathrm{Rot}_{\frac{\pi}{2}}$ on vector fields, which rotates each T_pM by $\frac{\pi}{2}$ in the positive direction, is covariantly parallel: i.e. $\nabla \mathrm{Rot}_{\frac{\pi}{2}} = 0$. This simply means that for any vector fields $U, V, \mathrm{Rot}_{\frac{\pi}{2}}(\nabla_U V) = \nabla_U \mathrm{Rot}_{\frac{\pi}{2}} V$, which can be easily checked. This implies that for a vector field V

$$\operatorname{div} V = 0 \Leftrightarrow \nabla \mathrm{Rot}_{\frac{\pi}{2}} V \text{ is symmetric,}$$
$$(\text{i.e. } \langle \nabla_U \mathrm{Rot}_{\frac{\pi}{2}} V, \widetilde{U}\rangle = \langle \nabla_{\widetilde{U}} \mathrm{Rot}_{\frac{\pi}{2}} V, U\rangle \text{ for all vector fields } U, \widetilde{U})\,.$$

Associated to a vector field W is the dual one form ω defined by $\omega(U) = \langle U, W\rangle$. One can verify by elementary means that

$$\begin{aligned} d\omega = 0 &\Leftrightarrow \nabla W \text{ is symmetric (i.e. } \langle \nabla_U W, \widetilde{U}\rangle = \langle \nabla_{\widetilde{U}} W, U\rangle \\ &\qquad \text{for all vector fields } U, \widetilde{U}.) \\ &\Leftrightarrow W \text{ is locally the gradient of some differentiable function.} \end{aligned}$$

Combining these two equivalences and applying them to $W = \operatorname{grad} f$, where $f\colon \mathcal{S} \to \mathbf{R}$ is some smooth function, we have

$$\begin{aligned} \Delta f &= \operatorname{div}(\operatorname{grad} f) \\ &= 0 \Leftrightarrow \mathrm{Rot}_{\frac{\pi}{2}}(\operatorname{grad} f) \text{ is (locally) the gradient of some function } f^*\,. \end{aligned}$$

Smooth functions $f\colon \mathcal{S} \to \mathbf{R}$ satisfying $\Delta f = 0$ are called *harmonic functions.* For a harmonic function f, the locally defined functions f^*, where $\operatorname{grad} f^* = \mathrm{Rot}_{\frac{\pi}{2}}\,(\operatorname{grad} f)$ are called *conjugate* harmonic functions associated to f. Note that $df^* = -df \circ \mathrm{Rot}_{\frac{\pi}{2}}$. Because grad f and grad f^* are orthogonal and have the same length, the locally defined mapping given by (f, f^*) is conformal wherever $\operatorname{grad} f \neq 0$. Thus, $f + i\, f^*$ is a local conformal coordinate in a neighborhood of any point $p \in \mathcal{S}$ where $df_p \neq 0$.

Suppose our surface $\mathcal{S}$ is given as an oriented immersed surface in $\mathbf{R}^3$. Let X denote its position vector; we think of X as an immersion $X\colon \mathcal{S} \to \mathbf{R}^3$. $\mathcal{S}$ inherits a Riemannian metric $\langle\cdot,\cdot\rangle$ from $\mathbf{R}^3$ and its Riemannian connection for vector fields can be considered to be the projection of differentiation D in $\mathbf{R}^3$; if U, V are vector fields on $\mathcal{S}, \nabla_U V$ is defined by the relation

$$DX(\nabla_U V) = [U(DX(V))]^T\,, \tag{2.4}$$

where $[\,]^T$ denotes projection onto $DX_p(T_p\mathcal{S})$. For $p \in \mathcal{S}$ let $N(p)$ denote the unit normal to $X(\mathcal{S})$ at $X(p)$. The endomorphism $\mathrm{Rot}_{\frac{\pi}{2}}$ on $T_p\mathcal{S}$ is just the pullback by DX of counter-clockwise rotation around $N(p)$ in $DX[T_p\mathcal{S}]$. We think of N as the Gauss mapping $N\colon \mathcal{S} \to S^2$ and denote its differential by DN. By identification of $T_p\mathcal{S}$ with $DX_p(T_p\mathcal{S})$ and $DX_p(T_p\mathcal{S})$ with $T_{N(p)}S^2$ by rigid translation in $\mathbf{R}^3$, we may think of DN_p as an endomorphism of $T_p\mathcal{S}$. As such it is called the *shape operator,* and denoted by S. The eigenvalues of

S are called the *principal curvatures*, and its eigenvectors are called *principal directions*. The average of the principal curvatures is the *mean curvature* and is denoted by H.

An important formula with strong consequences for minimal surfaces is:

$$\Delta X = -2HN . \tag{2.5}$$

By ΔX, we mean $(\Delta x_1, \Delta x_2, \Delta x_3)$ where (x_1, x_2, x_3) are the Euclidean coordinates in $\mathbf{R}^3$. We will verify (2.5) as follows. Writing $DX = (dx_1, dx_2, dx_3)$ and applying (2.1) coordinatewise, we have

$$\begin{aligned}(\nabla_U DX)(V) &= U(DX(V)) - DX(\nabla_U V) \\ &= U(DX(V)) - [U(DX(V))]^T \quad \text{(by 2.4)} \\ &= [U(DX(V))]^N \quad \text{(where } [\,]^N \text{ denotes projection} \\ &\qquad\qquad\qquad\qquad \text{onto } DX(T\mathcal{S})^{\perp}) \\ &= (U(DX(V)) \cdot N)N .\end{aligned}$$

The scalar factor of N in the last expression is ($\pm$, depending on your background) the Second Fundamental Form. Because $DX(V) \cdot N = 0$, $U(DX(V)) \cdot N = -DX(V) \cdot DN(U) = -\langle S(U), V \rangle$, where S is the shape operator. Thus

$$(\nabla_U DX)(V) = -\langle S(U), V \rangle N .$$

From (2.2) we conclude that $\Delta X = -(trS)N$, which is (2.5), as desired.

Remark 2.1. *Since the covariant derivative is most often developed on Riemannian manifolds without a given immersion, we indicate how the above definition is used for practical computations. For example, if W is a vector field in $\mathbf{R}^3$ we can consider its tangential component*

$$W^T := W - \langle W, N \rangle N$$

along the immersed surface $X(S)$. This also gives us a vector field V on the domain surface S such that $DX(V) = W^T$. Strictly speaking it is the covariant derivative of V that was defined, but one often allows abbreviations, e.g. writing W rather than $W \circ X$, in computations like:

$$\begin{aligned}\nabla_U(W^T) :&= DX(\nabla_U V) \\ &= (U(DX(V)))^T \\ &= (U(W - \langle W, N \rangle N))^T \\ &= (UW)^T - \langle W, N \rangle DN(U) .\end{aligned}$$

Minimal Surfaces

A minimal surface, $\mathcal{S}$, is one for which $H \equiv 0$; from (2.5) we can conclude that the coordinate functions of a minimal surface are harmonic. We also may conclude from the discussion above that if x is a coordinate function, $x + ix^*$

defines a conformal coordinate chart in a neighborhood of any $p \in \mathcal{S}$ where $dx \neq 0$. Since X is assumed to be an immersion, we conclude that at every $p \in \mathcal{S}$, at least one of the coordinate functions x_i satisfies $dx_{i|p} \neq 0$. Thus minimal surfaces inherit naturally, from their Euclidean coordinates, an atlas of holomorphic functions that define its Riemann surface structure.

Because $\operatorname{tr} S = 2H \equiv 0$ on a minimal surface $\mathcal{S}$, the Gauss mapping $N: \mathcal{S} \to S^2$ satisfies $\langle DN(U), DN(V)\rangle = \kappa^2 \langle U, V\rangle$, where $\pm\kappa$ are the principal curvatures, and also reverses orientation. We orient $S^2 \subset \mathbf{R}^3$ with the outward-pointing normal (so that *its* Gauss map is the identity and its principal curvatures equal to $+1$) and let σ be stereographic projection from the north pole to the complex plane considered to be the (x_1, x_2)-plane, also positively oriented. With these choices, σ is orientation-*reversing* and $g := \sigma \circ N: \mathcal{S} \to \mathbf{C} \cup \{\infty\}$ orientation-*preserving* and conformal, whenever $DN \neq 0$. Thus g is a *meromorphic function* on $\mathcal{S}$, considered now to be a Riemann surface.

We have shown that a minimal surface has a natural Riemann surface structure, with respect to which the Gauss map is meromorphic. Now we will use this information to realize minimal surfaces as the real parts of holomorphic curves. Recall that each coordinate function x gives rise, locally, to a conjugate harmonic function x^* and globally to a holomorphic differential form

$$dx + i dx^* = dx - i(dx \circ \operatorname{Rot}_{\frac{\pi}{2}}).$$

Associated to our minimal immersion $X = (x_1, x_2, x_3): S \to \mathbf{R}^3$ we also have the "conjugate" minimal immersion $X^* = (x_1^*, x_2^*, x_3^*)$ (which is well defined only on some covering $\widetilde{S}$ of S). The relation $dX^* = -dX \circ \operatorname{Rot}_{\frac{\pi}{2}}$ shows that X and X^* are locally immersions and that $\Psi := X + iX^*$ satisfies $d\Psi \circ \operatorname{Rot}_{\frac{\pi}{2}} = i \cdot d\Psi$. In particular Ψ is holomorphic and

$$X = \operatorname{Re} \Psi = \operatorname{Re} \int d\Psi = \operatorname{Re} \int (dX + i dX^*).$$

With respect to any local holomorphic parameter $z = u_1 + iu_2$ we have the usual connection with real differentiation:

$$\begin{aligned} d\Psi &= \Psi' dz \\ &= \left(\frac{\partial X}{\partial u_1} + i \frac{\partial X^*}{\partial u_1}\right) dz = \left(\frac{\partial X}{\partial u_1} - i \frac{\partial X}{\partial u_2}\right) dz. \end{aligned}$$

It is convenient to define for a real-valued differentiable function w

$$w_z = \frac{\partial w}{\partial z} := \frac{1}{2}\left(\frac{\partial w}{\partial u_1} - i \frac{\partial w}{\partial u_2}\right),$$

and extend this function linearly to complex-valued functions:

$$(v + iw)_z = v_z + i w_z.$$

If $f = v + iw$ is holomorphic, $f_z = f'$; the desirability of this identity is the reason for the factor of $\frac{1}{2}$. The Cauchy-Riemann equation for f is $v_z = iw_z$, and extending this operation coordinate-wise to $\Psi = X + iX^*$, we have

$$\begin{aligned}\Psi' = \Psi_z &= (X_z + iX_z^*)\\ &= 2X_z = \frac{\partial X}{\partial u_1} - i\frac{\partial X}{\partial u_2}\\ &= 2iX_z^* .\end{aligned}$$

We find an additional property by computing

$$(\Psi'^2) = 4X_z^2 = \left|\frac{\partial X}{\partial u_1}\right|^2 - \left|\frac{\partial X}{\partial u_2}\right|^2 - 2i\frac{\partial X}{\partial u_1}\cdot\frac{\partial X}{\partial u_2} = 0 ,$$

because all holomorphic reparametrizations are conformal. The image $\Psi(\widetilde{S})$ in $\mathbf{C}^3$ is therefore called a holomorphic null-curve. Conversly, we see that the real part of a holomorphic null-curve is *conformally* parametrized and, of course, harmonic with respect to the given parametrization. Since conformal changes of the metric do not change the harmonicity of functions, the real part of a holomorphic null-curve is also harmonic with respect to the induced Riemannian metric and hence minimal.

Having established a firm connection with complex analysis, we now tie in some more geometry by rewriting $d\Psi$ in terms of the Gauss map. This leads to a representation formulated by Enneper, Riemann and Weierstrass. Recall that the Gauss map became a meromorphic function after we distinguished a "vertical" direction for the stereographic projection; we take this to be the x_3-axis. Now we make full use of the fact that our surface is a complex curve: While quotients of one-forms on a surface are not functions, the quotients $\frac{d\Psi_1}{d\Psi_3}, \frac{d\Psi_2}{d\Psi_3}$ of holomorphic one-forms on the complex curve are functions, meromorphic functions on the underlying Riemann surface. Moreover, because of $X_z \cdot X_z = 0$, these functions are quadratically related:

$$\left(\frac{d\Psi_1}{d\Psi_3}\right)^2 + \left(\frac{d\Psi_2}{d\Psi_3}\right)^2 = -1 .$$

Finally, since

$$\frac{\partial X}{\partial u_1} = \operatorname{Re}\left(d\Psi\left(\frac{\partial}{\partial u_1}\right)\right) \quad \text{and}$$
$$\frac{\partial X}{\partial u_2} = \operatorname{Re}\left(d\Psi\left(\frac{\partial}{\partial u_2}\right)\right)$$

are orthogonal tangent vectors to the minimal surface, we must be able to compute the normal, i. e., the Gauss map g, from $\frac{d\Psi_1}{d\Psi_3}, \frac{d\Psi_2}{d\Psi_3}$. With a small trick, this procedure can be reversed. Put

$$f := \frac{-(d\Psi_1 + i d\Psi_2)}{d\Psi_3}$$
$$= \frac{d\Psi_3}{d\Psi_1 - i d\Psi_2} \quad \left(\text{because } \left(\frac{d\Psi_1}{d\Psi_3}\right)^2 + \left(\frac{d\Psi_2}{d\Psi_3}\right)^2 = -1\right).$$

Then we have

$$d\Psi = \left(\frac{1}{2}\left(\frac{1}{f} - f\right), \frac{i}{2}\left(\frac{1}{f} + f\right), 1\right) d\Psi_3 ,$$

which now depends only on one meromorphic function f and one holomorphic differential $d\Psi_3$. From now on, we will write $d\Psi_3$ as dh. Recall

$$dh := d\Psi_3 = dx_3 - i dx_3 \circ \mathrm{Rot}_{\frac{\pi}{2}} .$$

We know already that f must be closely related to the Gauss map. In fact, we find that inverse stereographic projection of f, namely

$$(2\,\mathrm{Re}\, f, 2\,\mathrm{Im}\, f, |f|^2 - 1)/(|f|^2 + 1)$$

is orthogonal to the tangent vectors $\mathrm{Re}(\frac{d\Psi}{d\Psi_3}), \mathrm{Im}(\frac{d\Psi}{d\Phi_3})$ since

$$\left\langle \left(\frac{1}{2}\left(\frac{1}{f} - f\right), \frac{i}{2}\left(\frac{1}{f} + f\right), 1\right), (2\,\mathrm{Re}\, f, 2\,\mathrm{Im}\, f, |f|^2 - 1) \right\rangle_{\mathbf{C}} = 0 \in \mathbf{C}.$$

This identifies f as the stereographic projection of the Gauss map of the minimal surface. We now state the "Weierstrass representation theorem" with emphasis on its global formulation due to Osserman [59].

Theorem 2.1. *Suppose S is a minimal surface in $\mathbf{R}^3$, M its Riemann surface, $g = \sigma \circ N$ the stereographic projection of its Gauss map. Then S may be represented (up to a translation) by the conformal immersion*

$$X(p) = \mathrm{Re} \int \Phi , \tag{2.6}$$

where

$$\Phi = (\phi_1, \phi_2, \phi_3) = \left((g^{-1} - g)\frac{dh}{2}, i(g^{-1} + g)\frac{dh}{2}, dh\right) . \tag{2.7}$$

Conversely, let M be a Riemann surface, $g: M \to \mathbf{C} \cup \{\infty\}$ a meromorphic function and dh a holomorphic one-form on M. Then (2.6) and (2.7) define a conformal minimal mapping of some covering of M into $\mathbf{R}^3$, which is regular provided the poles and zeros of g coincide with the zeros of dh. The mapping X is well-defined on M if and only if no component of Φ in (2.7) has a real period. That is

$$\mathrm{Period}_\alpha(\Phi) =: \mathrm{Re} \oint_\alpha \Phi = 0 \tag{2.8}$$

for all closed curves α on M.

We now compute the basic geometric quantities of M in terms of g and dh. The metric induced on M by X can be expressed as

$$ds^2 = \tfrac{1}{2}|\Phi|^2 = \tfrac{1}{4}(|g| + |g|^{-1})^2|dh|^2 , \tag{2.9}$$

The metric will be complete provided $\int_\delta ds = \infty$ for every divergent curve δ on M. The Gauss curvature of this metric may be computed as the ratio of the area stretching of dN to that of dX. Thus:

$$K = \frac{-4|dg|^2}{(|g| + |g|^{-1})^2|dh|^2} \cdot \frac{4}{(1 + |g|^2)^2} .$$

The last term is the conformal stretching of stereographic projection. The minus sign comes from the fact that stereographic projection reverses orientation. We may rewrite this as

$$K = \frac{-16}{(|g| + |g|^{-1})^4} \left|\frac{dg/g}{dh}\right|^2 . \tag{2.10}$$

To express the shape operator $S = DX^{-1} \cdot DN$ in terms of g and dh, we compute $X_{zz}dz^2$ and take the real part of the dot product with N:

$$X_{zz}dz^2 = \tfrac{1}{4}(X_{11} - X_{22} - 2iX_{12}) \cdot (du_1^2 - du_2^2 + 2idu_1du_2) .$$

Since in general $X_{ij} \cdot N = -X_i \cdot N_j = -\langle S(\frac{\partial}{\partial u_i}), \frac{\partial}{\partial u_j}\rangle := -s_{ij}$ and $X_{11} \cdot N = -X_{22} \cdot N$ because we are using conformal coordinates on a minimal surface:

$$-2\,\mathrm{Re}((X_{zz} \cdot N)dz^2) = (s_{11}du_1^2 + s_{22}du_2^2 + 2s_{12}du_1du_2) . \tag{2.11}$$

The right-hand side is the quadratic form associated with the shape operator S, expressed in terms of the conformal coordinates (u_1, u_2), where $z = u_1 + iu_2$. The left-hand-side of (2.11) can be written, using (2.7), as $-\mathrm{Re}(\Phi_z dz \cdot N)$. Differentiating (2.7) gives

$$\Phi_z dz = \frac{-1}{2}((g^{-1} + g), i(g^{-1} - g), 0)\frac{dg}{g}dh ,$$

and, using the fact that

$$N = \sigma^{-1} \circ g = \frac{2}{1 + |g|^2}\left(\mathrm{Re}\, g, \mathrm{Im}\, g, \frac{|g|^2 - 1}{2}\right) ,$$

we have from (2.11) that

$$\Sigma s_{ij} du_i du_j = \mathrm{Re}\left\{\frac{dg}{g}dh\right\} .$$

For a tangent vector $V = v_1\frac{\partial}{\partial u_1} + v_2\frac{\partial}{\partial u_2}$, we may express the Second Fundamental Form (2.11) as the real part of a holomorphic quadratic differential:

$$\langle S(V), V\rangle = \mathrm{Re}\left\{\frac{dg}{g}(V) \cdot dh(V)\right\} . \tag{2.12}$$

This formula allows us to conclude that a curve c on $X(M)$ is

$$\begin{aligned} \text{asymptotic} &\Longleftrightarrow \frac{dg}{g}(\dot{c}) \cdot dh(\dot{c}) \in i\mathbf{R}\,. \\ \text{principal} &\Longleftrightarrow \frac{dg}{g}(\dot{c}) \cdot dh(\dot{c}) \in \mathbf{R}\,. \end{aligned} \tag{2.13}$$

For a minimal surface X as in (2.7) and (2.8) with the Weierstrass data $\{g, dh\}$ on M, we make the following definition.

Definition 2.1. *The associate surfaces to X are the minimal immersions X_θ given by the Weierstrass data $\{g, e^{i\theta}dh\}$, $0 \leq \theta \leq \frac{\pi}{2}$, using* (2.6) *and* (2.7). *The conjugate surface X^* is equal to $-X_{\frac{\pi}{2}}$, where $X_{\frac{\pi}{2}}$ is the associate surface with data $\{g, idh\}$.*

Remark 2.1.1. *The Weierstrass data $\{e^{i\theta}g, dh\}$ produces the same minimal surface as $\{g, dh\}$ rotated by an angle θ around the vertical axis.*

One sees immediately that X_θ may not be well-defined on M because the period condition (2.8) is not necessarily satisfied. We have

$$X^* = -\operatorname{Im} \int \Phi\,,$$

and

$$X_\theta = (\cos\theta)X - (\sin\theta)X^*$$

locally. Clearly X_θ is well-defined on M, for all θ, $0 < \theta \leq \frac{\pi}{2}$, if and only if X^* is well-defined on M. Also $X^{**} = -X$. By definition, all members of the associate family share the same Gauss map and from (2.8) it follows that they are locally isometric.

The Schwarz Reflection Principle for Minimal Surfaces

A curve on any surface M in $\mathbf{R}^3$ is a straight line if and only if its geodesic curvature and its normal curvature vanish. A principal curve on M that is not a straight line has the property that it is a geodesic if and only if it lies in a plane orthogonal to the surface. Thus (2.13) can be used on geodesics to identify lines and planar principal curves. From (2.13) we see that principal geodesics of X correspond to straight lines of X^*, and vice versa. The Gauss map along a principal geodesic has values parallel to the plane of this curve, and the same values along the conjugate straight line – so that the line is orthogonal to the plane. The Schwarz reflection principle for minimal surfaces states that minimal surfaces with such curves must possess Euclidean symmetries.

If a minimal surface contains a line segment L, then it is symmetric under rotation by π about L. (If a minimal surface is bounded by a line segment L, it may be extended by rotation by π about L to a smooth minimal surface containing L in its interior.)

If a nonplanar minimal surface contains a principal geodesic – necessarily a planar curve – then it is symmetric under reflection in the plane of that curve. (If a minimal surface meets a plane orthogonally on its boundary the surface may be extended by reflection through the plane to a smooth minimal surface with this curve in its interior.)

We will prove this below. Finding a geodesic on a minimal surface satisfying one of the conditions of (2.13) is equivalent to finding a straight line or a planar geodesic, which with the Reflection Principle implies the existence of a Euclidean symmetry.

Suppose a minimal surface S contains a line segment L; without loss of generality we may assume that L is a portion of the x_3-axis containing $\vec{0}$. The surface S^* conjugate to S contains a planar geodesic in a horizontal plane. Because the coordinate functions of S are harmonic $z := x_3 + ix_3^*$ defines a conformal parameter in a neighborhood of L. To avoid confusion, we write $z = u_1 + iu_2$ and we may translate so that $(x_1, x_2, x_3)(0) = \vec{0}$. Then

$$x_3(u_1) = u_1, \ x_3^*(u_1) = 0$$

$$x_1(u_1) = x_2(u_1) = 0 .$$

Thus $x_3 + ix_3^*$ maps a line segment on the real axis to the real axis, while $x_1 + ix_1^*$ and $x_2 + ix_2^*$ map this same segment to the imaginary axis.

The Schwarz Reflection Principle for complex analytic functions states that if an analytic function f, defined in a neighborhood of a segment of the real axis, maps the segment of the real axis into a line $\ell \subset \mathbf{C}$ then $f(\overline{z}) = \rho \circ f(z)$, where ρ is reflection in ℓ. (Moreover if f is defined as a one-sided neighborhood of the real segment, the extension of f by $f(\overline{z}) := \rho \circ f(z)$ defines an extension of f analytic on a neighborhood of the real line segment).

Since reflection in the real (resp. imaginary) axis is $z \to \overline{z}$ (resp. $z \to -\overline{z}$), we may conclude that

$$x_3 + ix_3^*(\overline{z}) = \overline{x_3 + ix_3^*}$$

by definition, and

$$x_k + ix_k^*(\overline{z}) = -(\overline{x_k + ix_k^*}) \quad k = 1, 2 .$$

Hence

$$X(\overline{z}) := (x_1, x_2, x_3)(\overline{z}) = (-x_1, -x_2, x_3)(z)$$

$$X^*(\overline{z}) := (x_1^*, x_2^*, x_3^*)(\overline{z}) = (x_1^*, x_2^*, -x_3^*)(z) .$$

That is, S is symmetric under rotation about the x_3-axis, and S^* is symmetric under reflection in the (x_1, x_2)-plane, as claimed. (The statements in parentheses of the Schwarz Reflection Principle for minimal surfaces follow from the statements in parentheses for the classical reflection principle.) This completes the proof of the Schwarz Reflection Principle.

The Catenoid and the Helicoid

As a simple illustration, consider the catenoid, which can be represented as $M = \mathbf{C} - \{0\}$ with Weierstrass data $\{g, dh\} = \{z, \frac{dz}{z}\}$. See Figure 2.1.

The mapping $\rho_\theta(z) \to e^{i2\theta}\bar{z}$ is a reflection in the line L_θ given by $c(t) = te^{i\theta}$. Using (2.9) we see that ρ_θ is an isometry, and therefore its fixed point set $X(L_\theta)$ is a geodesic. The mapping (2.6) is well defined because there is only one homology cycle on which to check (2.8), namely that generated by the cycle $|z| = 1$, and:

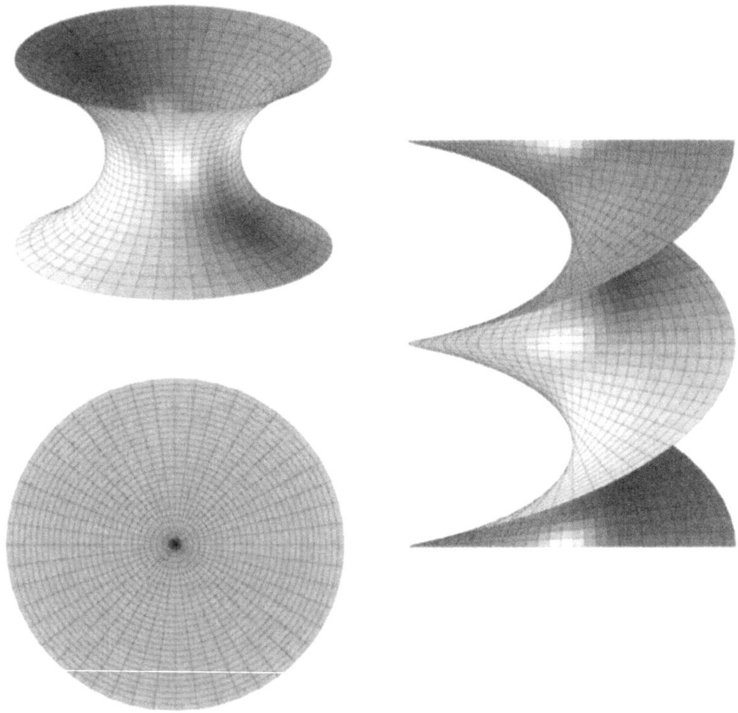

Fig. 2.1. The catenoid and helicoid. The catenoid is the image of $\mathbf{C} - \{0\}$ under the Weierstrass mapping with $g = z$ and $dh = cdz/z$, $c = 1$. The image of $\gamma := \{|z| = 1\}$ is the waist-circle of the catenoid (left). When $c = i$, the Weierstrass mapping is multivalued; its third coordinate has a vertical period of length 2π on γ (center). The multivalued mapping produces the helicoid, which can also be produced by reflection across the boundary lines of the image. For $c = e^{i\theta}, \theta \neq 0$, mod π, the image of γ is a helix and the full image surface is not embedded. For any $c, |c| = 1$, the full image surface is complete and has infinite total curvature unless $c = 1$, in which case we have the catenoid with total curvature equal to -4π

$$\begin{aligned}\text{Period}_{|z|=1}(\Phi) &= \text{Re}\int_{|z|=1}\left(\left(\frac{dz}{z^2}-dz\right), i\left(\frac{dz}{z^2}+dz\right), \frac{dz}{z}\right)\\ &= \text{Re}(0,0,2\pi i) = \vec{0}\,.\end{aligned}$$

Along rays $c(t) = te^{i\theta}, t > 0$, $\dot{c}(t) = e^{i\theta}$ and

$$\frac{dg}{g}(e^{i\theta})\cdot dh(e^{i\theta}) = \left(\frac{e^{i\theta}}{te^{i\theta}}\right)^2 = t^{-2} \in \mathbf{R}\,.$$

Hence each ray is a planar geodesic in a vertical plane containing the vector $(\cos\theta, \sin\theta, 0)$. Any two reflections must generate a rotation about a common axis and we can conclude, as we know, that the catenoid is a surface of rotation. Since

$$x_3(z) = \text{Re}\int_{z_0}^{z}\frac{dz}{z} = \ln\frac{|z|}{|z_0|}\,,$$

we see that the surface diverges to ∞ (resp. $-\infty$) in the vertical direction as $|z| \to \infty$ (resp. 0).

If we consider the conjugate surface X^* of the catenoid, we have Weierstrass data $\{z, idz/z\}$ on $\mathbf{C} - \{0\}$. But now

$$\text{Period}_{|z|=1}\{\Phi^*\} = \text{Re}\int_{|z|=1} i\Phi = \text{Re}(0,0,i\ln z) = (0,0,2\pi)\,.$$

Hence X^* is *not* well-defined on $\mathbf{C} - \{0\}$. It gives rise to a multi-valued mapping, the image of which we know to be fibred by horizontal lines and invariant under a vertical translation. This surface is, of course, the helicoid. Note that since the helicoid and catenoid are locally conjugate, they are locally isometric. But they are not congruent because the catenoid contains no lines. The Gauss map of the catenoid covers S^2 exactly once, omitting two points. Hence its total curvature is -4π. The helicoid is nonflat and periodic; hence it has infinite total curvature.

In the case of finite total curvature, $\int_M KdA > -\infty$, much more is known about the Weierstrass representation.

2.1 Finite Total Curvature

The representation Theorem 2.1 can be significantly sharpened when the minimal surface in question is complete and has finite total curvature.

Theorem 2.2 (Osserman, [48, 59]). *Let $X: M \to \mathbf{R}^3$ be a complete conformal minimal immersion with finite total curvature (not necessarily embedded). Then:*

(i) *M is conformally diffeomorphic to $\overline{M}_k - \{p_1, \ldots, p_r\}$ where $\overline{M}_k$ is a closed Riemann surface of genus k and $p_1 \ldots, p_r$ are points in $\overline{M}_k$, $r \geq 1$;*

(ii) *X is proper;*

(iii) *The Gauss map $N: M \to S^2$, which is meromorphic on M, extends to a meromorphic function on $\overline{M}_k$; the holomorphic one-form dh extends to a meromorphic one-form on $\overline{M}_k$.*

(iv) *The total curvature is an integer multiple of 4π and satisfies*

$$\int_M KdA \leq -4\pi(k + r - 1) = 2\pi(\chi(M) - r), \tag{2.14}$$

where k and r are integers defined in Statement (i), *and $\chi(M)$ is the Euler characteristic of M.*

Theorem 2.2 is a prescription for constructing finite total curvature examples. We will give an example of how to use it in Section 2.2.

In what follows, we will assume that $X: M \to \mathbf{R}^3$ is a complete, regular, conformal minimal immersion that has finite total curvature, and that $M = \overline{M}_k - \{p_1, \ldots, p_r\}$, where $\overline{M}_k$ is a compact, connected Riemann surface of genus k.

Definition 2.2. *Let D_j be a punctured neighborhood of $p_j \in \overline{M}_k$. We will refer to $X(D_j) = E_j$ as an end of M, and define, for fixed $R > 0$,*

$$S_{R,j} = \{q \in S^2 | Rq \in E_j\} \tag{2.15}$$

where S^2 denotes the unit sphere in $\mathbf{R}^3$.

Theorem 2.3 Gackstatter [24], Jorge-Meeks [42]. *Let $S_{R,j}$ be as defined in* (2.15).

(i) *For each $j = 1, \ldots, r$, $S_{R,j}$ converges smoothly, as $R \to \infty$, to a great circle, covered an integral number of times;*

(ii) *Let d_j be the multiplicity of the great circle $\lim_{R\to\infty} S_{R,j}$. Then*

$$\int_M KdA = 2\pi\left(2(1 - k) - r - \sum_{j=1}^{r} d_j\right) \tag{2.16}$$
$$= 2\pi\left(\chi(M) - \sum_{j=1}^{r} d_j\right).$$

Note that (2.16) gives an interpretation of the difference between the right- and left-hand sides of (2.14).

Definition 2.3. *The* total spinning *of S is the integer quantity*

$$n(S) = \sum_{j=1}^{r} d_j .$$

It is immediate from the definitions of $n(M)$ and d_j that $n(M) \geq r$, with equality if and only if each end has multiplicity one. This may be interpreted in terms of the end at p_j. We will say that M has an "embedded end at p_j" provided X embeds some punctured neighborhood of p_j. The end at p_j is embedded if and only if $d_j = 1$, as follows from the definition of d_j in Theorem 2.3 (ii).

Remark 2.2. *For later use in Section* 7 *we will need to estimate the degree, d, of the Gauss map of a complete minimal surface M in terms of the genus. We have from* (2.16)

$$d = \left(k + \frac{r}{2} - 1 + \frac{n(M)}{2}\right).$$

If $r \geq 2$ then clearly $d \geq k+1$. If $r = 1$, then $n(M)$ is just d_1 and must be odd, and $d_1 = 1$, if and only if the end of the surface is embedded. Since $\frac{r}{2} + \frac{n(M)}{2} \geq 1$ with equality if and only if $r = 1$ and $n(M) = d_1 = 1$, it follows that $d \geq k+1$ on a complete minimal surface, unless it has precisely one end, and that end is embedded. However such an exceptional minimal surface must be a plane, as we will show below. Hence on a complete nonplanar minimal surface

$$d \geq k+1.$$

Several arguments can be used to show that a complete immersed minimal surface M of finite total curvature with a single embedded end is a plane. The simplest uses the halfspace theorem (Theorem 2.4 *below in Section* 2.3.3*), which says that a complete, properly immersed minimal surface that lies in a halfspace must be a flat plane. Because we have finite total curvature and an embedded end, we can use Proposition* 2.3 (iv) *and Remark* 2.4 (i) *to conclude the same thing as follows. The end in question is asymptotic to the end of a plane or the end of a catenoid. In either case, the surface M lies in a halfspace. If the end is catenoidal it is clear that the plane defining the smallest halfspace containing M is tangent to M, violating the Maximum Principle. (See Section* 2.3.3.*) If the end of M is planar, then M lies in some slab. But again the boundary planes of the smallest slab are tangent to M, a contradiction unless the slab has zero thickness; i. e., M is a plane.*

2.2 The Example of Chen-Gackstatter

Chen and Gackstatter [10, 11] were the first to construct a complete, finite-total-curvature example by explicitly solving the period problem. In retrospect, and from our point of view, what they did was put a handle in Enneper's surface, making it a genus-one surface with one end, while preserving all of the symmetries and end behavior. See Figure 2.2. This was the first known complete minimal torus with finite total curvature and one end. We will present this construction, following Karcher [43], Barbosa and Colares [2] and Thayer [71, 70].

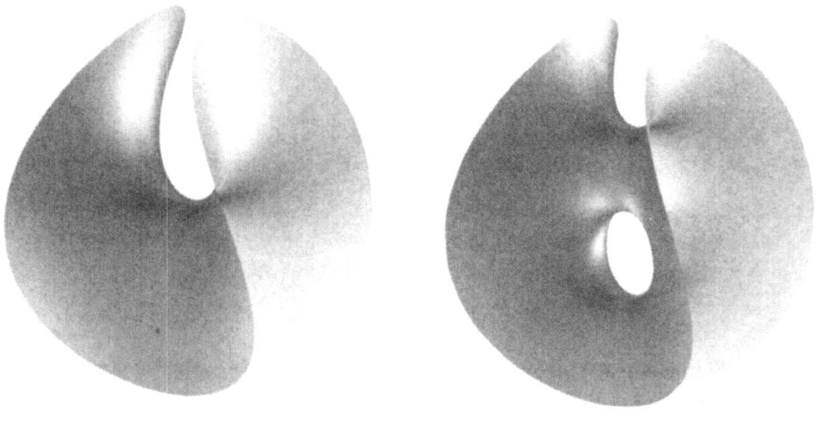

Fig. 2.2. Enneper's surface (left) and the Chen-Gackstatter Surface (right)

Enneper's surface is given by the data

$$g = z \qquad dh = zdz \tag{2.17}$$

on **C**, the sphere minus the point at infinity. There are no nontrivial closed cycles, so there is no period problem. Moreover, the metric induced by using (2.17) in (2.6) and (2.7) is given as in (2.9) by $ds = \frac{1}{2}(|z|^2 + 1)|dz|$ and is invariant under the maps $z \to e^{i2\theta}\overline{z}$, for any fixed θ. The line L_θ through $\vec{0}$ given by $e^{i\theta}t, \infty < t < \infty$, is the fixed-point set of this mapping and so is mapped to a geodesic on Enneper's surface. The normal curvature (multiplied by the length squared of the tangent vector) of this geodesic is given, according to (2.12) by $\text{Re}\{e^{2i\theta}\}$. By (2.13), L_θ is mapped to a straight line on Enneper's surface if $\theta = \pm\frac{\pi}{4}$, and to a planar geodesic if and only if $\theta = 0, \frac{\pi}{2}$. Assuming we start the integration in (2.6) at 0, this implies that Enneper's surface has the (x_1, x_3)- and the (x_2, x_3)-plane as reflective planes of symmetry, and contains the lines $x_1 \pm x_2 = x_3 = 0$.

The planar symmetry lines cut the surface into four congruent pieces. The symmetry lines themselves cross at $\vec{0}$ and the end point at infinity. In addition, the two straight lines cross at $\vec{0}$ and at the end, and the surface is invariant under rotation about these lines.

We will construct a genus-one surface S with the same symmetries and end behavior as Enneper's surface. From (2.16) we may conclude that the winding d_1 at the end of Enneper's surfaces equal to three. The two reflections generate a rotation, R, of order two about the vertical axis. By (2.16), the total curvature of this surface must be -8π, which is equivalent to saying that the degree of g is two. By the symmetry assumptions, the Gauss map is vertical at the end. By symmetry again, and the fact that the degree of g is two, all the vertical normals are fixed points of both reflections, which means they lie on the vertical axis. We place the intersection of the two lines at $\vec{0}$.

Because rotation about the lines on the surface reverses orientation, any axis point, p, of the surface (where the normal must be vertical) other than $\vec{0}$, is paired with $-p$, where g has the same value. There must be at least one such point p on the positive x_3-axis and we orient the surface so that $g(p) = \infty$. Since $g(\vec{0})$ is vertical, it follows that $g(0) = 0$ and $g = 0$ at the end. There can be no other points on the surface on the vertical axis where the Gauss map is vertical. We have

$$g(e) = 0\,, \quad g(\vec{0}) = 0\,, \quad g(p) = g(-p) = \infty\,, \quad p = (0,0,\alpha)\,,$$

where the end is at e on the torus $\overline{S}$.

Choose a triangulation of $\overline{S}$, with vertices including 0, $\pm p$ and e, which projects to a triangulation of $\overline{S}/R$. It will have half the number of faces and half the number of edges. But the number of vertices is $\frac{V}{2} + 2$, where V is the number of vertices in the triangulation of $\overline{S}$. Since $\overline{S}$ is a torus

$$\chi(\overline{S}/R) = \frac{\chi(\overline{S})}{2} + 2 = 2\,;$$

so $\overline{S}/R$ is a sphere.

Let $z\colon \overline{S} \to \overline{S}/R = S^2$ denote projection onto the quotient. The meromorphic function z has degree two and is branched at the points $\{\vec{0}, \pm p, e\} \subset \overline{S}$. Without loss of generality, we may assume that $z(\vec{0}) = 0$, $z(p) = 1$ and $z(e) = \infty$. Reflection K in either one of the vertical planes of symmetry induces the same order-two automorphism of S^2. This automorphism fixes $0, 1, \infty$ and $z(-p)$. We may conclude that this automorphism is complex conjugation $z \to \bar{z}$ and therefore $z(-p)$ is real. Rotation about either line in S induces an orientation reversing automorphism of S^2. This automorphism fixes 0 and ∞ while interchanging $\pm p$. Thus rotation induces $z \to -\bar{z}$ on S^2. Moreover, $z(-p) = -1$ and the two lines on S are projected by z onto the imaginary axis.

We now have the torus $\overline{S}$ presented as a double covering of S^2, branched at $0, \infty$ and ± 1. Therefore, the surface $\overline{S}$ is the Riemann surface

$$w^2 = z(z-1)\,(z+1)\,.$$

This is the square torus on which z is the Weierstrass $\mathcal{P}$-function, normalized to have a double pole at $e = (\infty, \infty)$, a double zero at $\vec{0} = (0,0)$, (and branch points at $\pm p = (\pm 1, 0)$). Introducing $\gamma = z/w$, we have

$$\gamma^2 = \frac{z}{(z-1)\,(z+1)}\,. \tag{2.18}$$

The function γ has degree equal to two, and has simple zeros at $\vec{0}$ and e, and simple poles at $\pm p$. Therefore $g = \rho\gamma$ for some nonzero complex constant ρ. The Gauss map is either purely real or purely imaginary along curves of S projecting onto the real axis in the z-plane. This is because these curves lie in the assumed symmetry planes $x_2 = 0$ and $x_1 = 0$. Because γ is real or imaginary over the real z-axis, it follows that ρ is either real or imaginary. After a rotation of $\mathbf{R}^3$, if necessary, we may assume ρ is real and positive. From Theorem 2.1 we know that the differential dh has simple zeros at 0 and $\pm p$ and no poles at any other finite points. At the end e, it must have a triple pole (producing end behavior like that of Enneper's surface). The one-form dz has precisely these zeros and poles. Hence, $dh = cdz$ for some nonzero complex constant c, which by scaling we may assume to be unitary. We will now show that c is real. To see this, observe that along lifts of the real axis $\frac{d\gamma}{\gamma}(\frac{\partial}{\partial t})$ is real as is $dz(\frac{\partial}{\partial t})$. But these curves are geodesic curvature lines so, by (2.13), c must be real. We may therefore choose $c = 1$ and conclude that

$$g = \rho\gamma \qquad dh = dz\,, \tag{2.19}$$

where z is the projection defined above and $\rho > 0$ is to be determined.

All that remains is to choose $\rho > 0$ so that (2.8) is satisfied. Clearly, the third component has no periods because z is well-defined on the square torus. Because e lies on the intersection of two vertical symmetry planes, the real period of X must be orthogonal to both symmetry planes, and hence zero, for any closed curve around e.

To determine the value of ρ that makes X in (2.6) well-defined, we exploit the symmetry of the situation. Consider the curve on the square torus that is the lift of the interval $[-1, 0]$ in the z-plane, on which γ is positive. (It is always real). This curve must be in the vertical plane $x_2 = 0$, assuming we start the integration from $(0, 0)$. We require that

$$\mathrm{Re}\int_0^{-1}(g^{-1} - g)dh = 0\,,$$

or

$$0 = \mathrm{Re}\int_0^{-1}\left(\frac{1}{\rho\gamma} - \rho\gamma\right)dz = \rho^{-1}\int_{-1}^{0}\sqrt{\frac{(t^2-1)}{t}}dt - \rho\int_{-1}^{0}\sqrt{\frac{t}{(t^2-1)}}dt\,.$$

Both of the integrals are convergent and positive. Choosing ρ to be the positive square root of the ratio of these integrals insures that this way of evaluating $-p = X((-1, 0))$ places it on the x_3-axis. Choice of negative square roots in the integrand will result in the same value of ρ, placing $X((1, 0))$ on the vertical axis. Thus X is well-defined on the torus and produces the genus-one Chen-Gackstatter example with the symmetry of Enneper's surface.

Remark 2.3. (i) *The only complete minimal surfaces in* $\mathbf{R}^3$ *with total curvature* -4π *are the Catenoid and Enneper's surface (Osserman* [58]*). Recently Lopez* [49] *proved that the unique genus-one complete minimal surface with total curvature* -8π *is the Chen-Gackstatter example. This result was also contained in the thesis of D. Bloss* [3].

(ii) *The Chen-Gackstatter genus-one construction has been generalized in two different ways. First, Karcher realized that the kth-order Enneper surface given by the Weierstrass data*

$$g(z) = z^k, dh = g(z)dz, z \in \mathbf{C},$$

could also be used as a model for the end behavior and symmetry of higher genus examples. Using the Lopez-Ros parameter (see Section 3.1, *Equation* (3.5)*) he was able to show that the periods could be killed for all* $k > 0$ [43]. *Second, one can add more handles by stacking them. Chen and Gackstatter did this themselves in the genus-two case* [11]. *Computationally, higher genus examples were found by E. Thayer, who used a computer to solve the period problem* (2.8). *Thayer also computed higher genus examples (up to* 35*) with higher winding orders* [71]. *N. do Espírito-Santo established existence for stacked handles in the genus-three case* [19]. *(See Remark* 4.2.*)*

2.3 Embeddedness and Finite Total Curvature: Necessary Conditions

We continue the general discussion of Section 2.1. Suppose $X: M \to \mathbf{R}^3$ has an embedded end at some $p \in \overline{M}_k - M$. Let $D - \{p\}$ be a punctured neighborhood of p on which X is one-to-one. We may assume, after a rotation if necessary, that the Gauss map, which extends to p, takes on the value $(0, 0, 1)$ at p. The two simplest examples are the plane and (either end of) the catenoid. Written as graphs over the (x_1, x_2)-plane, these ends have bounded and logarithmic growth respectively. This is in fact all that can happen.

Proposition 2.1 ([59, 67]). *Suppose $X: M \to \mathbf{R}^3$ has a complete embedded end $X: D - \{p\} \to \mathbf{R}^3$ of finite total curvature with the value of the extended Gauss map equal to $(0, 0, \pm 1)$ at p. Then, outside of a compact set, $X(D - \{p\})$ is the graph (over the exterior of a bounded domain in the (x_1, x_2)-plane) with the following asymptotic behavior:*

$$x_3(x_1, x_2) = \alpha \log \rho + \beta + \rho^{-2}(\gamma_1 x_1 + \gamma_2 x_2) + \mathcal{O}(\rho^{-2}) \tag{2.20}$$

for $\rho = (x_1^2 + x_2^2)^{1/2}$. Moreover the first two components ϕ_1, ϕ_2 of Φ in (2.7) have poles of order two at p and have no residues, while the third component $dh = \phi_3$ is either regular (which happens if and only if $\alpha = 0$ in (2.20)) or has a simple pole.

Proof. We will prove this proposition in the case that the Gauss map at the embedded end is one-to-one. In this case we may assume, without loss of generality, that $p = 0$ and $g(z) = z$ on D. (In general we may assume $g(z) = z^k$ on D. The reader is invited to carry out a similar computation, showing that $\alpha = 0$ when $k > 1$.)

From (2.6) and (2.7) we have

$$x_1(z) = \mathrm{Re} \int_{z_0}^{z} \phi_1 = \mathrm{Re} \int_{z_0}^{z} \frac{1}{2}(g^{-1} - g)dh = \mathrm{Re} \int_{z_0}^{z} \frac{1}{2}(z^{-1} - z)dh$$

$$x_2(z) = \mathrm{Re} \int_{z_0}^{z} \phi_2 = \mathrm{Re} \int_{z_0}^{z} \frac{i}{2}(g^{-1} + g)dh = \mathrm{Re} \int_{z_0}^{z} \frac{i}{2}(z^{-1} + z)dh$$

$$x_3(z) = \mathrm{Re} \int_{z_0}^{z} dh\,.$$

It is straightforward to compute

$$2(x_1 - ix_2) = \int_{z_0}^{z} g^{-1}dh - \overline{\int_{z_0}^{z} gdh} = \int_{z_0}^{z} z^{-1}dh - \overline{\int_{z_0}^{z} zdh}\,.$$

We will now look more closely at dh. Writing

$$dh = \left(\sum_{j=1}^{k} c_{-j} z^{-j} + c_0 + zw_1(z) \right) dz\,,$$

$w_1(z)$ holomorphic, we observe that $x_1 - ix_2$ is not well-defined if $c_0 \neq 0$. The first integral in the expression for $x_1 - ix_2$ would produce a term $c_0 \ln z$, which is not a well-defined function of z. For similar reasons (using the second integral) $c_{-2} = 0$. Knowing that $c_0 = 0$ allows us to conclude that dh has a pole at 0; otherwise the metric (2.9)

$$\frac{1}{2}(|g| + |g|^{-1})|dh| = \frac{1}{2}(|z| + |z|^{-1})|dh| = \frac{1}{2}(|z|^2 + 1)\frac{|dh|}{|z|}$$

is not complete at $|z| = 0$. We claim now that dh has a simple pole at $z = 0$. If not, let $c_{-k}z^{-k}, k \geq 3$, be the highest order term in the expansion. Then the dominant term in $\int zdh$ is $-(c_{-k}/k)z^{-k}, k \geq 3$. This means that on $|z| = \varepsilon$ small, $x_1 - ix_2(z)$ has self intersections. However, since we are assuming the end is embedded, this contradicts the fact that the normal is close to vertical near 0, so the projection of the image of a small circle $|z| = \varepsilon$ is one-to-one. Thus, $dh = (c_{-1}z^{-1} + zw_1(z))dz, w_1(z)$ holomorphic, as claimed. Because $x_3 = \mathrm{Re} \int_{z_0}^{z} dh$ is well-defined, c_{-1} must be real. Moreover, ϕ_1 and ϕ_2 have double poles without residues at $z = 0$.

We now can write

$$x_1 - ix_2 = -c_{-1}/z + c + zw_2(z) + \overline{\bar{z}w_3(z)}$$
$$x_3 = \text{const} + c_{-1} \ln |z| + \mathcal{O}(|z|^2)\,,$$

where $w_i(z)$ holomorphic and c is a constant. Because the end is complete, $c_1 \neq 0$. We compute

$$\rho^2 = |c_{-1}|^2/|z|^2 - 2\,\mathrm{Re}(\bar{c}_{-1}c/\bar{z}) + const + \mathcal{O}(|z|)$$

and $\mathrm{Re}(x_1 - ix_2) \cdot \bar{c}$, which we express as follows:

$$\mathrm{Re}(c_{-1}/z)\bar{c} = -\,\mathrm{Re}(x_1 - ix_2) \cdot \bar{c} + |c|^2 + \mathcal{O}(|z|)\,.$$

From the expression for ρ^2, one sees that $|z| = \mathcal{O}(\rho^{-1})$. Using this fact, and substitution of the second equation into the expression for ρ^2, gives after rearrangement:

$$\begin{aligned}|c_{-1}|^2/|z|^2 &= \rho^2 + 2\,\mathrm{Re}(x_1 - ix_2 \cdot \bar{c}) + const + \mathcal{O}(\rho^{-1})\\ &= \rho^2(1+\rho^{-2})(2\,\mathrm{Re}(x_1 - ix_2 \cdot c) + \mathcal{O}(\rho^2)\,.\end{aligned}$$

Taking the logarithm of both sides and using the fact that $\log(1+a) = a + \mathcal{O}(a^2)$, we get

$$\ln|z| = \ln|c_{-1}| - \ln\rho - \frac{\mathrm{Re}(x_1 - ix_2)\cdot c}{\rho^2} + \mathcal{O}(\rho^{-2})\,.$$

Substitution in the equation above for $x_3(z)$ yields:

$$x_3(z) = const - c_{-1}\ln\rho - \rho^{-2}c_{-1}\,\mathrm{Re}(x_1 - ix_2)\cdot c + \mathcal{O}(\rho^{-2}), \qquad (2.20')$$

which gives (2.20).

Definition 2.4. *Consider an embedded end of a complete minimal surface of finite total curvature. The end is said to be a flat or planar end if* $\alpha = 0$ *in Equation* (2.20), *and is a catenoid end otherwise. The constant* α *is called the logarithmic growth of the end.*

Remark 2.4. *From* (2.20′)*:*

(i) *It is evident that an embedded finite total curvature end is asymptotic to an actual catenoid end* ($h = \alpha\log r + \beta$) *or to a plane* ($h \equiv \beta$)*;*
(ii) *It also follows from Equations* (2.20′) *that the logarithmic growth of a complete embedded end, parameterized on a punctured disk* $D - \{p\}$*, is equal to* $-c_1$*, where* c_1 *is the residue of* dh *at* p*. If the end is parameterized by the exterior of a disk* ($p = \infty$) *then the logarithmic growth is equal to the residue of* dh *at infinity;*
(iii) *In the proof of Proposition* 2.1 *we showed that if the extended Gauss map is unbranched at an embedded end of finite total curvature, that end is a catenoid end. Since the only other geometric possibility is to have a flat end, it follows that branching of the extended Gauss map at an embedded end of finite total curvature is equivalent to the flatness of that end.*

Proposition 2.2 (Gackstatter [24], Jorge-Meeks [42]).

(i) *All of the ends of* M *are embedded* $\Leftrightarrow$ $n(M) = r$ $\Leftrightarrow$ *equality holds in* (2.14)*:*

$$\int K dA = -4\pi(k + r - 1)\,. \qquad (2.21)$$

(ii) *If no two ends of M intersect then, after a rotation, the Gauss map N satisfies $N(p_i) = (0, 0, \pm 1), i = 1, \ldots, r$.*

Remark 2.5. *If $X\colon M \to \mathbf{R}^3$ is an embedding, then the conditions of Proposition 2.2 must hold. However, these conditions are not sufficient to imply that X is an embedding. See the discussion and examples presented in Section 5.*

Proof. (i) follows immediately from Theorem 2.3 (ii), formula (2.16) and the observation made after Definition 2.3 that $d_j = 1$ when an end is embedded. Statement (ii) follows from Remark 2.4, because two ends that are of planar- or catenoid-type must intersect if their limit normals are not parallel.

Because formula (2.21) is important and we did not present the proof of Theorem 2.3 we will give a direct proof of it here. Formula (2.21) is clearly true for the flat plane ($k = 0, r = 1, K \equiv 0$). Hence we may assume $K < 0$ except at isolated points. M is by assumption a compact Riemann surface of genus k, from which r points, $p_1, \ldots p_r$, have been removed. Choose closed curves $\widetilde{\beta}_j$ that bound small disks D_j centered at p_j. Since the ends are assumed to be embedded, Remark 2.4,i) implies that they are asymptotic to half-catenoids or planes. This means that a sequence of curves homotopic to $\widetilde{\beta}_j$, whose lengths approach a minimum for all such curves, cannot diverge. Because $K < 0$ except at isolated points of M, a *unique* length-minimizing geodesic β_j exists in the homotopy class of $\widetilde{\beta}_j$ ([27]). Each geodesic bounds an end representative that is topologically an annulus. By intersecting a planar end with a large sphere or a catenoid end with a plane orthogonal to the limit normal and sufficiently far out, we may approximate the annular end E_j by a finite annulus bounded on one side by the geodesic β_j and on the other by a closed curve c. Moreover the closed curve c can be chosen, according to Proposition 2.1 and formula (2.20), to be of the form $c_\rho(\theta)$, $0 \leq \theta \leq 2\pi$, with

$$c_\rho(\theta) = (\rho \cos\theta, \rho \sin\theta, \alpha \log \rho + \rho^{-2}(\gamma_1 \cos\theta + \gamma_2 \sin\theta) + \mathcal{O}(\rho^{-2}))\,.$$

Here $\rho > 0$ can be made as large as desired, and γ_1, γ_2 are fixed. Increasing ρ corresponds to pushing $c_\rho(\theta)$ farther out on the end; as $\rho \to \infty$, $c_\rho(\theta)$ diverges. It is straightforward to see from the equation for $c_\rho(\theta)$ that its total curvature converges quickly to 2π. Using the Gauss-Bonnet formula,[5] we have

$$\int_{E_j} K dA \cong 2\pi\chi(E_j) - \int_{\beta_j} k_g ds - \int_c k_g ds \cong -\int_c \kappa ds\,,$$

since the Euler characteristic of an annulus is zero, β_j is a geodesic, c is asymptotically a circle and the end is planar. Hence, the total curvature of the end bounded by β_j is equal to -2π.

We next show that the β_j are pairwise disjoint. If two touch but do not cross, then of course they must coincide, which implies that S is the union of two ends with a common boundary. The surface S is an annulus ($g = 0, r = 2$)

[5] Here k_g and k denote the geodesic curvature and curvature in $\mathbf{R}^3$, respectively.

with total curvature -4π, so (2.21) is satisfied in this case. (By a result of Osserman [58], the only such surface is the catenoid. See also Lemma 3.1 in Section 3.1.) If two β_j cross, then a segment of one together with a segment of the other bounds a region Σ of the surface. Since Σ is simultaneously inside of two ends it is easy to see that Σ is simply connected. This leads to a contradiction of the fact that each α_j is the unique geodesic in its homotopy class. (Alternatively, we could use the Gauss-Bonnet formula again to get a contradiction; the Euler characteristic of Σ is 1 and its boundary consists of two geodesics meeting at two vertices whose exterior angles are less than π in absolute value:

$$2\pi = 2\pi\chi(\Sigma) = \int_\Sigma KdA + \Sigma \text{ exterior angles } < \int_\Sigma KdA + 2\pi\,.$$

This clearly contradicts the fact that $K < 0$ almost everywhere on Σ.) Hence the β_j are disjoint.

Now remove the r ends from S by cutting along the geodesics β_j $j = 1, \dots r$. This leaves a compact surface $\widehat{M}$ of genus k, bounded by r closed geodesics. A third (and final) application of Gauss-Bonnet gives

$$2\pi(2-2k-r) = 2\pi\chi(\widehat{M}) = \int_{\widehat{M}} KdA\,.$$

Combining this with the previously established fact that each end has total curvature -2π gives

$$\int_M KdA = \int_{\widehat{M}} KdA + r(-2\pi) = -4\pi(k+r-1)\,,$$

which is (2.21). See [9] for more details.

If S is embedded and complete with finite total curvature, then we know by Proposition 2.1 that outside of a sufficiently large compact set of $\mathbf{R}^3$, S is asymptotic to a finite number of half-catenoids and planes, which may be assumed to have the same vertical limit normal (Proposition 2.2 (ii)). If S is not a plane it must have *at least two* catenoid ends, one with positive, the other with negative logarithmic growth. This follows from the Halfspace Theorem (Theorem 2.4, which will be proved in Section 2.3.3). It states that a complete properly immersed minimal surface without boundary in $\mathbf{R}^3$, which is not a flat plane, cannot lie in any halfspace. If S has only one catenoid end, it would be in a halfspace yet not be a plane, a contradiction.

Since we are assuming that S is connected, the fact that S is properly embedded implies that $\mathbf{R}^3 - S$ consists of precisely two components. Outside of a sufficiently large compact set, the ends of S are stacked and thus ordered from top to bottom. This also means that the limit normals $(0, 0 \pm 1)$ alternate from one end to the next and that the logarithmic growth rates are also ordered: If the ends correspond to points $\{p_1 \dots p_r\} \subset \overline{M}$, and α_j is the logarithmic growth rate of the jth end:

$$\alpha_1 \le \alpha_2 \cdots \le \alpha_r\,, \qquad \alpha_1\alpha_r < 0\,. \tag{2.22}$$

2.3.1 Flux

On an oriented surface S, with boundary, let ν denote the outward-pointing unit-normal vector field on ∂S. For any vector field W that is C^1 with compact support on S, the divergence theorem states that

$$\int_S (\mathrm{div}_S\, W) dA = \int_{\partial S} \langle W, \nu \rangle ds\,.$$

If $S \subset \mathbf{R}^3$ and W is a vector field on $\mathbf{R}^3$, we may restrict W to S and consider its tangential component $W^T := W - \langle W, N\rangle N$. Suppose W is a constant vector field and S a minimal surface. Then W^T is the gradient of a harmonic function, the restriction to S of a linear function in $\mathbf{R}^3$; therefore $\mathrm{div}_S W^T = 0$. It follows from the divergence theorem that

$$\int_\gamma \langle W, \nu\rangle ds = \langle W, \int_\gamma \nu ds\rangle$$

for all constant vector fields W, i.e. $\int_{\partial S} \nu ds = 0$. Thus, for closed curves γ, $\int_\gamma \nu(s) ds$ is an homology invariant.

Definition 2.5. *The flux along γ is the vector quantity*

$$\mathrm{Flux}\,([\gamma]) := \int_\gamma \nu(s) ds\,.$$

More generally, if W is a Killing field and S a minimal surface, we still have $\mathrm{div}_S\, W^T = 0$. (This can be seen as follows: For any tangent vector U, and any vector field: $W \in \mathbf{R}^3$

$$\begin{aligned}\nabla_U W^T &= [UW^T]^T = [UW - U(\langle W, N\rangle N)]^T \\ &= [UW]^T - \langle W, N\rangle DN(U)\,.\end{aligned}$$

Since W is Killing, $\langle (UW)^T, U\rangle = \langle UW, U\rangle = 0$. Since S is minimal, $tr_S DN = 0$. Hence $\mathrm{div}_S\, W^T = tr_S \nabla W^T = tr_S [DW]^T - \langle W, N\rangle tr_S DN = 0$, as claimed. It follows in an analogous manner that $\int_\gamma \langle W, \nu\rangle ds$ is an homology invariant. This will be used in the next section.

We remark that the flux can also be defined by the same equation for any closed curve γ contained in S, using $\nu(s) = \mathrm{Rot}_{\frac{\pi}{2}}(d\gamma/ds)$. This coincides with our definition of flux when γ is a boundary curve of S and is easily seen to be homology-invariant.

Proposition 2.3. *Suppose $X: M \to \mathbf{R}^3$ is a complete conformal minimal immersion given by $X = \mathrm{Re} \int \Phi$ as in* (2.6), (2.7). *Suppose $\gamma = X(\widetilde{\gamma}), \widetilde{\gamma} \subset M$ is a closed curve. Let X^* denote the conjugate surface (see Definition* 2.1*), which is a (possibly multi-valued) conformal immersion $X^*: M \to \mathbf{R}^3$:*

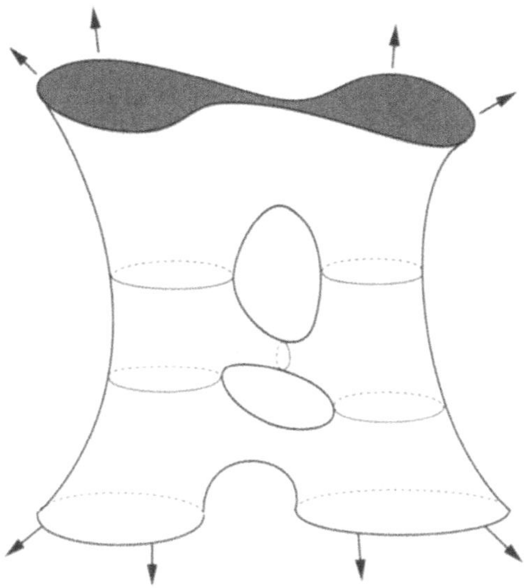

Fig. 2.3. Flux. Flux on a curve is computed by integrating the outward-pointing unit normal vector on a cycle. The total flux of the boundary of a minimal surface is zero

(i) $\mathrm{Flux}([\gamma]) = -\,\mathrm{Period}_{\widetilde{\gamma}}\, X^*$;

(ii) *If $\widetilde{\gamma}$ is in the homology class of a puncture, p,*

$$\mathrm{Flux}([\gamma]) = -2\pi\, \mathrm{Residue}_p\, \Phi\,;$$

(iii) *If $\widetilde{\gamma}$ is in the homology class of a puncture p, which represents a vertical embedded end of finite total curvature,*

$$\mathrm{Flux}([\gamma]) = (0, 0, 2\pi\alpha)\,, \tag{2.23}$$

where α is the logarithmic growth of the end. In particular, the flux of a homology class representing a vertical, embedded finite-total-curvature end is always vertical and is zero if and only if the end is flat;

(iv) *If $X(M)$ is an embedding with finite total curvature and logarithmic growth rates $\alpha_1, \ldots \alpha_r$ at the ends, then*

$$\sum_1^r \alpha_j = 0\,. \tag{2.24}$$

Proof. A constant vector field T in $\mathbf{R}^3$ is the gradient of a linear function f. The tangential component T^T of T restricted to $\mathcal{S} := X(M)$ is the (Riemannian) gradient of $f|_{\mathcal{S}}$, which is harmonic, according to (2.5), because

X is a minimal immersion. Moreover, $\mathrm{Rot}_{\frac{\pi}{2}} T^T = \mathrm{grad}\, f^*$, where f^* is the locally defined harmonic conjugate of f. (See the discussion between (2.3) and (2.4).) Now we compute

$$\begin{aligned} T \cdot \mathrm{Flux}([\gamma]) &= \int_\gamma \langle T^T, \nu \rangle ds = \int_\gamma \langle \mathrm{Rot}_{\frac{\pi}{2}} T^T, \mathrm{Rot}_{\frac{\pi}{2}} \nu \rangle ds \\ &= \int_\gamma \langle \mathrm{grad}\, f^*, \frac{-d\gamma}{ds} \rangle ds = - \int_\gamma df^* . \end{aligned}$$

Since this is true for *any* constant vector field T in $\mathbf{R}^3$, it follows that $\mathrm{Flux}([\gamma]) = -\mathrm{Period}_{\tilde{\gamma}} X^*$, which is (i). From this, (ii) follows immediately since

$$\int_{\tilde{\gamma}} \Phi = \mathrm{Re} \int_{\tilde{\gamma}} \Phi + i\,\mathrm{Im} \int_{\tilde{\gamma}} \Phi = \mathrm{Period}_{\tilde{\gamma}} X + i\ \mathrm{Period}_{\tilde{\gamma}} X^* = -i\ \mathrm{Flux}([\gamma]) . \tag{2.25}$$

To establish (iii) we use Remark 2.4(ii), together with statement (ii) above and the fact that the third component of Φ is dh. We have that the third component of $\mathrm{Flux}([\gamma])$, where $[\gamma]$ is the homotopy class of a puncture p, must equal $-2\pi\, \mathrm{Residue}_p\, dh = 2\pi\alpha$, where α is the logarithmic growth of the end at p. According to Proposition 2.1, the first two components of Φ have no residues at p.

If $X(M)$ has finite total curvature then according to Theorem 2.2(iii), dh is globally defined, with poles only at the ends $p_1 \ldots p_r$. Hence, using statement (iii)

$$0 = \sum_{i=1}^{r} \mathrm{Residue}_{p_i}\, dh = 2\pi \left(\sum_{i=1}^{r} \alpha_i \right) .$$

This establishes statement (iv). □

Remark 2.6. *If we consider the normal component of the Killing field, $W^N =: w^N N$, where $w^N = \langle W, N \rangle$, a calculation shows that*

$$-\Delta_S w^N + 2K w^N = 0 ,$$

where K is the Gauss curvature of S. That is, $w^N N$ is a Jacobi-field. From a geometric point of view this is expected, since a Killing field is an infinitesimal isometry, and the Jacobi equation is derived from the formula for the second variation of area of a minimal surface. This is considered in more detail in Section 7.

2.3.2 Torque

Let $R_\mathbf{u}$ be the Killing field associated with counter-clockwise rotation about the axis $\ell_\mathbf{u}$ in the direction $\mathbf{u} \in \mathbf{R}^3$. From the identity $(U \wedge V) \cdot W = det(U, V, W)$, for vectors U, V, W in $\mathbf{R}^3$, we have

$$(X \wedge \nu) \cdot \mathbf{u} = (\mathbf{u} \wedge X) \cdot \nu = R_\mathbf{u} \cdot \nu , \tag{2.26}$$

where X is the position vector of a minimal surface S and ν is the outward-pointing normal to a component γ of ∂S. Because $R_{\mathbf{u}}$ is a Killing field, $\int_\gamma R_{\mathbf{u}} \cdot \nu$ is a homology invariant and $\int_{\partial S} R_{\mathbf{u}} \cdot \nu = 0$. (See the discussion after Definition 2.5.) This motivates:

Definition 2.6.

$$\mathrm{Torque}_0(\gamma) = \int_\gamma X \wedge \nu \,.$$

Torque was introduced by Kusner in [45, 46]. From (2.26) it follows that the component of torque in the $\mathbf{u}$ direction is $\int_\gamma R_{\mathbf{u}} \cdot \nu$. If we move the origin from 0 to $W \in \mathbf{R}^3$ and let $\widehat{X}$ be the position vector measured from W, then $\widehat{X} = X - W$ and we can compute

$$\begin{aligned} \mathrm{Torque}_W(\gamma) &= \mathrm{Torque}_0(\gamma) - W \wedge \int_\gamma \nu \\ &= \mathrm{Torque}_0(\gamma) - W \wedge \mathrm{Flux}(\gamma) \,. \end{aligned} \tag{2.27}$$

It follows that if $\mathrm{Flux}(\gamma) = 0$, the torque of γ does not depend upon the base-point of X. It also follows that the torque does not change if we move the base point parallel to the flux. If E is an embedded end we may speak of the torque of E; it is the torque of a closed curve in the homology class of the ends. From Proposition 2.1 the following geometric proposition can be deduced.

Proposition 2.4. *If E is a vertical catenoid end, the torque of E is a horizontal vector. If E is a vertical flat end at which the degree of the Gauss map is at least three, the torque of E vanishes.*

In the case of a vertical catenoid end, we know from Proposition 2.3(iii) that the flux is a vertical vector of the form $(0, 0, 2\pi\alpha)$ where α is the logarithmic growth. We can conclude from (2.27) that there is a unique horizontal vector $W = (w_1, w_2, 0)$ for which the Torque of E is vertical. From Proposition 2.4, it follows that there is a unique vertical line on which the torque vanishes.

Definition 2.7. *For a vertical catenoid end E, the (unique) vertical line ℓ_E on which the torque vanishes is called the* axis *of E.*

The torque of E measured from a base point displaced by W from ℓ_E is

$$\begin{aligned} \mathrm{Torque}_W(E) &= W \wedge (0, 0, 2\pi\alpha) \\ &= 2\pi\alpha(w_2, -w_1, 0) \,. \end{aligned} \tag{2.28}$$

If an embedded finite-total-curvature minimal surface, whose catenoid ends, $E_1 \ldots E_n$, have logarithmic growth rates α_i, has either no flat ends or all flat ends of order three or greater, the total torque of all the ends must sum to zero. For any base point, let W_i be the horizontal displacement of $\ell_{E_i}, i = 1 \dots n$. Then

$$\begin{aligned} 0 &= \Sigma \operatorname{Torque}_{-W_i}(E_i) = 2\pi \operatorname{Rot}_{\frac{\pi}{2}}(\Sigma_{i=1}^n \alpha_i W_i), \quad \text{or} \\ &\Sigma \alpha_i W_i = 0\,. \end{aligned} \tag{2.29}$$

Remark 2.7. *If a surface has only two catenoid ends E_1, E_2, and either no flat ends or all flat ends of order three or greater, then placing the base point on ℓ_{E_1} forces $W_1 = 0$ and by* (2.29) *we must have $W_2 = 0$. That is, the catenoid ends E_1 and E_2 have the* same axis. *Pascal Romon [personal communication] points out that if there are three catenoid ends (and the same hypothesis about flat ends) their axes are coplanar.*

We may assume, without loss of generality that there are two ends with positive logarithmic growth. If we label these ends E_1 and E_2 and the third end E_3 and choose a base point on E_3, then it follows from (2.29) *that $W_1 = -\frac{\alpha_2}{\alpha_1} W_2$. Thus the axes ℓ_{E_1}, ℓ_{E_2} and ℓ_{E_3} are coplanar and, because $\frac{\alpha_2}{\alpha_1} < 0$, the axis ℓ_{E_3} with negative logarithmic growth lies between the two with positive growth. We note that on all known embedded examples, flat ends of order two do not occur. It is not known if they can occur.*

2.3.3 The Halfspace Theorem

We present here a proof of

Theorem 2.4 (The Halfspace Theorem for Minimal Surfaces, [40]). *A complete, properly immersed, nonplanar minimal surface in $\mathbf{R}^3$ is not contained in any halfspace.*

Our proof requires the use of the Maximum Principle for Minimal Surfaces: If S_1 and S_2 are two connected minimal surfaces with a point p in common, near which S_1 lies on one side of S_2, then a neighborhood of p in S_1 coincides with a neighborhood of p in S_2. This implies immediately that the analytic continuations of S_1 and S_2 coincide. In particular, if S_1 and S_2 are both complete, $S_1 = S_2$. We refer the reader to [17] for proofs. In the simple case that S_2 is a plane, the Maximum Principle states that a minimal surface $S = S_1$ that lies locally on one side of its tangent plane $T_pS = S_2$ must in fact be planar. This is easy to prove from the Weierstrass representation, Theorem 2.1. Without loss of generality, we may assume that $T_pS = \{x_3 = 0\}$ and that z is a local coordinate chart chosen so that $X(0) = p$. The Gauss map of S is vertical at p, so that $g(0) = \sigma \circ N(p) = 0, \infty$. Either g is constant, in which case S is flat and is a subset of $T_pS = \{x_3 = 0\}$, or 0 is an isolated pole or zero of g of order $k \geq 1$. In the latter case,

$$dh = z^k w(z) dz\,,$$

where $w(z)$ is holomorphic and $c = w(0) \neq 0$. Then

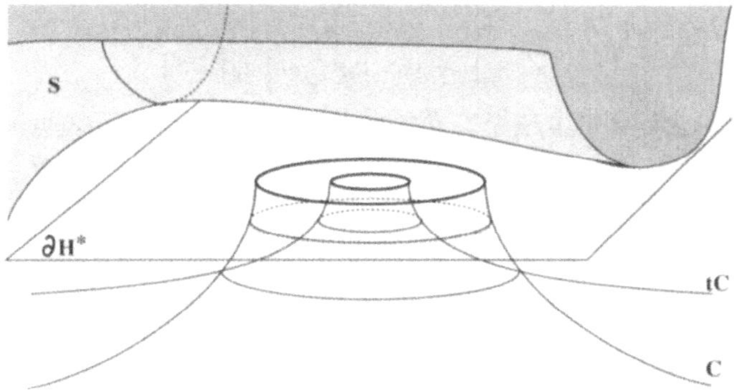

Fig. 2.4. The vertically translated half-catenoid C in the proof of Theorem 2.4

$$x_3(z) = \operatorname{Re} \int_0^z dh = \operatorname{Re} \int_0^z z^k w(z) dz = \operatorname{Re} \left(\frac{cz^{k+1}}{k+1} \right) + \mathcal{O}(|z|^{k+2}) .$$

Clearly $x_3(z)$ changes sign $2(k+1)$ times in any small neighborhood of $z = 0$. Hence S does not lie on one side of $T_pS = \{x_3 = 0\}$ near p.

Proof of Theorem 2.4. By assumption, S is contained in some halfspace H. The intersection, H^*, of the closed halfspaces containing S and having boundary parallel to ∂H is the smallest closed halfspace containing S. We will rotate and translate S so that $H^* = \{x_3 \geq 0\}$, and hence $\partial H^* = \{x_3 = 0\}$. If $S \cap \{x_3 = 0\} \neq \emptyset$, then the simple case of the Maximum Principle implies S is a plane. If S is not a plane $S \cap \{x_3 = 0\} = \emptyset$, and since S is properly immersed, S has no limit points in $\{x_3 = 0\}$. This means each point of $\{x_3 = 0\}$ has positive distance from S.

Consider a catenoid with axis parallel to the x_3-axis and waist circle in $\{x_3 = 0\}$. Let C denote the half-catenoid in $\{x_3 \leq 0\}$. Because x_3 is proper (and unbounded) on C it follows that we can vertically translate C so that the waist circle is in $\{x_3 > 0\}$ but the *solid* half-catenoid inside C is disjoint from S. For convenience in the rest of the proof, we move our coordinate system so that $\partial H^* = \{x_3 = -\varepsilon\}$ and the waist-circle of C lies in $\{x_3 = 0\}$. Also, we will now think of C as the boundary for the solid half-catenoid; i. e., we add to C the disk in $\{x_3 = 0\}$ bounded by the waist-circle.

Let tC denote the rescaling of C by a factor of $t > 0$. We will say that *tC lies below S* if S lies in the closure of the component of $\mathbf{R}^3$ above tC. (Similarly we may define what it means for a plane to lie below S.) For $t \leq 1$, the waist-circles of tC lie on C, so their distance from S is uniformly bounded away from 0. If $S \cap tC \neq \emptyset$, then the Maximum Principle implies $S = tC$, and S, being complete, must be a catenoid; this is a contradiction, since S lies in H^*. We may conclude that if tC is below S, then tC and S have no points in common. Furthermore, since $tC \cap H^*$ is compact and $S \subset H^*$, it follows that every tC lying below S actually has finite positive distance from S. In particular the set

$$T := \{t \in (0,1] : tC \text{ is below } S\}$$

is open in $(0,1]$. But if $\{t_n\} \subset T$ and $\lim_{n\to\infty} t_n = \tau > 0$, then τC is a half catenoid below S; if not, some $t_n C$ must not be below S. Hence T is closed, too. This means that $T = (0,1]$, so all tC are below S, $0 < t \leq 1$.

Note that the half-catenoids tC converge, as $t \to 0$, to the plane $\{x_3 = 0\}$, which is strictly inside $H^* = \{x_3 \geq -\varepsilon\}$. Since all the tC are below S, the limit plane $\{x_3 = 0\}$ is also below S. This contradicts the definition of H^* as the intersection of all the closed halfspaces containing S and having boundary parallel to ∂H. This contradiction completes the proof. □

Remark 2.8. *A slight modification of the same proof can be used to prove the following variant of Theorem* 2.4*: Suppose S is a properly immersed nonplanar minimal surface with compact boundary ∂S. If H is a halfspace containing S, and H^* a smallest halfspace containing S, with ∂H^* parallel to ∂H, then $\partial H^* \cap \partial S \neq \emptyset$.*

2.4 Summary of the Necessary Conditions for Existence of Complete Embedded Minimal Surfaces with Finite Total Curvature

We conclude this section by gathering together the results of our discussion to formulate a sharpened version of Theorem 2.2 for *embedded* complete, nonplanar minimal surfaces of finite total curvature.

Proposition 2.5. *Let $X: M \to \mathbf{R}^3$ be a complete, nonplanar, conformal minimal embedding with finite total curvature. Then:*

(i) *M is conformally diffeomorphic to $\overline{M}_k - \{p_1, \ldots, p_r\}$ where $\overline{M}_k$ is a closed Riemann surface of genus k and $p_1, \ldots, p_r$ are points in $\overline{M}_k, r \geq 2$;*

(ii) *X is proper. A punctured neighborhood of each p_i is mapped by X onto an end of $S = X(M)$ that is asymptotic to either a plane or a half-catenoid.*

(iii) *The Gauss map $N: M \to S^2$ extends to a meromorphic function on $\overline{M}_k$. All the normals on $\{p_1, \ldots, p_r\}$ must be parallel and after a rotation if necessary we may assume that $N(p_i) = (0, 0 \pm 1)$, $i = 1, \ldots r$. The holomorphic one form dh, the complex differential of the height function, extends meromorphically to $\overline{M}_k$.*

(iv) *The total curvature of M is $-4\pi \cdot \text{degree } g = -4\pi(k + r - 1)$, where $g = \sigma \circ N$, and $\sigma: S^2 \to \mathbf{C} \cup \{\infty\}$ is stereographic projection.*

(v) *The ends are naturally ordered by height from top to bottom. In this ordering, say $p_1 \ldots p_r$, the vertical unit normals at the ends alternate and the logarithmic growth rates α_j are ordered from biggest to smallest. Furthermore,*

$$\sum_{j=1}^{r} \alpha_j = 0 \,. \tag{2.30}$$

If S is not a plane, $r \geq 2$ and $\alpha_1 \cdot \alpha_r < 0$.

(vi) *At a catenoid end, g has a simple pole or zero, while at a flat end the pole or zero of g has higher order. The height differential dh has a simple pole at a catenoid end. At a planar end where g has a zero or pole of order m, dh has a zero of order $|m| - 2 \geq 0$. The logarithmic growth rate is equal to minus the residue of dh at the puncture corresponding to the end.*

To construct a complete, embedded minimal surface of finite total curvature with the Enneper-Weierstrass-Riemann representation ((2.6) (2.7) of Theorem 2.1), Proposition 2.5 dictates necessary conditions on the choice of $\overline{M}_k$, $\{p_1 \dots p_r\} \subset \overline{M}_k$, $g = \sigma \circ N$ and dh. However, in order for (2.7) to be single-valued on $M_k = \overline{M}_k - \{p_1 \dots p_r\}$, it is necessary that (2.8) be satisfied:

$$\mathrm{Period}_\alpha(\Phi) = \mathrm{Re} \int_\alpha \left((g^{-1} - g)\frac{dh}{2}, i(g^{-1} + g)\frac{dh}{2}, dh \right) = 0 \tag{2.31}$$

for all closed curves $\alpha \subset M_k$. (See Figure 3.1.) If Statement (iii) of Proposition 2.5 is satisfied, the resulting surface must have embedded parallel ends, which implies that outside of some compact set in $\mathbf{R}^3$, $X(M_k)$ is embedded. However, X may not be an embedding. See the examples in Section 5.

3. Examples with Restricted Topology: Existence and Rigidity

In this section, we present the few uniqueness results that are presently known. They take the following form: if the topological type is restricted or some other geometric property is specified, then the surface is uniquely determined. (See also Theorem 3.4, below.)

Theorem 3.1. *Let S be a complete, embedded minimal surface of finite total curvature.*

(1) *If S has one end, it is a flat plane;*
(2) *(The Lopez-Ros punctured-sphere Theorem)* (§ 3.1). *If S has genus zero, it is the catenoid or the flat plane;*
(3) *(The R. Schoen catenoid characterization)* [67]. *If S has two ends, it is the catenoid;*
(4) *If S has three ends and a symmetry group of order at least $4(k+1)$, where k is the genus, then S is the surface described in Theorem 3.2.*
(5) *(Costa's thrice-punctured-torus Theorem)* [16]. *If S has genus one and three ends, it is one of the surfaces $M_{1,x}$ in Theorem 3.3.*

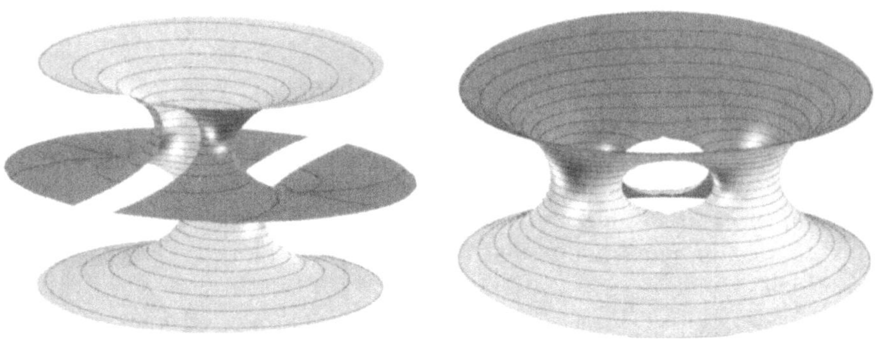

Fig. 3.1. Attempts to produce nonexistent minimal surfaces.

Left: According to the Lopez-Ros punctured sphere theorem (Theorem 3.1(2)), a complete embedded minimal surface with genus zero and finite total curvature is either the plane or the catenoid. Nonetheless, it is possible to write down Weierstrass data that would produce such an example, provided the period problem (2.8) were solvable (which it is not). One can do this with a high degree of symmetry. Under the assumptions that the desired example has three parallel vertical ends – top and bottom catenoidal, and the middle one flat – one vertical plane of symmetry and a horizontal line diverging into the flat end, the resulting Weierstrass data is

$$g = r(z-1)(z+1)^{-1}, \quad dh = dz$$

on $\mathbf{C}-\{\pm 1\}$ with $r>0$. The gap in the image indicates the failure of the Weierstrass-Enneper mapping to be single-valued. The value of r chosen here is 1.
Right: According to the Schoen catenoid characterization (Theorem 3.1(3)) the catenoid is the only complete embedded minimal surface with two ends and finite total curvature. Nevertheless, we can find symmetric Weierstrass data that fails only in that the period problem is not solvable. Under the assumptions that the desired example has genus one, two vertical planes of reflective symmetry and one horizontal plane of reflective symmetry, the Weierstrass data is $\{g, dh\}$, where

$$g^2 = (1-z)(r-z)((1+z)(r+z))^{-1}, \quad dh = (z^2-1)^{-1}dz\,,$$

$r > 1$, on the rectangular torus determined by the equation relating g and z. The image looks like a catenoid through which someone has tried to drill a tunnel from both sides. The tunnels don't meet up, as one sees here (for $r = 2.5$). By Theorem 3.1(3), it is futile to try other values of r.

Statement (1) follows from Theorem 2.4 or Proposition 2.5 (v). Statement (4) is proved in [36]. We will give a full proof of (2) in Section 3.1.

Corollary 3.1. *The only complete, embedded minimal surfaces with* $\int K dA \geq -8\pi$ *are the plane and the catenoid (with total curvature* 0 *and* -4π *respectively).*

Proof. From Proposition 2.2, it follows that the total curvature is an integer multiple of -4π. If the total curvature is zero, we must have the plane.

If it is -4π, then by a result of Osserman [58] it must be the catenoid. (See also Lemma 3.1 in Section 3.1.)

We need to show that total curvature -8π is not possible. In that case we would have, by Proposition 2.2, (i), that $k + r = 3$. But the various cases of Theorem 3.1 exclude this possibility. □

Proposition 3.1. *Let $X: M \to \mathbf{R}^3$ be a connected complete minimal immersion with finite total curvature. Let $n(M)$ be as in Definition 2.3. Then for any $p \in \mathbf{R}^3$, the number of points in $X^{-1}(p)$ is at most $n(M) - 1$, with the sole exception of the case when $X(M)$ is the flat plane.*

Corollary 3.2. *Let $X: M \to \mathbf{R}^3$ be as in Theorem 3.1. If $n(M) = 2, X(M)$ is the catenoid.*

Proof of Proposition 3.1. We will use a version of the monotonicity formula to prove this proposition: *If p is a point on a complete immersed minimal surface and $A(r)$ is the area of the minimal surface inside a Euclidean ball of radius r, then $A(r)/\pi r^2$ is a nondecreasing function of r, which is strictly increasing unless the minimal surface is a collection of planes.*

Suppose that $p \in X(M)$ and $X^{-1}(p)$ contains n points. Assume that $X(M)$ is not a plane. (It cannot be a collection of planes because we assume that $X(M)$ is connected.) Hence, for small r, $A(r)/\pi r^2 > n$. But as a consequence of Theorem 2.3, $A(r)/\pi r^2$ converges to $n(M)$ as $r \to \infty$. From monotonicity we conclude that $n < n(M)$. If $X(M)$ is a flat plane, $n = 1 = n(M)$. □

Proof of Corollary 3.2. It follows from Proposition 3.1 that X is an embedding and since $n(M) = 2$, it is not the plane. It must therefore have two embedded ends. From Theorem 3.1(3), it must be the catenoid. □

According to Corollary 3.1, the next smallest possible total curvature for an embedded example is -12π; that is, $k + r = 4$. By Theorem 3.1, this can only happen when $k = 1$ and $r = 3$, and in fact it does.

Theorem 3.2 ([35, 36]). *For every $k \geq 2$, there exists a complete properly embedded minimal surface of genus $k-1$ with three annular ends. After suitable rotation and translation, the example of genus $k - 1$, which we will call M_k, has the following properties:*

1. *M_k has one flat end between its top and bottom catenoid ends. The flat end is asymptotic to the (x_1, x_2)-plane.*
2. *M_k intersects the (x_1, x_2)-plane in k straight lines, which meet at equal angles at the origin. Removal of the k lines disconnects M_k. What remains is, topologically, the union of two open annuli;*
3. *The intersection of M_k with any plane parallel (but not equal) to the (x_1, x_2)-plane is a single Jordan curve;*
4. *The symmetry group of M_k is the dihedral group with $4k$ elements generated by reflection in k vertical planes of symmetry meeting in the x_3-axis, and rotation about one of the lines on the surface in the (x_1, x_2)-plane.*

Removal of the intersection of M_k with the (x_1, x_2)-plane and the vertical symmetry planes disconnects M_k into $4k$ congruent pieces, each a graph.

5. *M_k is the unique properly embedded minimal surface of genus $k-1$ with three ends, finite total curvature, and a symmetry group containing $4k$ or more elements.*

As will follow from our presentation in Section 4, the Riemann surface $\overline{M}_k$ is given by

$$w^k = -\left(\tfrac{1}{2}\right) z^{k-1}(z-1)\,(z+1)\,. \tag{3.1}$$

The catenoid ends are located at $(z, w) = (\pm 1, 0)$, and the flat end is at (∞, ∞). The Gauss map and the complex differential of the height function are given by

$$g = \rho\, w\,, \qquad dh = \frac{dz}{(z-1)\,(z+1)} \tag{3.2}$$

where ρ is a constant determined by the necessity of satisfying the period condition (2.8) in Theorem 2.1. The symmetry of the surface is used to show that there is only one period condition. (See Figure 3.3, left-hand column, for pictures of the M_k.)

Remark 3.1. *The example with $k = 2$ was found by Celso Costa in* 1982. *He proved it was complete and had embedded ends of the specified type. See Costa* [14, 15] *and Hoffman* [28, 29].

The surfaces M_k each lie in a family of embedded minimal surfaces.

Theorem 3.3 ([30], and Section 4). *For every $k \geq 2$, there exists a one parameter family, $M_{k,x}, x \geq 1$, of embedded minimal surfaces of genus $k-1$ and finite total curvature. The surfaces $M_{k,1}$ are precisely the surfaces M_k of Theorem* 3.2. *The surfaces $M_{k,x}, x > 1$ have all three ends of catenoid type and a symmetry group generated by k vertical planes of reflectional symmetry. The Riemann surface $\overline{M}_{k,x}$ is given by*

$$w^k = -c_x z^{k-1}(z-x)(z+x^{-1})\,, \qquad c_x = (x+x^{-1})^{-1}\,. \tag{3.3}$$

The catenoid ends are located at $(z, w) = (x, 0), (-x^{-1}, 0)$ and (∞, ∞). The Gauss map and differential of the height function are given by

$$g = \frac{\rho w}{mz+1}\,, \qquad dh = \frac{z(m+z^{-1})dz}{(z-x)(z+x^{-1})}\,, \tag{3.4}$$

where ρ and m are constants determined by the period conditions. (When $x = 1$, $m(1) = 0$, and (3.3) *and* (3.4) *give* (3.1) *and* (3.2) *for the surfaces $M_k = M_{k,1}$.)*

Remark 3.2. *The surfaces $M_{k,1}$ of Theorem* 3.3 *are the surfaces M_k of Theorem* 3.2. *In Section* 4, *we find it more convenient to work with the function $u = z/w$ instead of w. The formulae* (3.3) *and* (3.4) *are presented (e.g. in* (4.12)*) in terms of u and z. See* (4.4) *where we perform the conversion, into a (u, z) expression, of the Riemann surface equation* (3.3).

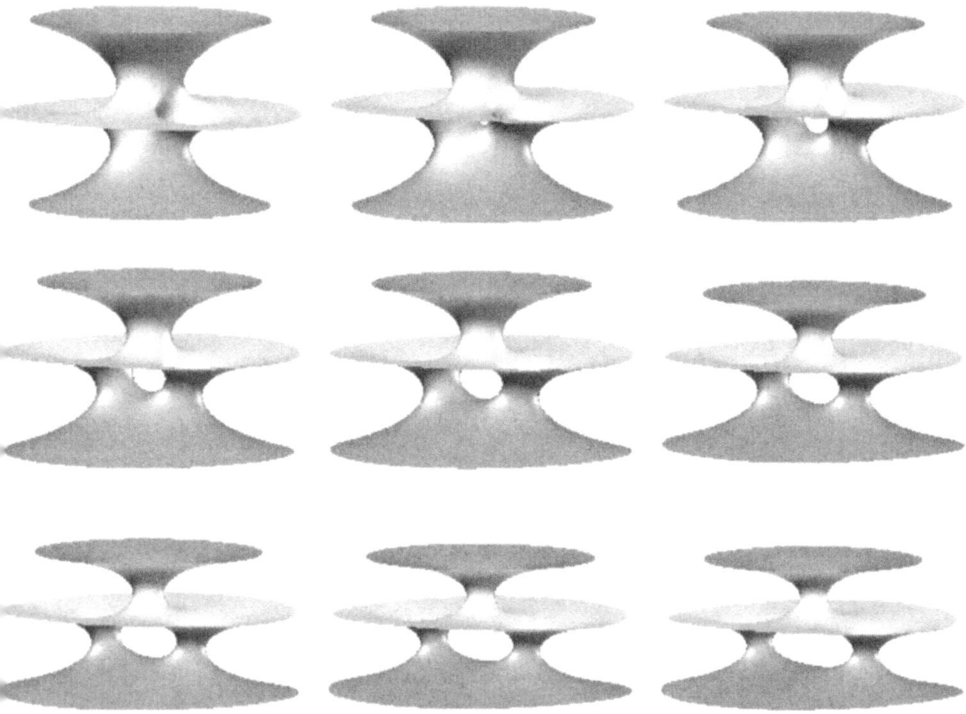

Fig. 3.2. Costa's surface (*upper left*) and a few members of the family of her deformations, as described in Theorem 3.3

In Section 5, we will discuss examples with more than three ends, mention some other results about the structure of the "space" of embedded finite total curvature examples, and present some questions and conjectures. In Section 4 we describe in detail the ideas behind the construction of the surfaces $M_{k,x}$ of Theorem 3.3. As we pointed out in the introduction, these are the only higher-genus examples that have been fully analyzed.

3.1 Complete Embedded Minimal Surfaces of Finite Total Curvature and Genus Zero: the Lopez-Ros Theorem

In this section we prove the second statement of Theorem 3.1, which is Theorem 3.5 below. Recall that for a closed curve $\gamma \subset S$, its flux is defined in Section 2.3.1 as

$$\mathrm{Flux}([\gamma]) = \int_\gamma \mathrm{Rot}_{\frac{\pi}{2}} \left(\frac{d\gamma}{ds}\right) ds\,.$$

Fig. 3.3. Higher-genus embedded minimal surfaces. The surfaces in the column on the left are the surfaces $M_k, k = 3, 4, 6$ described in Theorem 3.2, with genus 2 (top), 3 (*middle*) and 5 (*bottom*). The middle end in each of these surfaces is flat. In each row, the other two surfaces are deformations of the left-most one. These are the surfaces described in Theorem 3.3. Their middle ends are catenoidal

Flux $([\gamma])$ depends only on the homology class of γ. We say that S *has vertical flux* provided Flux $([\gamma])$ is a vertical vector for all closed $\gamma \subset S$.

Our presentation follows that of Perez and Ros [60], whose approach yields the following theorem.

Theorem 3.4. *Let S be a complete, embedded minimal surface with finite total curvature and vertical flux. Then S is the catenoid or the plane.*

As usual, we may realize S by a conformal minimal embedding $X: M \to \mathbf{R}^3$, where $M = \overline{M} - \{p_1 \dots p_r\}$, and rotate S in $\mathbf{R}^3$ so that $g = 0$, or ∞ at the ends $\{p_1 \dots p_r\}$.

If S is complete and embedded, with genus zero, the only closed curves one has to consider are those associated to the ends $\{p_1 \dots p_r\}$. In Proposition 2.3ii) we established that the flux of the homology class of p_j is equal to the vertical vector $(0,0,2\pi\alpha_j)$, where α_j is the logarithmic growth of the end at p_j. Hence we have as an immediate consequence of Theorem 3.4:

Theorem 3.5 ([51]. *Let S be a complete minimal surface with genus zero and finite total curvature. Then S is the catenoid or the plane.*

The proof of Theorem 3.4 exploits what has come to be called the Lopez-Ros deformation. If $\{g, dh\}$ is the Weierstrass data of S, we define on $\overline{M}$ the data

$$g_\lambda = \lambda g \qquad dh_\lambda = dh\,, \tag{3.5}$$

for any $\lambda > 0$. Notice that the zeros and poles of $\{g_\lambda, dh_\lambda\}$ are the same as those of $\{g, dh\}$. The deformation we consider is given by the Weierstrass representation (2.6), (2.7):

$$X_\lambda := \mathrm{Re}\int \Phi_\lambda := \mathrm{Re}\int \big(\frac{1}{2}(\lambda^{-1}g^{-1} - \lambda g), \frac{i}{2}(\lambda^{-1}g^{-1} + \lambda g), 1\big)dh\,. \tag{3.6}$$

Note that X_λ may be multi-valued. The metric and curvature of X_λ are given by (2.9) (2.10) as

$$\begin{aligned} ds_\lambda &= (\lambda^{-1}|g|^{-1} + \lambda|g|)|dh| \\ K_\lambda &= \frac{-16}{(\lambda^{-1}|g|^{-1} + \lambda|g|)^4}\,\frac{|dg/g|^2}{|dh|^2}\,. \end{aligned} \tag{3.7}$$

We will prove three propositions about the Lopez-Ros deformation that allow us to prove Theorem 3.4. The propositions are stated here and then the proof of Theorem 3.4 is presented. The proofs of the propositions follow.

Proposition 3.2. *Let $X\colon M \to \mathbf{R}^3$ be a conformal minimal immersion:*

(i) *X is complete if and only if X_λ is complete for all $\lambda > 0$;*
(ii) *If X_λ is single-valued, the total curvature of X is finite if and only if the total curvature of X_λ is finite for all $\lambda > 0$;*
(iii) *The immersions X_λ are single-valued for all $\lambda > 0$ if and only if X has vertical flux.*

Proposition 3.3. *Let $X\colon M \to \mathbf{R}^3$ be a complete, embedded minimal surface of finite total curvature, for which X_λ is single-valued for all $\lambda > 0$. Then X_λ is an embedding for all $\lambda > 0$.*

Proposition 3.4. *Suppose $X\colon M \to \mathbf{R}^3$ is a conformal minimal immersion.*

(i) *If for some $p \in M, N(p) = (0, 0 \pm 1)$ (i. e., $g(p) = 0$ or $g(p) = \infty$) then, on every neighborhood of p, X_λ is* not *an embedding, for λ sufficiently large.*
(ii) *Suppose $X\colon M = \{z |\ 0 < |z| < \varepsilon\} \to \mathbf{R}^3$ is a conformal embedding representing a flat end. Then X_λ is not an embedding, for λ sufficiently large, unless X represents the end of a flat plane.*

Proof of Theorem 3.4. Let $X\colon M \to \mathbf{R}^3$ be a conformal minimal embedding with $S = X(M)$. The vertical flux condition guarantees that the Lopez-Ros deformation X_λ, defined by (3.5) and (3.6), is single-valued for all $\lambda > 0$ (Proposition 3.2 above). Because X is a complete embedding with finite total curvature, the same is true for $X_\lambda, \lambda > 0$ (Propositions 3.2, and 3.3 above). But this means that S has no vertical points (points where $g = 0, \infty$) and no flat ends, unless S is itself a flat plane (Proposition 3.4 above).

Assume that S is not a plane.

Because S is embedded and has finite total curvature, $M = \overline{M} - \{p_1 \dots p_r\}$, where $\overline{M}$ is compact. The ends at $p \in \{p_1, \dots p_r\}$ are either planar or catenoid-type. The zeros of dh can occur only at vertical points of g or at flat ends, while dh has simple poles at the catenoid ends. (See Proposition 2.5 in Section 2.3.4.) From the previous paragraph we can conclude that dh has no zeros. (Because S is not a plane, $dh \not\equiv 0$.) Thus $\chi(\overline{M}) = 2$, i.e. $\overline{M}$ is the sphere and $r = 2$. We may conclude from Lemma 3.1 that S is the catenoid. □

Lemma 3.1. *Suppose S is a complete, embedded minimal surface with total curvature -4π and two ends. Then S is a catenoid.*

This follows from (2.21). Proposition 2.5(i) and Theorem 3.1(3)). It also follows from Osserman's characterization of complete minimal suraces of total curvature -4π [58]. We give a direct proof here in order to make this section as self-contained as possible.

Proof of Lemma 3.1. By Proposition 2.5(iv) S must have genus zero. Because $g = 0, \infty$ at the ends, and the values must alternate (see Proposition 2.5), we may assume, without loss of generality, that $\overline{M} - \{p_1, p_2\} = \mathbf{C} - \{0\}$ and $g(0) = 0, g(\infty) = \infty$. Since g has no other poles or zeros, it follows that $g(z) = az, a \neq 0$. Without loss of generality, we may assume that a is real, because rotation of S about a vertical axis changes g by a unitary multiplicative factor. Because dh has a simple pole at 0 and at ∞ and no other zeros or poles, $dh = c\,dz/z$ for some $c \neq 0$. Letting $\zeta = az$ we have $g(\zeta) = \zeta$ and $dh = c\,d\zeta/\zeta$. Clearly, dh has a real period on $|\zeta| = 1$ if c is not real. Thus our surface is the catenoid as described in Section 2. (See Figure 2.1.)

We now present proofs of the three propositions.

Proof of Proposition 3.2. From (3.7) it is evident that for each $\lambda > 0$

$$\underline{c}\,ds \leq ds_\lambda \leq \overline{c}\,ds$$
$$\overline{c}^{\,-4}|K| \leq |K_\lambda| \leq \underline{c}^{\,-4}|K|,$$

where $\overline{c} = \max\{\lambda^{-1}, \lambda\}$ and $\underline{c} = \min\{\lambda^{-1}, \lambda\}$. From these inequalities, statements (i) and (ii) follow directly. To prove statement (iii), we use (2.25) from Section 2.3.1. For any closed curve $\gamma = X(\widetilde{\gamma}), \widetilde{\gamma} \subset M$, we have

$$\int_{\widetilde{\gamma}} \Phi = \operatorname{Re} \int_{\widetilde{\gamma}} \Phi + i \operatorname{Im} \int_{\widetilde{\gamma}} \Phi = \operatorname{Period}_\gamma(X) - i \operatorname{Flux}_X([\gamma]) = -i \operatorname{Flux}_X([\gamma])\,.$$

From this identity it is evident that the vertical flux condition is equivalent to the exactness of ϕ_1 and ϕ_2. Since $\phi_1 - i\phi_2 = g^{-1}dh$ and $-(\phi_1 + i\phi_2) = gdh$, exactness of ϕ_1 and ϕ_2 is equivalent to the exactness of $g^{-1}dh$ and gdh. But from (3.6), it is clear that $\mathrm{Period}_\gamma(X_\lambda) = \mathrm{Re}\int_\gamma \Phi_\lambda = 0$, for all $\lambda > 0$ and all $[\gamma]$, if and only if $g^{-1}dh$ and gdh are exact. □

Proof of Proposition 3.3. Since the Lopez-Ros deformation, (3.5) and (3.6), changes neither zeros nor poles of g and dh, the end types of X_λ at the ends p_j of $M = \overline{M} - \{p_1 \dots p_r\}$ are the same for all $\lambda > 0$. That is, flat ends remain flat, and catenoid-type ends remain catenoidal; even more, their logarithmic growth rates, $\alpha = -\,\mathrm{Residue}_p\, dh$, are independent of λ.

Fix $\lambda_0 > 0$ and consider two distinct ends p_i, p_j , $i \neq j$. Choose disjoint neighborhoods D_i, D_j of p_i and p_j so that $X(D_i)$ and $X(D_j)$ are end representatives. Suppose X_{λ_0} is an embedding. If these ends have different logarithmic growths, i. e., if $\alpha_i \neq \alpha_j$, the distance function from $X(D_i)$ to $X(D_j)$ is not only bounded away from zero but unbounded. This is because the ends are asymptotic to catenoid ends (or to a plane if one of the α's is zero) with different logarithmic growth rates. Since these growth rates are independent of λ, it follows that $X_\lambda(D_i) \cap X_\lambda(D_j) = \emptyset$ for λ sufficiently close to λ_0. In case $\alpha_i = \alpha_j$ and $X_{\lambda_0}(D_i)$ and $X_{\lambda_0}(D_j)$ are asymptotic to ends with the same growth rate, we appeal to the Maximum Principle at Infinity, proved in [52], which states that the distance between these embedded annular ends is bounded away from zero; i. e., they are *not* asymptotic at infinity. Thus, again, $X_\lambda(D_i) \cap X_\lambda(D_j) = \emptyset$ for λ sufficiently close to λ_0. We may conclude that, for λ sufficiently close to λ_0, X_λ is an embedding. Stated differently, what we have proved is that

$$L{:}= \{\lambda > 0 \mid X_\lambda \text{ is injective}\}$$

is an open set. It is nonempty since $1 \in L$.

We now wish to show that L is closed. Suppose $\{\lambda_k\}$ is a sequence in L that converges to λ_0, and suppose X_{λ_0} is *not* an embedding; i. e., $X_{\lambda_0}(q_1) = X_{\lambda_0}(q_2)$ for some points $q_1, q_2 \in M, q_1 \neq q_2$. Because X_{λ_k} converges, uniformly on compact subsets of M, to X_{λ_0} it follows that some neighborhood $\mathcal{O}_1$ of q_1 is mapped by X_{λ_0} to be on one side of the image of some neighborhood $\mathcal{O}_2$ of q_2. By the Maximum Principle, stated in Section 2.3.3, $X_{\lambda_0}(\mathcal{O}_1) = X_{\lambda_0}(\mathcal{O}_2)$ and by analyticity, $X_{\lambda_0}{:}\, M \to \mathbf{R}^3$ is a finite covering map, whose image $\mathcal{S}{:}= X_{\lambda_0}(M)$ is a complete, embedded minimal surface of finite total curvature in $\mathbf{R}^3$. We will show that X_{λ_0} is in fact one-to-one.

Given two disctinct punctures p_i, p_j, $i \neq j$, the quantity

$$\mathrm{Re}\int_{p_i}^{p_j} dh_\lambda = \mathrm{Re}\int_{p_i}^{p_j} dh$$

is a measure of the vertical distance between the two ends. Notice that it is independent of λ. Since X is an embedding, the absolute value of this integral is infinite if the logarithmic growth rates at p_i and p_j are different. If the logarithmic growth rates are the same, it is still nonzero, a consequence of the Maximum Principle at Infinity quoted above. The mapping X_{λ_0} takes a sufficiently small neighborhood of any puncture p_j onto an end of $\mathcal{S}$, and each end of $\mathcal{S}$ is the image of some neighborhood of some puncture point. Hence the number of ends of $\mathcal{S}$ is not greater than the number of ends of $X_{\lambda_0}(M)$, i. e. not greater than r. If neighborhoods of two distinct punctures, p_i, p_j, $i \neq j$, are mapped to the same end of $\mathcal{S}$, then

$$\mathrm{Re}\int_{p_i}^{p_j} dh = \mathrm{Re}\int_{p_i}^{p_j} dh_{\lambda_0} = 0\,,$$

which is impossible. Thus the number of ends of $\mathcal{S}$ equals r and, furthermore, we can find sufficiently small neighborhoods of p_j, $1 \leq i \leq r$, so that each neighborhood is mapped by X_{λ_0} onto a different end of $\mathcal{S}$. This means that X_{λ_0} is one-to-one near the punctures and hence one-to-one everywhere; i. e. X_{λ_0} is injective. □

Proof of Proposition 3.4. Suppose q is a point where the Gauss map is vertical. Without loss of generality, we may assume that $g(q) = 0$. Choose conformal coordinates near q so that q corresponds to $z = 0$ and $g(z) = z^k$ for some $k > 0$ on $D_r(0) := \{|z| < r\}$. Because we are assuming regularity, dh must have a zero of order k at 0, i. e.,

$$g(z) = z^k \qquad dh = z^k(a + zf(z))dz\,,$$

where a is some nonzero complex constant and $f(z)$ is holomorphic on $D_r(0)$. For each $\lambda > 0$, the change of variables $z = \lambda^{-\frac{1}{k}}\zeta$ from $D_{\lambda^{\frac{1}{k}}r}(0)$ to $D_r(0)$ allows us to express g_λ and dh_λ as follows:

$$g_\lambda(z(\zeta)) = \zeta^k \qquad dh_\lambda = dh = \lambda^{-(1+\frac{1}{k})}\zeta^k(a + \lambda^{-\frac{1}{k}}\zeta f(\lambda^{-\frac{1}{k}}\zeta))d\zeta \tag{3.8}$$

on $D_{\lambda^{\frac{1}{k}}r}(0)$.

Rescale dh_λ by a factor of $\lambda^{(1+\frac{1}{k})}$ so that $dh_\lambda = \zeta^k(a + \lambda^{-\frac{1}{k}}\zeta f(\lambda^{-\frac{1}{k}}\zeta))d\zeta$. The rescaled data produce immersions, which we will also call X_λ, that converge uniformly, on compact subsets of $\mathbf{C} = \lim_{\lambda\to\infty} D_{\lambda^{\frac{1}{k}}r}(0)$, to the immersion produced by the Weierstrass data

$$\widehat{g} = \zeta^k \qquad \widehat{dh} = a\zeta^k d\zeta\,. \tag{3.9}$$

We will call this immersion $\widehat{X}: \mathbf{C} \to \mathbf{R}^3$. The limit surface is a rescaled version of the k-Enneper surface defined and discussed in Remark 2.3(ii). This surface is clearly not *embedded.* (See Figure 3.4.) Since rescaling does not create or destroy self-intersections, it follows that X_λ is not injective for λ sufficiently large. This completes the proof of the first statement of the proposition.

We now prove the second statement. Because a planar end has zero flux, X_λ is well-defined for all $\lambda > 0$, according to Proposition 3.2. Because it has finite total curvature, the Weierstrass data extends to zero meromorphically.

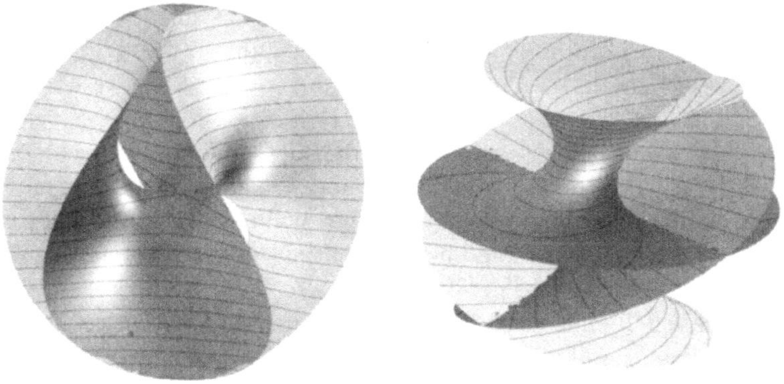

Fig. 3.4. The limit surfaces in the proof of Proposition 3.4 *Left:* The k-Enneper surface given by (3.9), with $k = 2$. *Right:* The limit surface given by (3.10), with $k = 2$. There is one horizontal flat end and one immersed end with winding number equal to three

Without loss of generality, we may assume that the limit normal at the end is $(0,0,-1)$, so $g(0) = 0$. Because the end is flat, the zero of g at $z = 0$ has order $k \geq 2$ and dh has a zero of order $k-2$ at $z = 0$. After a change of variables, if necessary, we may assume that

$$g(z) = z^k \qquad dh = z^k(az^{-2} + f(z))dz\,,$$

with $a \neq 0$ and $f(z)$ holomorphic on some disk $D_r(0) = \{|z| < r\}$. As in the proof of the first part of the proposition, we change variables by $z = \lambda^{-\frac{1}{k}}\zeta$ and rescale – this time by $\lambda^{1-\frac{1}{k}}$ – so that

$$g_\lambda(\zeta) = \zeta^k \qquad dh = \zeta^k(a\zeta^{-2} + \lambda^{-\frac{2}{k}} f(\lambda^{-\frac{1}{k}}\zeta))d\zeta$$

on $D'_{\lambda^{\frac{1}{k}}r}(0)$. The associated immersions converge, uniformly on compact subsets of $\mathbf{C} - \{0\} = \lim_{\lambda\to\infty} D'_{\lambda^{\frac{1}{k}}r}(0)$, to an immersion $\widehat{X}\colon \mathbf{C} - \{0\} \to \mathbf{R}^3$ given by the Weierstrass data

$$\widehat{g}(\zeta) = \zeta^k \qquad \widehat{dh} = a\zeta^{k-2}d\zeta\,. \tag{3.10}$$

This immersion is complete and has a flat end at 0. However, it has a nonflat end at infinity, and that end is not embedded. (See Figure 3.4.) It follows that X_λ is not one-to-one for $\lambda > 0$ sufficiently large. □

4. Construction of the Deformation Family with Three Ends

We wish to construct complete, embedded minimal surfaces of genus $k-1, k \geq 2$ with three catenoid ends and k vertical planes of symmetry intersecting in a common vertical line. They will be the surfaces $M_{k,x}$ of Theorem 3.3. These surfaces should be deformations of the examples M_k of Costa ($k=2$) and Hoffman-Meeks [35, 36] ($k \geq 3$), given in Theorem 3.2, which have two catenoid ends and a middle flat end. There will be, for each genus, a one parameter family of such deformations. We begin by locating the zeros and poles of $g = \sigma \circ N$, the stereographic projection of the Gauss map. We will then show that the conformal types of the examples are quite restricted.

Assume we have such a surface S for some fixed $k \geq 2$. Recall from Theorem 2.2 that the conformal type of S is that of a compact Riemann surface punctured in three points (one for each end). Each of these ends is fixed by each of the reflections. Thus the closure of the fixed-point set of each reflective isometry contains all the puncture points. We express this by saying that: "Each of the k symmetry planes passes through each of the three punctures."

Because the surface is properly embedded, it divides $\mathbf{R}^3$ into two regions. The vertical line of intersection of the symmetry planes may be taken to be the x_3-axis. The symmetry planes make equal angles of π/k, otherwise the surface would have more symmetry planes by Schwarz reflection. The top and bottom ends must be catenoid-type. (This follows from Proposition 2.5. See also the discussion after Theorem 2.4.) Since these ends have the same limit-normals (because there are an odd number of ends and their normals must alternate between $(0,0,+1)$ and $(0,0,-1)$), it follows that, for $x_3 > 0$ sufficiently large, the x_3-axis is in one component of $\mathbf{R}^3 - S$, and in the other component for $x_3 < 0$, sufficiently negative. Thus the x_3-axis passes through the surface an odd number of times.

We will show that it passes through *exactly once*. The reflections in vertical planes generate a cyclic rotation group of symmetries, of order k, about the x_3-axis. Because S is embedded, at any point p of intersection of S and the x_3-axis, S must have a vertical normal vector. The order of the zero or pole of g at p is one less than the number of curves in $S \cap T_pM$. If S has a rotational symmetry of order k, this means that the zero or pole has order at least $k-1$. On the other hand, it follows from Proposition 2.2 or Proposition 2.4 that

$$\operatorname{degree} g = \operatorname{genus} S + 2 = k + 1\,. \tag{4.1}$$

If there is more than one axis point, then the top-most and the bottom-most points have the same orientation as the catenoid ends. Orient S so that $g \circ N$ has zeros at these points. Then the number of zeros of g is at least $2(k-1)+2$, which is strictly greater than $k+1$ for $k \geq 2$. This contradicts (4.1). Hence there is a unique axial point where g has a zero of order at least $k-1$. But since g has a zero at each of the extreme catenoid ends, this makes $k+1$ zeros. We can conclude from (4.1) that the axial point is a zero of g of order exactly $k-1$, and there are no other zeros.

We can now locate the poles of g. If the middle end is flat, then at the level of the end, say $x_3 = c$, the plane $\{x_3 = c\}$ must intersect S in $2kj$ divergent curves, $j \geq 1$. Using this, we can easily show that g has a pole of order $kj+1$. By (4.1) we can conclude that $j = 1$ and that there are no other poles.

On the other hand, if the middle end is of catenoid-type, the Gauss map has a simple pole there, leaving k more poles to be found at finite points of S. The planes of symmetry divide S into $2k$ symmetric pieces. If a pole of the Gauss map is found in the interior of one of these regions, there must be at least $2k$ of them. This is a clear contradiction. The sole remaining possibility is to have the poles occurring in the symmetry planes, one on every other of the $2k$ halfplanes.

We will now make a sketch of the portion of S in a wedge between successive halfplanes. Each symmetry line from the saddle must diverge to an end with the same normal as at the saddle, i. e. the top or the bottom end. Because the order of the Gauss map at the saddle is $k+1$, adjacent curvature lines must diverge to *different* ends.

The picture suggests that this piece of surface is simply connected. Clearly this surface piece cannot have positive genus, else S would have genus in excess of $2k$, a contradiction.

To show it is simply connected, we will compute its Euler characteristic. By (4.1), its total curvature is $-4\pi(k+1)/2k$. By taking a sufficiently large piece of the surface we see that each end contributes $\pi + \pi/k$ to the total curvature of the boundary. The first term comes from two right angles at vertices, while the second term comes from the fact that the surface is asymptotic to a catenoid (or plane) and we are seeing a π/k-arc of a circle, which has essentially no normal curvature. All the other boundaries are geodesics. Adding in a contribution of $\pi - \pi/k$ at the single axis point, we have from the Gauss-Bonnet formula

$$\begin{aligned} 2\pi\chi &= \int K\,dA + \int k_g + \sum_1^4 \text{vertex angles} \\ &= \frac{-4\pi(k+1)}{2k} + 0 + 3\left(\pi + \frac{\pi}{k}\right) + \left(\pi - \frac{\pi}{k}\right) \\ &= 2\pi\,. \end{aligned}$$

Thus the Euler characteristic of the piece of S in the wedge is 1 and it is simply-connected. We will now show that the sketch is qualitatively correct. There is but one component of the boundary and if we start at the top catenoid end, (considered now in the compactification as a vertex), neither edge emanating from this vertex can go to the bottom catenoid end. This is because the surface is embedded and the top and bottom ends have the same orientation. Hence one edge runs from the top end to the middle end, while the other runs to the

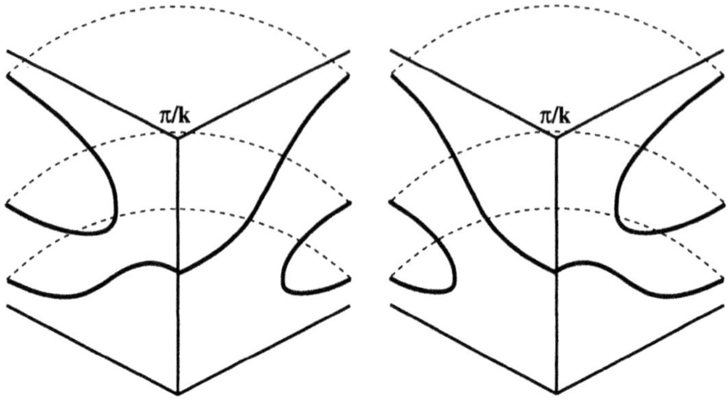

Fig. 4.1. Successive wedges of S

axis point. After a Euclidean motion, we may assume that if the middle end is a catenoid end, then its logarthmic growth is positive, i. e. it goes up. Thus the two edges emanating from the top catenoid end must look qualitatively like Figure 4.1.

4.1 Hidden Conformal Symmetries

We wish to determine the underlying conformal structure of the surfaces described above. In the previous section, we established that the compactification of the example of genus $k-1$, $k \geq 2$, assumed to have k vertical planes of symmetry, can be decomposed into $2k$ geodesic 4-gons, all of whose vertex angles equal π/k. All the 4-gons have the same vertices. The Riemann surface must be chosen to have conformal involutions corresponding to the reflections. If we choose a metric of constant curvature, these conformal involutions are isometries. In the *constant curvature model* our surface is made up of congruent geodesic 4-gons with all angles equal to π/k (rectangles in the case $k = 2$, hyperbolic 4-gons when $k > 2$) and with common vertices. For such 4-gons we need to prove a standard fact from hyperbolic geometry. This will allow us to determine the Riemann surface structure explicitly.

In the case $k = 2$ of genus one, a Euclidean rectangle has two reflective symmetry lines and, conformally, there is a one-parameter family of rectangles. For the case $k > 2$ we will need the same information from hyperbolic geometry about *α-quadrangles*. By an α-quadrangle we mean a quadrilateral in the hyperbolic plane, all of whose interior angles are equal to α.

Proposition 4.1. *For every $\alpha \in (0, \frac{\pi}{2})$ there exists a 1-parameter of α-quadrangles. Every α-quadrangle has all the symmetries of a Euclidean rectangle and, for each α, there is a unique α-square, i. e. an α-quadrangle with all the symmetries of a square. (In particular its sides all have the same length.)*

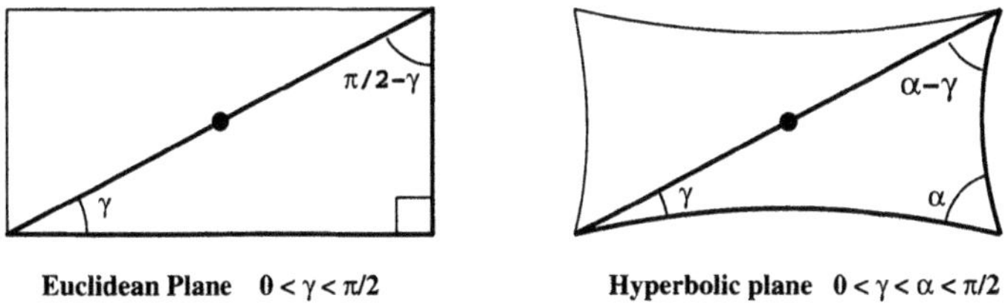

Fig. 4.2. Euclidean rectangles and hyperbolic α-quadrangles

Proposition 4.1 implies that the underlying Riemann surfaces of the minimal surfaces of interest to us have conformal symmetries that are not given by isometries of the minimal surfaces. This is important enough to prove.

Proof of Proposition 4.1. In the Euclidean plane we can parametrize the conformal family of rectangles by the angle $\gamma \in (0, \frac{\pi}{2})$ between the diagonal and an edge. For a given γ, take a triangle with angles $\gamma, \frac{\pi}{2}$ and $\frac{\pi}{2} - \gamma$, and rotate it about the midpoint of the edge opposite the right angle. The two triangles together give the rectangle associated to γ. In the hyperbolic plane: For any angles α, β, γ, the condition $\alpha + \beta + \gamma < \pi$ is necessary and sufficient for the existence of a triangle with these three angles, and, such a triangle is unique up to hyperbolic isometries. Given $\alpha \in (0, \frac{\pi}{2})$, for any $\gamma \in (0, \alpha)$, let $\beta := \alpha - \gamma$ and consider the α, β, γ-triangle. Rotate this triangle about the midpoint of the edge opposite α; the two triangles together form an α-quadrangle. For each fixed α, different values of $\gamma \in (0, \alpha)$ give conformally different α-quadrangles because the α, β, γ-triangles are different.

Next, we wish to show that any α-quadrangle has the symmetries of a rectangle. Take any edge, e_1, of the α-quadrangle, and let σ be the geodesic that is the perpendicular bisector of e_1. Reflection in σ maps e_1 into itself and interchanges the geodesics that extend the edges, e_0, e_2, adjacent to e_1. If the edge e_3 opposite e_1 is symmetric with respect to σ, we are done. If not, e_3 and its σ-reflection $\widetilde{e}_3$ intersect σ and meet the geodesic extensions of e_0 and e_2 at an angle equal to α. This produces two triangles with a common vertex at $e_3 \cap \widetilde{e}_3 \in \sigma$ and the same angle $\phi > 0$ there, whose other two angles are α and $\pi - \alpha$, a contradiction since the angle sum is $\pi + \phi > \pi$. Hence $\widetilde{e}_3 = e_3$ and the α-quadrangle has the required symmetry.

We wish to show that every α-quadrangle lies in the family constructed above. (Note that our proof that an α-quadrangle has reflectional symmetries did *not* assume that the quadrangle was in this family.) Given an α-quadrangle, the two symmetry geodesics cross at a point, which we will refer to as the *center*. Observe that 180° rotation about the center is a symmetry of the α-quadrangle; it is the composition of the two reflective symmetries. The geodesic join from any vertex to the center extends by this rotation to a diagonal of the α-quadrangle with the center as its midpoint. The diagonal divides the α-quadrangle into two $\alpha, \gamma, \alpha - \gamma$-triangles for some $\gamma \in (0, \alpha)$. Hence any α-quadrilateral lies in the family we constructed.

If our α-quadrangle was constructed as above with $\gamma = \alpha/2$, note that the diagonal produced from the edge of the $\alpha, \alpha/2, \alpha/2$ triangle (used to construct it) bisects the angles at its endpoints. It follows that reflection in this diagonal is a symmetry of this special α-quadrangle. This is the *α-square.* □

In the next section, we will build Riemann surfaces of genus $k-1$ from $2k$ congruent π/k-quadrangles. Like rectangular tori, they form noncompact 1-parameter families.

4.2 The Birdcage Model

From the previous two sections we can conclude that the conformal structure of our minimal surface of genus $k-1$ is given by $8k$ quadrilaterals in hyperbolic (Euclidean) space when $k > 2$ $(k = 2)$, with three right angles, one angle of π/k, and edges appropriately identified. In this section, which strictly speaking is not necessary for the next sections, we introduce an embedded surface in 3-space that realizes the conformal symmetries of our minimal surfaces in a geometrically evident manner. It has the same conformal type and will make the assembly of the surface out of quadrilaterals very clear.

Start with k symmetrically placed half-meridians on the unit sphere, $k \geq 2$. Let $S = S_k$ be a tubular neighborhood of these curves in $\mathbf{R}^3$. The thickness, $\varepsilon > 0$, of the tubes will correspond to a conformal parameter. The k symmetry planes of the meridians and the equatorial plane slice S into $4k$ geodesic 4-gons. Each 4-gon has two vertices on the polar axis with angle π/k and two vertices on the equator with angle $\pi/2$. For each quadrilateral we have from Gauss-Bonnet that

$$\int K dA = 2\pi - \left[2\left(\frac{\pi}{2}\right) + \frac{2\pi(k-1)}{k}\right],$$

or

$$\int K dA = \pi - \frac{2\pi(k-1)}{k} = -\pi + \frac{2\pi}{k}, \tag{4.2}$$

which implies that the total curvature of S is $-(4k-8)\pi = 2\pi(2-2(k-1))$. This verifies that the genus of S is $k-1$, which is evident from its construction. Note that the reflectional symmetries in the meridian planes generate a k-fold rotational symmetry, and this symmetry fixes the four polar points.

The birdcage surface has the additional isometry of reflection in the equatorial plane. There is another conformal automorphism, which amounts to inversion through the sphere of radius $\sqrt{1-\varepsilon^2}$. These two additional automorphisms restricted to the geodesic 4-gons produce the automorphisms corresponding to the symmetries of a Euclidean rectangle. In general, the composition of the equatorial reflection with the other conformal automorphism is an orientation-*preserving* conformal involution, the *hyperelliptic involution.* The quotient of the birdcage by this involution is a sphere; in the next section

Fig. 4.3. The birdcage model, $k = 3$, and the fundamental 4-gon

the quotient map is constructed by means of the Riemann Mapping Theorem (using a choice of identification of S^2 with $C \cup \infty$), and called the function u. When the genus is greater than 1, the branch points of u are the so-called *Weierstrass points* of the surface.

Remark 4.1. *The most symmetric of these Riemann surfaces are also the conformal models for one series of Lawson's minimal surfaces in* S^3 ([47]). *They can be found by extending Plateau solutions, since the symmetry-diagonals of the quadrilaterals, the ones from the* π/k*-angles, are great-circle arcs of length* $\pi/2$ *in* S^3.

The birdcage model serves as a convenient visualization of the decomposition of the Riemann surface into $8k$ 4-gons. We will now work directly with the hyperbolic (or Euclidean when $k = 2$) 4-gons having three right angles and one angle of π/k. Note that each of the 4-gons with all angles equal to π/k, decomposes into four 4-gons, each having three right angles and one angle π/k at one of the polar points.

Fig. 4.4. The birdcage decomposition The decomposition of the birdcage surface into hyperbolic polygons. Illustration for the case $k = 3$.

Upper left: The birdcage with one basic 4-gon darkened. This 4-gon has three right angles and one angle of π/k at an axis point.
Upper right: Four basic 4-gons darkened. Each one has an angle π/k at a different axis point. Taken together, they form a quadrilateral all of whose angles equal π/k.
Lower left: The birdcage decomposed into $2k$ π/k quadrilaterals.
Lower right: The birdcage decomposed into 4 right $2k$-gons, each centered at an axis point

In the next section we construct functions on the surface using this decomposition and the Riemann Mapping Theorem. This will allow us to make the transition from this hyperbolic picture to an algebraic description of our surfaces.

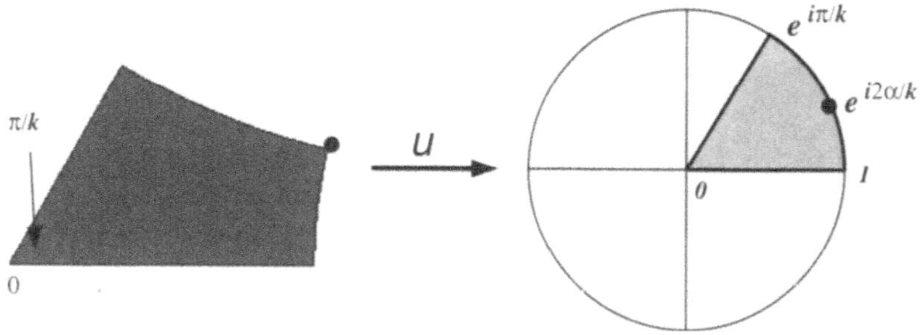

Fig. 4.5. The conformal mapping used to define u. The illustration here is for $k = 3$. The dotted vertex is the branch point of u. Its image is determined by the mapping, and is not assigned in advance

4.3 Meromorphic Functions Constructed by Conformal Mappings

Notice that each axis point is the center of a $2k$-gon, all of whose angles are right angles. These $2k$-gons are made up of $2k$ of the 4-gons with three right angles, the vertex with angle π/k lying at the polar point. Consider one of these small 4-gons.

We map the quadrilateral into the sector $0 \leq \theta \leq \pi/k$ on the unit disk considered as the closed hemisphere of S^2 by taking the central vertex with angle π/k into 0, its adjacent vertices to 1 and $e^{i\pi/k}$. This determines the image of the opposite vertex, which is a point on the unit circle. We will denote its value by $e^{2i\alpha/k}$, $0 < \alpha < \pi/2$. We will call this map u and extend it to S by Schwarz reflection. Note that at the midpoint of each of the four $2k$-gons, u has a simple pole or a simple zero, and u maps each of the four $2k$-gons onto a hemisphere. Thus, u is a degree-2 meromorphic function and we may think of S as a branched covering of S^2 with $2k$ simple branch points and branch values $e^{i(2\alpha+j\pi)/k}$.

In particular, S is *hyperelliptic*. In the case $k = 2$, u is clearly a degree-2 elliptic function. (In this case it follows directly from the construction of S that S is a rectangular torus.)

We observe that the symmetric construction of u allows us to choose α such that $\pi/4 < \alpha \leq \pi/2$. Given a choice of α between 0 and $\pi/4$, simply choose as fundamental domain an adjacent quadrilateral and adjust u by multiplication by $e^{\pi i/k}$.

4.3.1 The Function T and Its Relationship to u

Fix the same quadrilateral on which u is defined as in the previous section. We map this domain onto a quarter-disk, specifying that the vertex opposite the π/k-vertex goes to the origin and the other two $\pi/2$-vertices are mapped

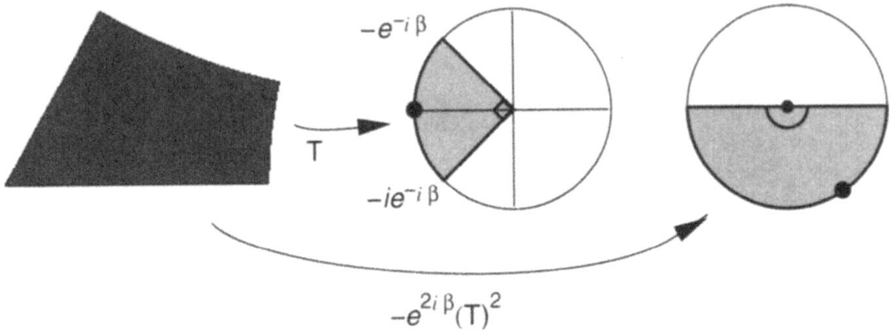

Fig. 4.6

to the unit circle. This determines the mapping up to a rotation in **C**, i. e. up to a unitary multiplicative factor. We rotate so that the π/k-vertex is mapped to -1. We call this mapping T and we label (as in the diagram above) the other two vertices $-e^{-i\beta}$ and $-ie^{-i\beta}$, where the first one is in the upper half-plane.

We will show that, in fact, $\beta = \alpha$, where α is the angle defined by the mapping u. To do this we will construct, from T and u, the same conformal map to the lower semi-disk. First of all $(T)^2$ maps our quadrilateral to a semi-disk, which we rotate by $-e^{2i\beta}$ so that $-e^{-i\beta} \to -1, -ie^{-i\beta} \to +1$, of course $0 \to 0$, and $-1 \to -e^{2i\beta}$. In particular, $-e^{2i\beta}(T)^2$ takes the π/k-vertex to $-e^{+2i\beta}$. Turning now to u, note that u^k maps the quadrangle to the upper semi-disk. Composition of u^k with the linear fractional transformation $\nu(\zeta) = \frac{\zeta - e^{2i\alpha}}{1-\zeta e^{2i\alpha}}$ not only takes the upper semi-disk to the lower semi-disk, but also has the property that $\nu(u^k) = -e^{2i\beta}(T)^2$ on the three vertices opposite the π/k-vertex. Thus $\nu(u^k) = -e^{2i\beta}(T)^2$ everywhere, in particular at the π/k-vertex itself; i. e.

$$-e^{+2i\alpha} = \nu(0) = -e^{2i\beta}(-1)^2 = -e^{2i\beta}\,.$$

Since α and β were both chosen in the range $(0, \frac{\pi}{2})$, $\alpha = \beta$ and

$$\frac{u^k - e^{2i\alpha}}{1 - u^k e^{2i\alpha}} = -e^{2i\alpha}(T)^2\,.$$

The above equation actually should be considered as the relation between the coordinate maps T and u, and is in fact an equation for our surface.

4.4 The Function z and the Equation for the Riemann Surface in Terms of z and u

We prefer to modify T by a Möbius transformation so that: its zeros and poles are more closely related to what we will need in the Weierstrass representation; and it is related to u by an equation with real coefficients. Accordingly let $M(\zeta) = \frac{i(\zeta+1)}{1-\zeta}$ and define

$$z = M \circ T .$$

Note that $T = \frac{z-i}{z+i}$ and that M takes -1 to 0, 0 to i and 1 to ∞. Since T takes on the values $-1, 0, -ie^{-i\alpha}, -e^{-i\alpha}$ at the vertices of the fundamental 4-gon, we know that z takes on the values 0 and i at two of the vertices. A calculation shows that z takes on the value $\cot(\frac{\pi}{4} + \frac{\alpha}{2})$ at the vertex where T takes on the value $-ie^{-i\alpha}$.[6]

We note that z extends to the $2k$-gon with a zero of order k at the midpoint, and covers the hemisphere $k/2$ times. Extending to the rest of S by Schwarz reflection produces a function of degree k whose zeros and poles (all of order k) occur at the special polar points. We chose the branch values of u in order to relate u to z in a particularly simple way. We actually have this relationship already because we know that

$$\frac{u^k - e^{2i\alpha}}{1 - u^k e^{2i\alpha}} = -e^{2i\alpha}T^2 = -e^{2i\alpha}\left(\frac{z-i}{z+i}\right)^2 .$$

Manipulation of this expression will yield (4.3) below. We prefer to derive (4.3) by matching the zeros and poles of z and u. Consider $Z := -z + z^{-1}$. From the identity

$$\cot\alpha - \tan\alpha = 2\cot 2\alpha$$

and the values of z and u given in the previous diagrams, we can easily determine the values of the functions $z^{-1} - z$, and u^{-k} at vertices. Notice that the poles and zeros of $Z + \cot\alpha - \tan\alpha$ match those of u^{-k} and $(\cot\alpha + \tan\alpha)u^{-k}$ equals Z at the points where $u^{-k} = \pm 1$. Hence

$$z^{-1} - z + \cot\alpha - \tan\alpha = (\cot\alpha + \tan\alpha)u^{-k} .$$

---

[6] Let $M(\zeta) = \dfrac{i(\zeta+1)}{1-\zeta}$. Since $M(e^{i\theta}) = i \cdot \frac{e^{i\theta}+1}{1-e^{i\theta}} = -\frac{e^{i\theta/2}+e^{-i\theta/2}}{(e^{i\theta/2}-e^{-i\theta/2})/i} = -\frac{\cos\theta/2}{\sin\theta/2} = -\cot\theta/2$, it follows that

$$M(-e^{-i\alpha}) = M(e^{i(\pi-\alpha)}) = -\cot\left(\frac{\pi}{2} - \frac{\alpha}{2}\right) = -\tan\frac{\alpha}{2} ,$$
$$M\left(e^{i(-\frac{\pi}{2}-\alpha)}\right) = -\cot\left(-\frac{\pi}{4} - \frac{\alpha}{2}\right) = \cot\left(\frac{\pi}{4} + \frac{\alpha}{2}\right) .$$

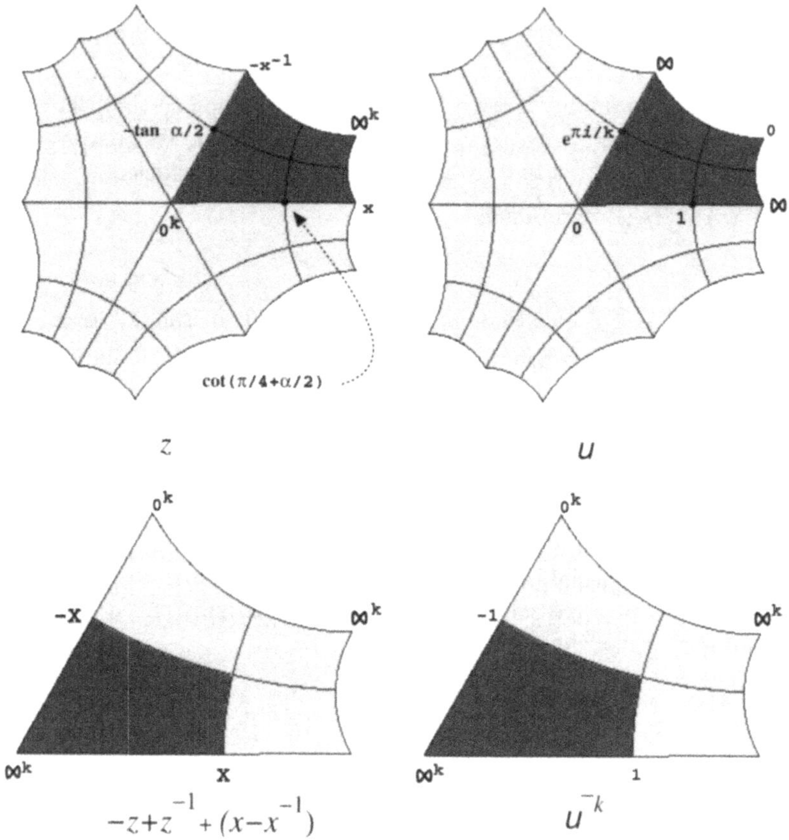

Fig. 4.7. The functions z and u

This produces an equation for the surface S, which we write in two equivalent ways:

$$
\begin{aligned}
&z^{-1} - z = (x + x^{-1})u^{-k} - (x - x^{-1}) \\
&z^{-1}(z - x)(z + x^{-1}) = -(x + x^{-1})u^{-k}, \quad \text{where } x = \cot\alpha\,.
\end{aligned} \tag{4.3}
$$

Here, x may be considered a conformal parameter for our family of Riemann surfaces, $0 < x \leq 1$.

Differentiate this equation to get

$$(z + z^{-1})\frac{dz}{z} = k(x + x^{-1})u^{-k}\frac{du}{u}$$

We now observe that branch points of u occur at points where $z = \pm i$, as we know by construction. By construction or by calculation from Equation (4.3) we can conclude that the branch values of u are of the form $e^{\pm 2i\alpha/k}\omega$, where

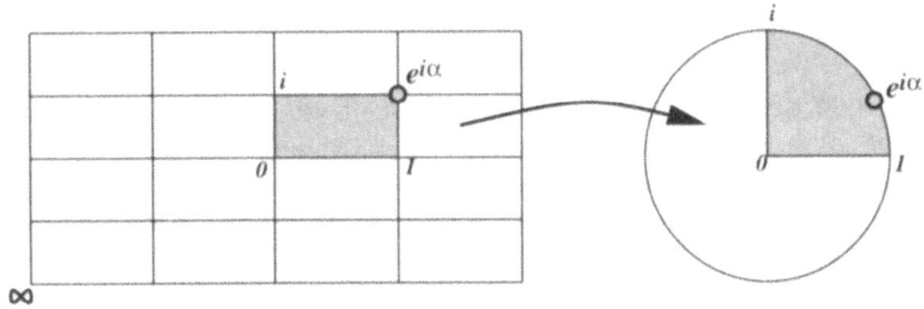

Fig. 4.8. When $k = 2$, the Riemann surface is a rectangular torus. The fundamental domain constitutes $1/16th$ of the surface. Values of u are indicated on the surface. The mapping is specified by fixing the values – $0, 1$, and i at three vertices. It is branched at the fourth vertex; – its value is determined by the mapping, which is in turn determined by the proportions of the rectangle

ω is a kth root of unity. At points where $z \neq 0, \infty$, poles of u are branch points of z, with branch values $x = \cot\alpha$ or $-x^{-1} = -\tan\alpha$.

Remark 4.2. *In Section 3, we wrote the above equations in a different and, for historical reasons, more familiar form. Multiply both sides of* (4.3) *by* z^k *and define* $w := z/u$. *Then we may write* (4.3) *as*

$$\left(\frac{z}{u}\right)^k = w^k = -X^{-1}z^{k-1}(z-x)\,(z+x^{-1}) \qquad X = (x+x^{-1})\,, \tag{4.4}$$

and $x, 0 < x \leq 1$, *parametrizes our family of Riemann surfaces. In case* $k = 2$, *we have*

$$w^2 = -X^{-1}z(z-x)\,(z+x^{-1})\,, \qquad x \geq 1 \tag{4.5}$$

and we recognize the equation of a rectangular torus. The function z *is thus seen to be a geometrically normalized form of the Weierstrass* $\wp$*-function:* $z = c_1(\wp - c_2)$, *where* c_2 *is such that* z *has a double zero at the center of one of the 4 gons.*

We also note that on the Riemann surfaces (4.4) *one can define data for Chen-Gackstatter-type surfaces (see Section 2.2) with higher symmetry using*

$$g = \rho\cdot\left(\frac{w}{z}\right)^k = \rho\cdot u^{-k},$$
$$dh = i(z+z^{-1}-2)\frac{dz}{z}\,.$$

(See Karcher [43]*.) This construction has been extended by Thayer* [71].

4.5 The Weierstrass Data

We now determine the Weierstrass data for our minimal surface of genus $k-1, k \geq 2$, with three vertical catenoid ends and k symmetrically placed vertical planes of reflective symmetry. We know that the underlying Riemann surface will be S, as presented in Equation (4.3) or (4.4). Note that we have a 1-parameter family depending on $\pi/4 \leq \alpha < \pi/2$ (or equivalently $x^{-1} = \tan\alpha \geq 1$).

In terms of our decomposition of S, we will look at the region between two adjacent symmetry halfplanes. This corresponds to four basic quadrilaterals joined together to form a π/k-quadrangle with vertices consisting of the polar points of the birdcage where the vertex angle is π/k. The top and bottom ends are located at points where $u = \infty$. The middle end is placed where $z = \infty$, which is where $u = 0$.

We will specify the Gauss map here by noticing that the values of g must be 0 or ∞ at the ends. (See Figure 4.9 and Table 4.1) Since there is no branching of the Gauss map at a catenoid end, these are simple zeros and poles. By contrast, at the saddle we must have at least k curves of intersection with the horizontal tangent plane, which implies that the branching of g at the saddle has order at least $k-1$. Orient the surface so that $g = 0$ at the saddle. We know that the degree of g must be equal to $k+1$, and since $g = 0$ at two other catenoid ends, we can conclude that g has a zero of order $k-1$ at the saddle. Referring to Table 4.1, we note that

$$g_1 := \frac{u^{-1}z}{mz+1} \tag{4.6}$$

has the correct zero/pole structure at the saddle and the ends, while having a simple pole at an edge-point in between. Here, m is a free parameter, placing the finite vertical normal at the points where $z = -m^{-1}$. Placement of the vertical normal as it is in the picture requires us to choose $-\infty < (-m^{-1}) < -1/x$, or

$$0 < m < x = \cot\alpha\,. \tag{4.7}$$

Otherwise, the finite vertical normal would be positioned elsewhere. Fixing m determines the Gauss map up to a real multiplicative factor. (Multiplying by a unitary complex number produces a rotation of the surface.)

$$g = \rho g_1 \qquad \rho > 0\,. \tag{4.8}$$

We are now in a position to determine the complex differential dh of the height function. It must have a simple pole at each catenoid end, a simple zero at the (simple) finite pole of g and a zero of order $k-1$ at the saddle. Direct calculation of the zeros and poles of $\frac{dz}{z}$, u^k and $mz+1$ is straightforward. (See Figure 4.9.) We may conclude that

$$dh = cu^k(mz+1)\frac{dz}{z}\,, \tag{4.9}$$

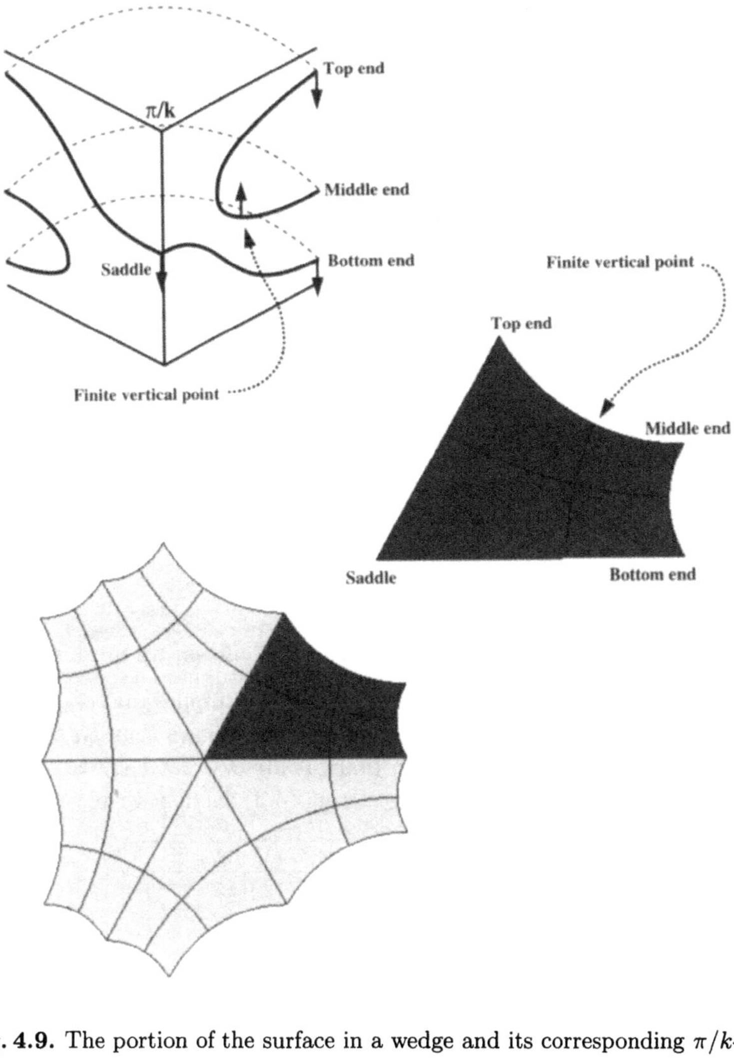

Fig. 4.9. The portion of the surface in a wedge and its corresponding π/k-gon

where c is a nonzero complex multiplicative constant. We wish to show that this constant must be chosen to be real. First of all we note that the curves on the boundary of the fundamental quadrilateral correspond to planar geodesics (in vertical planes) along which the Gauss map must take values in a meridian circle. Thus $\frac{dg}{g}$ is real on the tangent vectors of these curves. On the other hand, the quadratic differential associated with the second fundamental form of a minimal surface is given as in (2.12) by

$$\langle S(V)), V\rangle = \operatorname{Re}\left(\frac{dg(V)}{g} \cdot dh(V)\right). \tag{4.10}$$

Table 4.1

$(z,w) =$	$(0,0)$	$(x,0)$	$(-x^{-1},0)$	(∞,∞)	$z=-m^{-1}$
	Saddle	Bottom End	Top End	Middle End	Finite Vertical Points
g	0^{k-1}	0	0	∞	∞
u^{-1}	∞	0	0	∞	–
z	0^k	x	$-x^{-1}$	∞^k	$-m^{-1}$
$mz+1$	–	–	–	∞^k	0
dh	0^{k-1}	∞	∞	∞	0
u^k	0^k	∞^k	∞^k	0^k	–
dz/z	∞	0^{k-1}	0^{k-1}	∞	–
du/u	∞	∞	∞	∞	–
$\frac{mz+1}{z+z^{-1}}$	0^k	–	–	–	0

Since these planar geodesics are also principal curves, we must have $\frac{dg}{g}(V) \cdot dh(V)$ real along these edges. (See 2.14.) But this implies that dh is real along these edges. From our definitions of u and z, it follows that z and du/u are real along these curves. Hence from (4.9), it follows that c is real. We rescale so that this constant is equal to one. Using (4.3) we may write:

$$dh = u^k(mz+1)\frac{dz}{z} = \frac{-(x+x^{-1})(mz+1)dz}{(z-x)(z+x^{-1})}\,. \tag{4.11}$$

We will use both expressions for dh.

Summary. We summarize what we have achieved.

The Riemann surface and Weierstrass data are as in (4.4), (4.6)–(4.8), (4.11):

$(x+x^{-1})u^{-k} = z^{-1} - z + (x - x^{-1}) = -z^{-1}(z-x)(z+x^{-1})$, $x = \cot\alpha$, $\pi/4 \le \alpha < \pi/2$, minus the points $(z,u) = (x,\infty), (-x^{-1},\infty)(\infty,0)$. The end at $(\infty,0)$ is the middle end.

$$\begin{aligned} g &= \rho g_1 = \rho\frac{u^{-1}z}{mz+1} \qquad \rho>0, \quad 0 \le m < x\,; \\ dh &= u^k(mz+1)\frac{dz}{z} = \frac{-(x+x^{-1})(mz+1)dz}{(z-x)(z+x^{-1})} \end{aligned} \tag{4.12}$$

This data will produce in the Weierstrass Representation

$$X(p) = (x_1, x_2, x_3)(p) = \mathrm{Re}\int_{p_0}^{p} \Phi\,,$$

$$\Phi = \left(\frac{1}{2}(g^{-1} - g), \frac{i}{2}(g^{-1} + g), 1\right) dh\,,$$

a (possibly multivalued) minimal immersion into $\mathbf{R}^3$. From the development in Sections 4.1 to 4.5, it is clear that if the surfaces $M_{k,x}$ of Theorem 3.3 (also described at the beginning of Section 4) exist, they must have a Weierstrass Representation in the form above, for appropriate choices of constants k, x, m and ρ. All Weierstrass data of the above form produce the symmetry of the surfaces that we require. To see this, first note that the conformal automorphisms

$$f(z,u) := (\overline{z}, \overline{u})$$

$$\tau(z,u) := (z, \varepsilon u) \qquad \varepsilon = e^{-\frac{2\pi i}{k}}$$

fix $p_0 := (0,0)$ and the ends $(x,\infty), (-x^{-1},\infty)$ and $(\infty, 0)$. We choose to integrate from this p_0 in the expression for $X(p)$, and observe that we may write

$$x_1 - ix_2(p) = \int_{p_0}^{p} g^{-1}dh - \overline{\int_{p_0}^{p} gdh}$$

$$x_3(p) = \mathrm{Re}\int_{p_0}^{p} dh\,.$$

Since $f^*(dh) = \overline{dh}$ and $\tau^*(dh) = dh$, it is evident that $x_3(\tau(p)) = x_3(f(p)) = x_3(p)$. Because $f^*g = \overline{g}$ and $\tau^*g = \varepsilon g$, it follows that

$$x_1 - ix_2(f(p)) = \overline{x_1 - ix_2(p)} \qquad = x_1 + ix_2(p)$$

$$x_1 - ix_2(\tau^j(p)) = \overline{\varepsilon}^j(x_1 - ix_2(p)) = 1 \leq j < k\,.$$

Thus f induces reflection in the (x_1, x_3)-plane, while τ^j induces rotation by $2\pi j/k$ about the vertical axis. (We could have used the Schwarz Reflection Principle to prove that f induces a reflection.)

4.6 The Logarithmic Growth Rates

We proceed to determine the logarithmic growth rates of the catenoid ends. From Proposition 2.5(vi), we know that this growth rate is equal to minus the residue of dh at the end. At the bottom end, where $z = x$, and at the top end, where $z = -x^{-1}$, dz is regular, while at the middle end, where $z = \infty$, dz/z has a simple pole with residue -1. From (4.12) we can read off the residue of dh at $z = x$ and $z = -x^{-1}$:

$$\begin{aligned}\mathrm{Residue}(dh)_{z=x} &= \frac{-(x+x^{-1})(mx+1)}{x+x^{-1}} = -(mx+1)\\ \mathrm{Residue}(dh)_{z=-x^{-1}} &= \frac{-(x+x^{-1})(-mx^{-1}+1)}{(-x^{-1}-x)} = -\frac{m}{x}+1\,.\end{aligned} \tag{4.13}$$

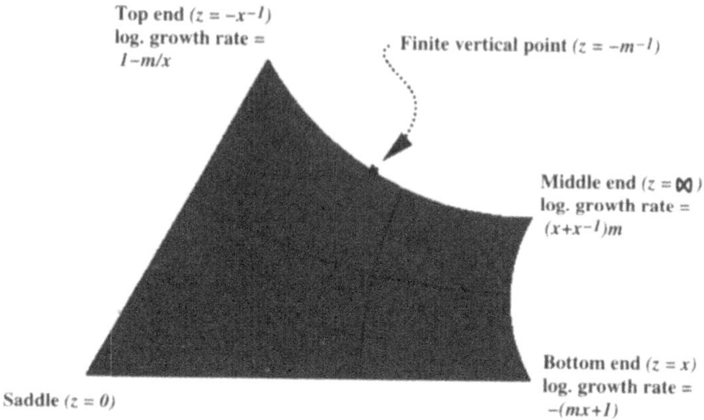

Fig. 4.10. Logarithmic growth rates at the ends

Because the residues must sum to zero (and dh has poles only at the three ends) we get for the middle end

$$\text{Residue}(dh)_{z=\infty} = m(x + x^{-1}) .$$

We could also compute this directly by rewriting the expression for dh in (4.12) in terms of z and expressing it in terms of z^{-1}.

$$dh = \frac{(x - x^{-1})(m + z^{-1})}{z^{-2} + (x - x^{-1})z^{-1} - 1} \frac{dz^{-1}}{z^{-1}} .$$

The logarithmic growth rates are the negatives of these residues: With $m \in [0, x)$ arbitrary, we have $-(mx+1) < 0 < m(x+x^{-1})$. However, the "middle" end can grow faster than the "top" end, in which case the surface would not be embedded. To avoid this we have the growth rate condition

$$m(x + x^{-1}) < -mx^{-1} + 1 . \tag{4.14}$$

4.7 The Period and Embeddedness Problems for the Surfaces $M_{k,x}$

We must show that the surfaces $M_{k,x}$ in Theorem 3.3 are in fact well defined by the mapping $X: \Sigma \to \mathbf{R}^3$ in the Weierstrass representation, (2.6) and (2.7) on the Riemann surface Σ given by (4.12). That is, according to (2.8), we must show that

$$\text{Period}_\alpha(\Phi) = \text{Re} \oint_\gamma \Phi = \text{Re} \oint_\gamma (\phi_1, \phi_2, \phi_3) = \vec{0} \tag{4.15}$$

for all closed curves γ on the punctured Riemann surface. These formulae are restated in the summary at the end of Section 4.5, where we proved that the immersions must have the required Euclidean symmetry. The built-in symmetry of the mapping is crucial. Consider a fundamental quadrilateral as in Figure 4.7. In Figure 4.11, we label the saddle by S and the top, middle and bottom ends by T, M and B, respectively. Opposite sides of the quadrilateral are mapped into parallel vertical planes by $X = \operatorname{Re}\int \Phi$. We require that these planes coincide. After a rotation, we can assume that the curve $X(\widehat{ST})$ lies in the $x_2 = 0$ plane. For $X(\widehat{BM})$ to lie in the same plane, we need only check that

$$\operatorname{Re}\oint_{\gamma_1} \phi_2 = 0\,, \tag{4.16}$$

where γ is the symmetry line of the quadrilateral between edges ST and BM.

The curves $X(\widehat{SB})$ and $X(\widehat{TM})$ lie in planes parallel to the plane $x_2 \cos\pi/k = x_1 \sin\pi/k$. They lie in the same plane if and only if

$$\operatorname{Re}\oint_{\gamma_2} (\phi_2 \cos\pi/k - \phi_1 \sin\pi/k) = 0\,. \tag{4.17}$$

Notice that the symmetry forces $\operatorname{Re}\int_\alpha \phi_3 = 0$ for every closed curve α on the punctured Riemann surface; we get no conditions from the third differential.

There is no geometric reason why the second period condition (4.17) should be more complicated than the first. In fact it is sufficient to look in detail at the first condition (4.16) and, at the end, handle the second condition by a parameter substitution. To see this, consider the following expressions for ϕ_1 and ϕ_2.

$$\begin{aligned} 2\rho\phi_1 &= (g_1^{-1} - \rho^2 g_1)dh = \left(u\left(\frac{1+mz}{z}\right)^2 - \rho^2 u^{-1}\right) u^k dz \\ 2\rho\phi_2 &= i(g_1^{-1} + \rho^2 g_1)dh = i\left(u\left(\frac{1+mz}{z}\right)^2 + \rho^2 u^{-1}\right) u^k dz\,. \end{aligned} \tag{4.18}$$

If we substitute, in the expression for $\phi_2, \widetilde{u} = e^{i\pi/k}u, \widetilde{z} = -z$ and $\widetilde{m} = -m$ and define $\widetilde{\phi}_2 = \phi_2(\widetilde{u}, \widetilde{z}, \widetilde{m}, \rho)$, we have $\widetilde{\phi}_2 = \phi_2 \cos\pi/k - \phi_1 \sin\pi/k$. Thus the second period condition (4.17) can be written in the form

$$\operatorname{Re}\oint_{\gamma_2} \widetilde{\phi}_2 = 0\,. \tag{4.17$'$}$$

Now note that $\widetilde{u}$ and $\widetilde{z}$ satisfy the same Riemann surface equation as u and z, provided we substitute $\widetilde{x} = 1/x$ for x:

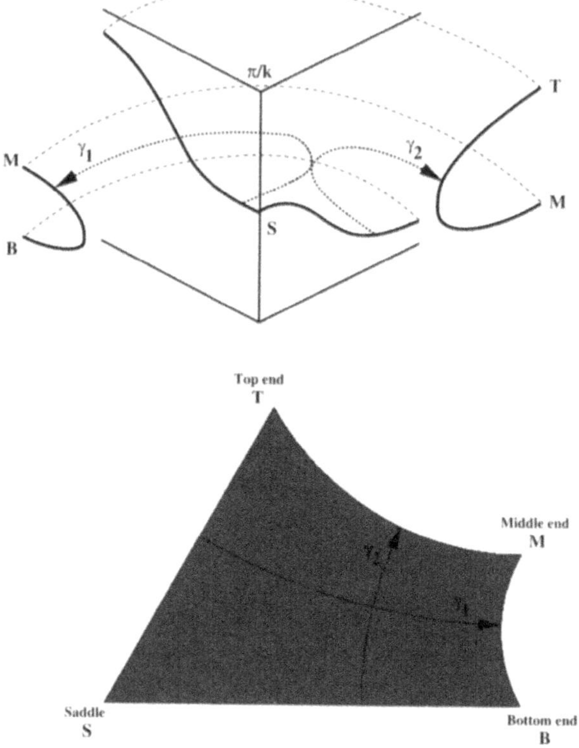

Fig. 4.11. The period problem

$$(\widetilde{x}+\widetilde{x}^{-1})\widetilde{u}^{-k} = \left(x+\frac{1}{x}\right)(-1)u^{-k} = -(z^{-1}-z)-(x-x^{-1})$$
$$= (\widetilde{z}^{-1}-\widetilde{z})+(\widetilde{x}-\widetilde{x}^{-1})\,.$$

Notice that $\widetilde{x} = x^{-1} = \tan\alpha = \cot(\widetilde{\alpha})$, where $\widetilde{\alpha} = \pi/2-\alpha$. Since we will explicitly express (4.16) in terms of m and α, we will be able to express (4.17) by the substitution $m \to \widetilde{m} = -m$ and $\alpha \to \widetilde{\alpha} = \pi/2-\alpha$. Using (4.18) we may express (4.16) and (4.17) in terms of the integrals

$$Q_j = Q_j(m,\alpha) := \int_{\gamma_j}\left(\frac{1+mz}{z}\right)^2 u^{k+1}dz,$$
$$C_j = C_j(\alpha) := \int_{\gamma_j} u^{k-1}dz\,,$$

$j = 1, 2$; namely,

$$Q_1(m,\alpha) = \rho^2 C_1(\alpha),$$
$$Q_2(\widetilde{m},\widetilde{\alpha}) = \rho^2 C_2(\widetilde{\alpha}) \qquad \widetilde{m} = -m\,, \quad \widetilde{\alpha} = \frac{\pi}{2} - \alpha\,. \tag{4.19}$$

For each α, we want to find (ρ, m) satisfying these conditions, which express the period problem. Note that Q_j is quadratic in m, while, C_j depends on α only. Each condition in (4.19) determines $\rho^2 = \rho^2(m,\alpha)$, and it is easy to check that $\rho^2 > 0$. The two conditions agree if and only if

$$\frac{Q_1(m,\alpha)}{Q_2(\widetilde{m},\widetilde{\alpha})} = \frac{C_1(\alpha)}{C_2(\widetilde{\alpha})} \quad \text{(Compatibility condition for } m = m(\alpha)). \tag{4.20}$$

Our strategy is to reduce this condition to a linear equation in $(m - m^{-1})$, whose obvious solvability shows that, for each α, we can find $m(\alpha), \rho(\alpha)$ solving the period problem (4.19). We then need to show that the logarithmic growth rates are correctly ordered. That is, the growth rate at the middle end (the one corresponding to the puncture $(z,u) = (\infty, 0)$) lies between the growth rate at the other two ends. From (4.13) and (4.14), this is equivalent to

$$2\tan\alpha + \cot\alpha = 2x^{-1} + x < m^{-1} \quad \text{(Growth rate condition for } m = m(\alpha)). \tag{4.21}$$

Proposition 4.2. *Fix $k \geq 2$. For every $\pi/4 \leq \alpha < \pi/2$, there exists a unique non-negative $m(\alpha) < \cot\alpha$ that satisfies the compatibility condition* (4.20). *Moreover $m(\alpha)$ satisfies the growth condition* (4.21) *and is a continuous function of α.*

We will give a complete proof of this Proposition in the case $k > 2$. For the torus case, $k = 2$, the arguments are a bit more difficult. (See [30].) As indicated above, existence of the surfaces $M_{k,x}$ in Theorem 3.3 follows from Proposition 4.2. When $\alpha = \frac{\pi}{4}, m(\alpha) = m(\frac{\pi}{4}) = 0$, and the middle end is flat; these are the surfaces $M_{k,1}$ of Theorem 3.3, which correspond to the surfaces M_k of Theorem 3.2. These surfaces M_k are embedded. Thus each $M_{k,x}$ lies in a continuous family of examples whose first member is embedded. The embeddedness of the surfaces in the $M_{k,x}$ is a consequence of the growth condition (4.21) and the following result from [30].

Proposition 4.3. *Suppose $h_t\colon N \to \mathbf{R}^3$, $0 \leq t \leq A$ is a continuous one-parameter family of complete minimal surfaces of finite total curvature. Suppose that for all t, the ends of $h_t(N) = N_t$ are all vertical, and that h_0 is an embedding but h_A is not. Let $T = \sup\{t | h_t$ is an embedding.$\}$ Then h_T is an embedding if $T < A$, and if one orders the ends by height, there are at least two ends with the same logarithmic growth.*

4.8 The Details of the Solution of the Period Problem, I. Simplification of the Integrals

Recall from (4.3) (or (4.12)) that the Riemann surface equation is

$$z^{-1} - z = (x + x^{-1})u^{-k} - (x - x^{-1}). \tag{4.22}$$

Differentiation gives us a relation between the differentials dz and du:

$$(z^{-1} + z)\frac{dz}{z} = k(x + x^{-1})u^{-k}\frac{du}{u}. \tag{4.23}$$

Two differential forms have the same periods, or are homologous, $w \sim \eta$, if they differ by the differential of a function: $w = \eta + df$. For example:

$$g_1 dh = u^{k-1}dz \sim -(k-1)zu^{k-2}du.$$

This gives us explicit integral expressions for C_1 and C_2 defined in the previous section.

Lemma 4.1

$$\begin{aligned} C_1 &= \operatorname{Re}\int_{\gamma_1} ig_1 dh \\ &= \frac{2}{k}(k-1)(x + x^{-1})\int_0^{\alpha}\sqrt{\cos^2\phi - \cos^2\alpha}\,\cos\left(\left(1 - \frac{2}{k}\right)\phi\right)d\phi \\ C_2 &= \frac{2}{k}(k-1)(x + x^{-1})\int_0^{\widetilde{\alpha}}\sqrt{\cos^2\phi - \cos^2\widetilde{\alpha}}\,\cos\left(\left(1 - \frac{2}{k}\right)\phi\right)d\phi. \end{aligned}$$

Proof. The path γ_1 is symmetric with respect to a branch point of u; therefore u has the same values at opposite points and du changes sign at the branch point. Moreover z has the value i at this point and is symmetric under 180° rotation, i. e. its values at opposite points are $z, -1/z$. Hence

$$\int_{\gamma_1} 2g_1 dh = -(k-1)\int_{\gamma_1}(z + z^{-1})u^{k-1}\frac{du}{u}.$$

We can eliminate $z + z^{-1}$ as follows:

$$\begin{aligned} \left(\frac{z + 1/z}{x + 1/x}\right)^2 &= \left(\frac{1/z - z}{x + 1/x}\right)^2 + \frac{4}{(x + 1/x)^2} \\ &= \left(u^{-k} - \frac{x - 1/x}{x + 1/x}\right)^2 + \frac{4}{(x + x^{-1})^2}. \end{aligned}$$

Recall $x = \cot\alpha$. Hence $(x - x^{-1})(x + x^{-1})^{-1} = \cos 2\alpha$; with this

$$\left(\frac{z + 1/z}{x + 1/x}\right)^2 = u^{-k}\cdot(u^k + u^{-k} - 2\cos 2\alpha). \tag{4.24}$$

Furthermore, along the path γ_1 we can take u as coordinate function on the Riemann surface; since u is unitary along γ_1 we have

$$u = e^{i\phi}, u^k + u^{-k} = 2\cos k\phi, \quad 0 \le \phi \le \frac{2\alpha}{k}$$

$$du/u = id\phi\,.$$

We insert this and (4.24) in the last integral and get

$$\operatorname{Re}\int_{\gamma_1} ig_1 dh = (k-1)(x+x^{-1})\int_0^{2\alpha/k} \sqrt{2\cos k\phi - 2\cos 2\alpha}\,\cos\left(\left(\frac{k}{2}-1\right)\phi\right)d\phi\,.$$

We substitute $t = k\phi$ and use $\cos 2\alpha = 2\cos^2\alpha - 1$ to derive the expression for C_1 in the lemma. Using the symmetric treatment discussed above, we get C_2 if we replace α by $\widetilde{\alpha} := \frac{\pi}{2} - \alpha$. □

Now we turn to $Q_1(m)$. We have at first

$$\frac{1}{g_1}dh = \left(m^2 + \frac{1}{z^2} + \frac{2m}{z}\right)u^{k+1}dz\,,$$

so that $\int_{\gamma_1}\frac{1}{g_1}dh$ is given in terms of three different period integrals on the Riemann surface. We have to reduce these integrals to the two in the previous lemma and to two others before we can discuss them. First, multiply the differential relation (4.23) by $u^{k+\ell}$ to get

$$u^{k+\ell}(1+z^{-2})dz = k(x+x^{-1})u^\ell \cdot \frac{du}{u} \underset{[\ell\neq 0]}{\sim} 0\,.$$

In particular,

$$\frac{1}{z^2}u^{k+1}dz \sim -u^{k+1}dz\,.$$

Hence we achieved a first reduction to two integrals per path:

$$\frac{1}{g_1}dh \sim \left(m^2 - 1 + \frac{2m}{z}\right)u^{k+1}dz\,.$$

We deal with $u^{k+1}dz \sim -(k+1)u^{k+1}zdu/u$ as before by using the symmetry along γ_1:

$$\int_{\gamma_1} 2u^{k+1}dz = -\int_{\gamma_1}(k+1)u^{k+1}\left(z+\frac{1}{z}\right)\frac{du}{u}\,.$$

Again $(z+1/z)$ is eliminated using (4.24) to get almost the same integral as in the lemma:

$$2\operatorname{Re}\int_{\gamma_1} iu^{k+1}dz = \frac{4(k+1)}{k}(x+x^{-1})\int_0^\alpha \sqrt{\cos^2\phi - \cos^2\alpha}\cdot\cos\left(\left(1+\frac{2}{k}\right)\phi\right)d\phi\,. \tag{4.25}$$

The remaining integral $\int u^{k+1}dz/z$ can be rewritten as a linear combination of the integrals we already have. We multiply the differential relation (4.23) to get

$$\left(z+\frac{1}{z}\right)u^{k+1}dz = k(x+x^{-1})zdu$$

and we multiply the surface equation (4.22) to get

$$(\frac{1}{z}-z)u^{k+1}dz = (x+x^{-1})udz + (x^{-1}-x)u^{k+1}dz\,.$$

Before we add these two equations, we use $udz \sim -zdu$. Then

$$\frac{2}{z}u^{k+1}dz \sim (k-1)(x+x^{-1})zdu + (x^{-1}-x)u^{k+1}dz\,.$$

The second differential on the right has been handled in (4.25). The remaining one is treated with path-symmetry as before:

$$\int_{\gamma_1} 2zdu = \int_{\gamma_1}\left(z+\frac{1}{z}\right)du\,,$$

and $(z+1/z)$ is eliminated with (4.24). We obtain

$$2\operatorname{Re}\int_{\gamma} izdu = -\frac{4}{k}(x+x^{-1})\int_0^{\alpha}\sqrt{\cos^2\phi-\cos^2\alpha}\,\cos((-1+2/k)\phi)d\phi\,. \tag{4.26}$$

This reduction to the same integral as in the lemma allows us to finish the period discussion. (That this should happen seemed at first an undeserved piece of good luck. On reflection, it is simple to see why this happens, at least in the case of genus one: If one has two differential forms without residues and with linearly independent periods– as is the case in the situation at hand– then the periods of *any* differential form without residues are linear combinations of the periods of the first two.) First shorten the notation by defining

$$\begin{aligned} E_{\mp}(\gamma) &:= \int_0^{\gamma}\sqrt{\cos^2\phi-\cos^2\gamma}\cdot\cos((1\mp 2/k)\phi)d\phi\,,\\ c &:= \frac{2}{k}(x+x^{-1})\,. \end{aligned} \tag{4.27}$$

In these terms the lemma says

$$C_1 = c\cdot(k-1)\cdot E_-(\alpha),\quad C_2 = c\cdot(k-1)\cdot E_-(\widetilde{\alpha})\,.$$

The simplifications concentrated in (4.25), (4.26) deal with all the terms in

$$\frac{1}{g_1}dh \sim \left(m^2-1+\frac{2m}{z}\right)u^{k+1}dz$$

to give

$$\begin{aligned} Q_1(m) := \operatorname{Re} \int_{\gamma_1} -i\frac{dh}{g_1} &= c \cdot \left[(1-m^2)(k+1)E_+(\alpha) + (\cot\alpha - \tan\alpha)\cdot m \cdot (k+1)E_+(\alpha)\right. \\ &\qquad \left. + (\cot\alpha + \tan\alpha)\cdot m\cdot (k-1)E_-(\alpha)\right]. \end{aligned}$$

The substitution $\alpha \to \widetilde{\alpha} := \pi/2 - \alpha$, $m \to \widetilde{m} := -m$ gives

$$\begin{aligned} Q_2(m) = c \cdot \left[(1-m^2)(k+1)E_+(\widetilde{\alpha}) + (\cot\alpha - \tan\alpha)\cdot m \cdot (k+1)E_+(\widetilde{\alpha})\right. \\ \left. - (\cot\alpha + \tan\alpha)\cdot m\cdot (k-1)E_-(\widetilde{\alpha})\right]. \end{aligned}$$

The compatibility condition (4.20) is now fractional linear in $M := \frac{1}{m} - m$ and of the form

$$\frac{M + a + b\cdot q}{M + a - b\cdot \widetilde{q}} = \frac{q}{\widetilde{q}},$$

namely:

$$\frac{(\frac{1}{m} - m) + (\cot\alpha - \tan\alpha) + (\cot\alpha + \tan\alpha)\frac{k-1}{k+1}\cdot\frac{E_-(\alpha)}{E_+(\alpha)}}{(\frac{1}{m} - m) + (\cot\alpha - \tan\alpha) - (\cot\alpha + \tan\alpha)\frac{k-1}{k+1}\cdot\frac{E_-(\widetilde{\alpha})}{E_+(\widetilde{\alpha})}} = \frac{E_-(\alpha)/E_+(\alpha)}{E_-(\widetilde{\alpha})/E_+(\widetilde{\alpha})}$$

This simplifies to

$$\left(\left(\frac{1}{m} - m\right) + (\cot\alpha - \tan\alpha)\right)\cdot\frac{k+1}{k-1}\cdot\left(\frac{E_+(\widetilde{\alpha})}{E_-(\widetilde{\alpha})} - \frac{E_+(\alpha)}{E_-(\alpha)}\right) = 2\cdot(\cot\alpha + \tan\alpha). \tag{4.28}$$

4.9 The Details of the Solution of the Period Problem, II. The Monotonicity Lemma

We will prove the following

Lemma 4.2 (The Monotonicity Lemma). *Fix $k \geq 2$. The function $\gamma \to \frac{E_+(\gamma)}{E_-(\gamma)}$ is strictly decreasing on $(0, \frac{\pi}{2})$, with*

$$\lim_{\gamma\to 0}\frac{E_+(\gamma)}{E_-(\gamma)} = 1\,, \quad \lim_{\gamma\to\frac{\pi}{2}}\frac{E_+(\gamma)}{E_-(\gamma)} = \frac{k-1}{k+1}\,.$$

The integrals $E_\pm(\gamma)$ are defined in (4.27).

Before presenting the proof, we will show how the Lemma is used to prove Proposition 4.2. Recall from Section 4.3 that the Riemann surfaces determined by the conformal parameter α, with $\alpha > \pi/4$, are obtained from those with $\alpha < \pi/4$ by exchanging generators. The choice of $\alpha > \pi/4$ was made to be compatible with the location of the point where the normal is vertical (i. e.

where $z = -m^{-1}$) on the planar geodesic symmetry curve *between* the end-punctures at the points where $z = -x^{-1} = -\tan\alpha$ and $z = \infty$. To see this, let $\alpha > \pi/4$. Then the Monotonicity Lemma implies

$$\frac{2}{k+1} = \frac{E_+}{E_-}(0) - \frac{E_+}{E_-}(\frac{\pi}{2}) \geq \frac{E_+(\tilde{\alpha})}{E_-(\tilde{\alpha})} - \frac{E_+(\alpha)}{E_-(\alpha)} > 0\,.$$

With this, it is clear that (4.28) is *always* solvable for $m^{-1} - m$ and that (for $\alpha > \pi/4$)

$$m^{-1} - m > (k-1)(\cot\alpha + \tan\alpha) + (\tan\alpha - \cot\alpha) > 0 \tag{4.29}$$

$$m^{-1} - \tan\alpha + (\cot\alpha - m) > (k-1)(\cot\alpha + \tan\alpha) > 0\,.$$

This shows that $M := m^{-1} - m > 0$, that $m = -\frac{M}{2} + \sqrt{1 + \frac{M^2}{4}} < 1/m$, and that $m < \cot\alpha$. Therefore the positive solution $m = m(\alpha)$ of (4.28) (which is our compatibility condition (4.20)), puts the vertical normal on the symmetry line defined by $-\infty < z < -\tan\alpha = -x^{-1}$, that runs between the punctures at $z = -\tan\alpha$ and $z = \infty$. Thus we have established the first part of Proposition 4.2 for any $k \geq 2$. For $k > 2$, the inequality (4.29) has as an immediate consequence the inequality

$$m^{-1} > 2\tan\alpha + \cot\alpha = 2x^{-1} + x\,,$$

which is the growth rate condition (4.21). This concludes the proof of Proposition 4.2, when $k > 2$.

Remark 4.3. *The case $k = 2$, the genus=1 case, turned out to be more difficult for us than the higher genus cases. In addition to the Monotonicity Lemma, we proved, for $k = 2$:*

$$\frac{E_+(\tilde{\alpha})}{E_-(\tilde{\alpha})} - \frac{E_+(\alpha)}{E_-(\alpha)} \leq \frac{2}{3}\sin^2\alpha\,.$$

If we insert this in the compatibility condition (4.28), *and use the identity* $\sin^{-2}\alpha = \cot^2\alpha + 1$, *we get, instead of* (4.29),

$$\begin{aligned} m^{-1} - m &> \frac{\cot\alpha + \tan\alpha}{\sin^{-2}\alpha} - \cot\alpha + \tan\alpha \\ &= \cot^3\alpha + \tan\alpha[\cot^2\alpha + 2] \\ &= \cot^3\alpha + \cot\alpha + 2\tan\alpha \\ &> 2\tan\alpha + \cot\alpha = 2x^{-1} + x\,, \end{aligned}$$

which implies the growth rate condition (4.21) *in the case* $k = 2$.

4.9.1 Proof of the Monotonicity Lemma

To establish the claim

$$0 \leq \alpha \leq \gamma \leq \frac{\pi}{2} \Rightarrow \frac{E_+}{E_-}(\alpha) \geq \frac{E_+}{E_-}(\gamma)$$

it is equivalent to prove

$$E_-(\alpha) \cdot (E_-(\gamma) - E_+(\gamma)) - E_-(\gamma) \cdot (E_-(\alpha) - E_+(\alpha)) \geq 0 \tag{4.30}$$

and sufficient to prove

$$E_-(\alpha) \cdot (E_- - E_+)'(\alpha) - E'_-(\alpha) \cdot (E_- - E_+(\alpha)) \geq 0 \,. \tag{4.31}$$

From the definition of these integrals in (4.27) we have

$$E'_-(\alpha) = \sin\alpha\cos\alpha \int_0^\alpha \frac{1}{\sqrt{\cos^2\phi - \cos^2\alpha}} \cos\left(\left(1 - \frac{2}{k}\right)\phi\right) d\phi$$

and using $\cos(\alpha - \beta) - \cos(\alpha + \beta) = 2\sin\alpha\sin\beta$ we have

$$(E_- - E_+)(\alpha) = \int_0^\alpha \sqrt{\cos^2\phi - \cos^2\alpha} \cdot 2\sin\phi \cdot \sin\left(\frac{2}{k}\phi\right) d\phi \,,$$

$$(E_- - E_+)'(\alpha) = \sin\alpha\cos\alpha \int_0^\alpha \frac{2\sin\phi \cdot \sin(\frac{2}{k}\phi)}{\sqrt{\cos^2\phi - \cos^2\alpha}} d\phi \,.$$

Now we define decreasing functions f_1, g_1 and increasing functions f_2, g_2 with choices of constants $c_i > 0$ such that $\int_0^\alpha g_i(\phi)d\phi = 1$:

$$\begin{aligned}
&0 \leq f_1(\phi) := \sqrt{\cos^2\phi - \cos^2\alpha} \\
&0 \leq g_1(\phi) := c_1 \cdot \cos\left(\left(1 - \frac{2}{k}\right)\phi\right) && \text{(decreasing)} \\
&0 \leq f_2(\phi) := 1/\sqrt{\cos^2\phi - \cos^2\alpha} \\
&0 \leq g_2(\phi) := c_2 \cdot 2\sin\phi \cdot \sin\left(\frac{2}{k}\phi\right) && \text{(increasing)}\,.
\end{aligned}$$

With this notation (4.31) is equivalent to

$$\int_0^\alpha f_1 \cdot g_1 d\phi \cdot \int_0^\alpha f_2 \cdot g_2 d\phi \geq \int_0^\alpha f_1 \cdot g_2 d\phi \cdot \int_0^\alpha f_2 \cdot g_1 d\phi$$

and this inequality is implied by

$$\int_0^\alpha f_1 \cdot (g_1 - g_2) d\phi \geq 0 \,, \qquad \int_0^\alpha -f_2 \cdot (g_1 - g_2) d\phi \geq 0 \,.$$

Since $g_1 - g_2$ is decreasing, with $\int_0^\alpha (g_1 - g_2) = 0$, there exists a $\phi^* \in (0, \alpha)$ such that

$$\phi \le \phi^* \Rightarrow (g_1 - g_2)(\phi) \ge 0\,,$$
$$f_1(\phi) \ge f_1(\phi^*) \ge 0 \text{ and } - f_2(\phi^*) \le -f_2(\phi) \le 0$$
$$\phi^* \le \phi \Rightarrow (g_1 - g_2)(\phi) \le 0$$
$$0 \le f_1(\phi) \le f_1(\phi^*) \text{ and } - f_2(\phi) \le -f_2(\phi^*) \le 0\,.$$

Therefore

$$\int_0^\alpha f_1(\phi)(g_1 - g_2)(\phi)d\phi \ge \int_0^\alpha f_1(\phi^*)(g_1 - g_2)(\phi)d\phi = 0$$
$$\int_0^\alpha -f_2(\phi)(g_1 - g_2)(\phi)d\phi \ge \int_0^\alpha -f_2(\phi^*)(g_1 - g_2) = 0\,.$$

□

5. The Structure of the Space of Examples

We begin this section by giving a list of the constructions to date of complete, embedded minimal surfaces with finite total curvature. None of these examples have been as completely analyzed as those of Theorem 3.3. For some, a complete existence proof independent of computation has yet to be found. For others there are existence proofs that amount to computing, with error estimates, the degree of the period map. All of these examples have the property that they are part of deformation families of surfaces, which begin with all but two of the ends flat and continue through surfaces all of whose ends are catenoidal. Most of the families, in contrast to the surfaces $M_{k,x}$ of Theorem 3.3 eventually contain surfaces that are not embedded because the growth rates of the ends change their order and a lower end overtakes a higher one.

(1) Four ended examples with $k \ge 2$ vertical planes of symmetry, one horizontal plane of symmetry, genus $2(k-1)$, two flat ends and two catenoid ends [74]. (See Figure 5.1.)
(2) A one parameter family deforming the surfaces above through surfaces with four catenoid ends. Eventually these surfaces cease to be embedded. [73, 4, 74, 75].
(3) The surfaces in 2) with symmetric tunnels through their waist planes [73, 4]. These surfaces have $k \ge 2$ vertical planes of symmetry and genus $3(k-1)$. The begins with a surface with two flat ends and moves through surfaces with all four ends catenoidal. Eventually the surfaces are not embedded. (See Figure 5.2.)
(4) Five ended examples. [5]. (See Figure 5.3.)

These examples have in common the fact that they were discovered first with the help of insights guided by and guiding computation. The software MESH, developed by James T. Hoffman et al. ([8, 41]) together with various

Fig. 5.1. Four-ended embedded minimal surfaces of genus $2k, k > 0$

Fig. 5.2. Four-ended embedded minimal surfaces. *Left:* The surface described in 3) with $k = 1$. *Right:* Deformation of the four-ended surface on the left, now with all ends catenoidal. The surface is no longer embedded

numerical routines for killing periods (finding simultaneous zeros of several real-valued functions of several variables) has been crucial to this research. This was also the case for the three-ended examples of Theorem 3.3. One has to produce Weierstrass data by methods analogous to those presented in Section 4.1–4.4. The main obstacle to constructing more complex examples with methods like those in Section 4 is the difficulty in killing the periods. Even numerically, for n large, finding the simultaneous zero of n functions on a domain in $\mathbf{R}^n$ is time-consuming and unstable for functions (the periods given by Weierstrass integrals) that themselves are approximated by numerical integration.

The state of knowledge at present is rather poor. It is not known whether these surfaces can be perturbed further to reduce their symmetry. There is at present no general implicit-function-theorem-type result for the period mapping. Some important theoretical progress has been made by Wohlgemuth [74, 75]. The existence of certain of these surfaces (the most symmetric ones

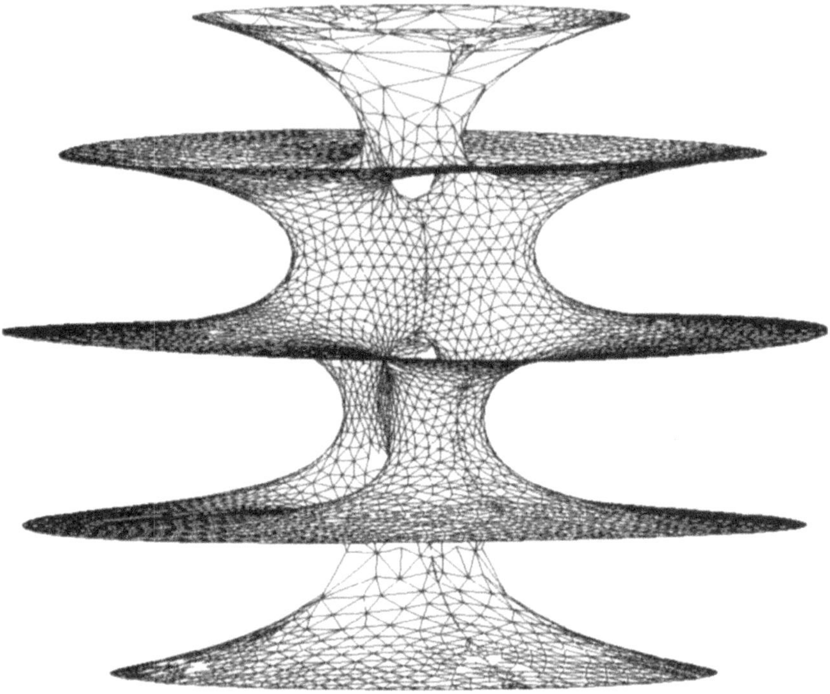

Fig. 5.3. Five-ended embedded minimal surface of genus 3, discovered by Boix and Wohlgemuth. See (4) in the list above

with flat ends) can be proved by theoretical means involving finding desingularizing limits of unstable solutions to the Plateau Problem. (See, for example [34].) However, while this is done for the surfaces of Theorem 3.2, this procedure has never produced an example not previously discovered by the method outlined above.

5.1 The Space of Complete, Embedded Minimal Surfaces of Finite Total Curvature

Let $\mathcal{M}$ be the space of complete, embedded minimal surfaces of finite total curvature, normalized so that all the ends have vertical normals and so that the maximum absolute value of the Gauss curvature is equal to one and occurs at the origin. (Note that the plane is not in $\mathcal{M}$.)

Definition 5.1. *The subspace* $\mathcal{M}_{i,j} \subset \mathcal{M}, i \geq 0, j \geq 2$, *consists of all surfaces in* $\mathcal{M}$ *of genus* i *with* j *ends. We define*

$$\mathcal{M}^*_{k,r} = \bigcup_{i\leq k,\, j\leq r} \mathcal{M}_{i,j}\,.$$

We consider two elements of $\mathcal{M}$ to be the same if they differ by a Euclidean motion. From (2.21) it follows that the total curvature of an example in $\mathcal{M}_{ij}$ is equal to $-4\pi(1-(i+j))$. This means that the subspace of $\mathcal{M}$ with total curvature at least $-4\pi n$ is a union of a finite number of $\mathcal{M}^*_{k,r}$. In this terminology, we may restate some of Theorem 3.1 and Theorem 3.4 as follows:

(i) $\mathcal{M}^*_{o,r}$ consists of one surface, the catenoid [51];
(ii) $\mathcal{M}^*_{k,2}$ consists of one surface, the catenoid [67];
(iii) $\mathcal{M}_{1,3}$ consists of the surfaces $M_{2,x}$ in Theorem 3.3, which is naturally a half-open interval $[1,\infty)$ [16].

Note from this last item that the spaces $\mathcal{M}_{ij}$ are not necessarily compact. However it can be shown that in $\mathcal{M}$, which we consider with the compact-open topology, every divergent sequence of surfaces $M_{k,x}$ has a subsequence that converges to the catenoid [30]. Thus while $\mathcal{M}_{1,3}$ is not compact, $\mathcal{M}^*_{1,3}$ is naturally a closed interval and is compact. (Its end points are Costa's surface and the catenoid.) The following result generalizes this fact.

Theorem 5.1 [(30)]. *$\mathcal{M}^*_{k,r}$ is compact. Specifically, every sequence of surfaces in $\mathcal{M}^*_{p,k}$ possesses a subsequence that converges smoothly on compact subsets of $\mathbf{R}^3$ to a surface in $\mathcal{M}^*_{p,k}$.*

This theorem is of limited utility because at the present time it is not known, in any general fashion, which of the spaces $\mathcal{M}_{i,j}$ are empty and which are not.

5.2 Some Questions and Conjectures

One could imagine that the Costa example is constructed in the following heuristic manner. Take a catenoid and intersect it by its waist plane. Remove a tubular neighborhood of the intersection circle and smoothly join up the four boundary circles by a genus one minimal surface. The higher-genus examples $M_k, k\geq 3$ can be thought of as improved approximations.

Theorem 5.2 ([38]).

(i) *As normalized in Theorem 3.2, the surfaces $M_k, k\geq 2$, possess a subsequence that converges as $k\to\infty$ to the union of the plane and the catenoid. The convergence is smooth away from the intersection circle.*
(ii) *Considered as surfaces in $\mathcal{M}$ (i. e., normalized so that the maximum value of K is equal to 1 and occurs at the origin), the sequence $\{M_k\}$ possesses a subsequence that converges to Scherk's singly-periodic surface.*

Fig. 5.4. One of a family of complete minimal surfaces of genus zero with two parallel, embedded catenoid ends and one embedded flat end, at which the Gauss map has an order-two branch point. The surface is not embedded because the flat end is not parallel to the catenoid ends

It has been conjectured by Hoffman and Meeks that for a complete, embedded, nonplanar minimal surface of finite total curvature,

$$k \geq r - 2\,, \tag{5.1}$$

where k is the genus and r is the number of ends. Theorem 3.1, Statements 2 and 3, results of Lopez-Ros and of Schoen, can be interpreted as saying that equality holds in (5.1) when either $k = 0$ or $r = 2$. The examples of Theorems 3.2 and 3.3, as well as all other known examples satisfy this inequality. Inequality 5.1 may be stated as a conjecture

Conjecture 5.1.

$$\mathcal{M}_{k,r} \textit{ is empty if } k < r - 2\,.$$

The heuristic idea behind the conjecture is that to increase the number of ends (r) by adding a flat end and then looking for a Costa style surface like the surface M_k, which is nearby and embedded, forces the genus (k) to increase. Related to this conjecture is the question of whether the order of the Gauss map at a flat end of a complete, embedded minimal surface can be equal to two. According to Remark 2.3(iii) the order must be at least two, and in all known *embedded* examples of finite total curvature the order is at least three.

(See Remark 2.8.) However, there are complete embedded periodic minimal surfaces with flat ends where the Gauss map has degree two. The most famous of these is the example of Riemann [62, 63]. (See also [31], [37] and [39].) For

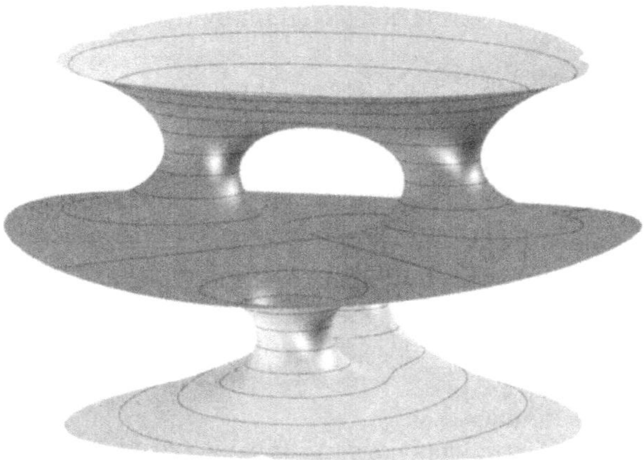

Fig. 5.5. The Horgan Surface. The surface depicted here with genus two and three ends probably does not exist. The periods appear to be zero, but one of them is just very small. Computationally, it appears that this period always has the same sign and is never zero, although it becomes vanishingly small. However, in the limit, the surface degenerates

immersed minimal surfaces, it is possible to have such an end. Consider the Weierstrass data

$$g(z) = \rho((z-r)(z+r))^{-1}, \quad dh = (z^2 - r^2)(z^2-1)^{-2}dz, \quad r \neq 1,$$

on $S^2 - \{\pm 1, \infty\}$. The surface has sufficient symmetry (vertical plane of reflection and one horizontal line) to make the period problem one-dimensional. The period problem is solvable by a residue calculation, which determines the choice of r(for each choice of ρ within a prescribed range). The ends at ± 1 are catenoid ends and the end at $g(z) = \infty$, where $g(z)$ has a zero of order 2, is flat. However, while all the ends are separately embedded, they are not parallel; the surface is complete and immersed but not embedded.

Recall from Proposition 2.1) and Remark 2.4(i) that each embedded catenoid end of a complete embedded minimal surface of finite total curvature is asymptotic to the end of a specific vertical catenoid. There are always at least two such ends, according to Proposition 2.5(v) and the discussion following Theorem 2.4. Each such end has an axis determined by the term of order $\mathcal{O}(\rho^{-1})$ in (2.21) and made precise in Definition 2.7. Because all of the known examples can be constructed by symmetry methods like those used in Section 4, it follows that all of the axes of the catenoid ends of these examples coincide. One basic question is whether or not there exist examples whose catenoid ends have *different* axes. Such an example could have, at most, a symmetry group of order four, generated by a reflection through a plane containing all of the axes (if such a plane exists) and an order two rotation about a horizontal line (if such a line exists). If the axes were not coplanar, only the rotation could exist. Such a rotation would imply that the number of catenoid ends was even.

As observed in Remark 2.8, if there are three catenoid ends then their axes are coplanar and the end whose axis is in the middle has logarithmic growth with sign opposite that of the other two.

It is known that any intrinsic isometry of a complete *embedded* minimal surface of finite total curvature in $\mathbf{R}^3$ must extend to a symmetry of that surface; that is an isometry of $\mathbf{R}^3$ that leaves the surface invariant [13, 36]. If the surface is not embedded, this is not necessarily the case. The Enneper surface and its generalizations $g(z) = z^k, dh = g(z)dz, z \in \mathbf{C}$, have intrinsic isometric rotations $z \to uz$, $|u| = 1$, but only k of them extend to symmetries of the surface. (See Remark 2.3(ii)).

An even more basic problem concerns the question of whether or not there exists a complete embedded minimal surface with no symmetries except the identity. That all the known examples have significant symmetry groups is an artifact of their construction and says nothing about the general situation. It would be extremely surprising if the hypotheses of finite total curvature together with embeddedness implied the existence of a nontrivial symmetry.

A related question concerns the surfaces $M_{k,x}$ in Theorem 3.3. Costa's Theorem 3.1.5 states that for the genus-one examples ($k = 2$)

$$\mathcal{M}_{1,3} = \{M_{2,x} \mid 0 < x \leq 1\}. \tag{5.2}$$

It would be very interesting to know to what extent this is true for higher genus:

$$\mathcal{M}_{k,3} \stackrel{?}{=} \{M_{k+1,x} \mid 0 < x \leq 1\}. \tag{5.3}$$

It is not even known if this is true with the added hypothesis that the symmetry group be of order at least $2(k+1)$, as is the case for the surfaces $M_{k+1,x}$. Costa's proof of (5.2) uses the full force of elliptic function theory, and hence is restricted to genus one. A geometric proof of (5.2) would be extremely useful to have, in order to begin to understand what happens in the higher-genus, three-ended case.

6. Finite Total Curvature Versus Finite Topology

A surface has finite topology provided it is homeomorphic to a compact Riemann surface from which a finite number of points have been removed. As a consequence of Theorem 2.2, we know that a complete immersed minimal surface whose total curvature is finite must have finite topology. In fact, it is *conformal* to a punctured compact Riemann surface. In this section, we will discuss to what extent finite topology implies finite total curvature for complete properly embedded minimal surfaces.

The helicoid is a complete, embedded, simply-connected (genus zero and one end) minimal surface. Because it is periodic and not flat, its total curvature is infinite. This simple classical example shows that finite topology does not imply finite total curvature. Osserman has posed the question of whether or not the helicoid is the only nonplanar, complete, embedded, minimal surface of genus-zero with one end. According to Theorem 3.1 there is no such surface of finite total curvature of any genus. One could ask, more generally, if there are any complete embedded minimal surfaces of finite topology and one end, besides the plane and the helicoid. For example, does there exist such a surface that is morphologically the helicoid with a finite number of handles? (We have recently shown that the answer is yes. See Section 6.3.)

An end of a complete surface of finite topology is necessarily annular. That is, it is homeomorphic to a punctured disk. If it is conformal to a punctured disk and the total curvature is infinite, then the Gauss map must have an essential singularity. All known embedded annular ends are asymptotic to the end of a helicoid. No embedded end is known that is conformal to an annulus $r_1 \leq |z| < r_2$.

6.1 Complete, Properly-Immersed, Minimal Surfaces with More Than One End

Under the assumption that the surface has more than one end, much more is known.

Theorem 6.1 (Hoffman and Meeks [35]). *Let M be a complete properly embedded minimal surface in $\mathbf{R}^3$ with more than one topological end. Then at most two of the annular ends have infinite total curvature.*

Corollary 6.1 ([35]). *Let M be a complete embedded minimal surface in $\mathbf{R}^3$ with finite topology. Then all except two of the ends have finite total curvature and are asymptotic to planes or catenoids, all with parallel limit normals.*

This Corollary was strengthened by Fang and Meeks who proved

Theorem 6.2 (Fang and Meeks [21]). *Let M be a properly embedded minimal surface in $\mathbf{R}^3$ with two annular ends that have infinite total curvature. Then these ends lie in disjoint closed halfspaces and all other annular ends are flat ends parallel to the boundary of the halfspaces.*

Corollary 6.2. *Suppose M is a complete properly embedded minimal surface of finite topology. If M does not have finite total curvature, then there are three possibilities:*

(i) *M has a catenoid end and exactly one end with infinite total curvature that lies in a halfspace;*

(ii) *M has exactly one end (of infinite total curvature) and this end does not lie in a halfspace;*
(iii) *M has two ends of infinite total curvature that lie in disjoint closed halfspaces, and a finite number of flat ends in the slab between these halfspaces. There are no catenoid ends.*

In particular a complete properly embedded minimal surface of finite topology with more than one end can never have an end asymptotic to a helicoid end (because such an end never lies in a halfspace).

6.2 Complete Embedded Minimal Surfaces of Finite Topology and More Than One End

What sorts of infinite total curvature ends can appear on a complete embedded minimal surface with finite topology? Nitsche considered ends that were fibred by embedded Jordan curves in parallel planes. These are called *Nitsche ends* by Meeks and Rosenberg [53]. For concreteness we assume that such an end A lies in the halfspace $\{x_3 \geq 0\}$ with boundary a simple closed curve in $x_3 = 0$, and that $\{x_3 = t\} \cap A$ is a simple nonempty closed curve, for all $t \geq 0$. Nitsche proved

Proposition 6.1 ([57]). *A Nitsche end, all of whose level curves are star shaped, is a catenoidal end.*

In particular, it has finite total curvature. A Nitsche end necessarily has the conformal type of a punctured disk. Certainly, we may assume that the end is conformally $\{z\,|0 \leq \alpha < |z| \leq 1\}$. The height function $h(z) = x_3(z)$ is harmonic and so is essentially determined by its boundary values. Suppose $\alpha > 0$. Then $h = 0$ on $|z| = 1$ and, $h = \infty$ on $|z| = \alpha$. Since $|c_1 \ln|z||$ is also harmonic, and $|c_1 \ln|z||$ lies below h on the boundary of $\alpha < |z| \leq 1$, it follows that

$$h(z) > c_1 \ln|z| \qquad \alpha < |z| < 1$$

for any real negative c_1. This means $h(z) \equiv +\infty$ on the interior of the annulus, which is a contradiction. Hence we can take the disk to be $\{z|0 < z \leq 1\}$, and $h(z) = c_1 \ln|z|, c_1 < 0$. Thus dh is determined, and using the Weierstrass representation (2.6), (2.7), we can determine the end completely by the knowledge of its Gauss map. For the Nitsche end to have infinite total curvature, g must have an essential singularity. Thus finding a non-catenoidal Nitsche end is equivalent to finding an analytic function on the punctured unit disk with an essential singularity at the origin for which (2.6) (2.7) provide a well-defined embedding. Toubiana and Rosenberg have given examples that are well-defined but not embedded [66]. We also remark that on $\mathbf{C}, g(z) = \exp(F(z))$, and $dh = dz$, where $F(z)$ is entire, gives a complete, immersed, simply-connected minimal surface whose Gauss map has an essential singularity at infinity.

The importance of understanding Nitsche ends is underscored by the following strong result of Meeks and Rosenberg.

Theorem 6.3 ([53]). *Let M be a complete embedded minimal surface of finite topology and more than one end. Then any end of infinite total curvature is a Nitsche end. In particular M is conformal to a compact Riemann surface punctured in a finite number of points.*

Rosenberg conjectured that a Nitsche end must have finite total curvature and hence must be asymptotic to a catenoid. This is called the "generalized Nitsche Conjecture". A proof of the Nitsche conjecture would imply that finite topology and finite total curvature are equivalent for complete embedded minimal surfaces with at least two ends.

6.3 Complete Embedded Minimal Surfaces of Finite Topology with One End

In 1992-93, Hoffman, Karcher and Wei [32, 33] constructed a complete embedded minimal surface of finite topology and infinite total curvature.

Theorem 6.4. *There exists a complete embedded minimal surface, $\mathcal{H}e_1$, of genus-one with one end that is asymptotic to the end of a helicoid. The*

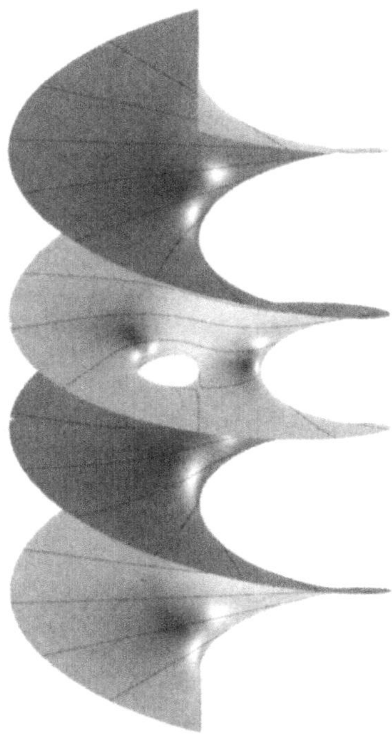

Fig. 6.1. The genus one helicoid, $\mathcal{H}e_1$

surface contains two lines, one vertical and corresponding to the axis of the helicoid, the other horizontal crossing the axial line. Rotation about these lines generates the full isometry group of $\mathcal{H}e_1$*, which is isomorphic to* $Z_2 \oplus Z_2$*.*

This is the first properly embedded minimal surface with infinite total curvature, whose quotient by symmetries does not have finite total curvature.

7. Stability and the Index of the Gauss Map

Recall that a minimal surface $S \subset \mathbf{R}^3$ is by definition a critical point of the area functional for compactly supported perturbations. Let $X: \Sigma \to \mathbf{R}^3$ be a conformal parametrization of S and $N: \Sigma \to S^2$ its Gauss map. If D is a compact subdomain of Σ, and $X_t: D \to \mathbf{R}^3$ is a normal variation of D with variation vector-field fN, $f \in C_0^\infty(D)$, the second derivative of area is given by

$$\frac{d^2}{dt^2}\,\mathrm{Area}(X_t) = \int_D (|\nabla f|^2 + 2Kf^2)dA =: Q_D(f,f)\,, \tag{7.1}$$

where K is the Gauss curvature. The second variation operator is defined to be

$$L =: -\Delta + 2K\,, \tag{7.2}$$

in terms of which

$$Q_D(f,f) = \int_D fLf\,dA\,.$$

The sign of Δ is determined by our definition in (2.3). The index of L on D, denoted $\mathrm{ind}(D)$, is defined to be the dimension of the maximal subspace of $C_0^\infty(D)$ on which Q_D is negative definite. A domain $X(D) \subset S$ is said to be *stable* if $\mathrm{ind}(D) = 0$, i.e., if (7.1) is nonnegative for all $f \in \mathbf{C}_0^\infty(D)$.

For a non-compact minimal surface $X(\Sigma) = S \subset \mathbf{R}^3$, the index (Σ) is defined to be the supremum of $\mathrm{ind}(D)$ taken over all compact $D \subset \Sigma$. The following theorem summarizes the results of several authors relating index to stability.

Theorem 7.1. *Let* $X: \Sigma \to \mathbf{R}^3$ *be a conformal minimal immersion of an orientable surface.*

(i) *(Schwarz* [68]*.) If the Gaussian image of* Σ *lies in a hemisphere,* S *is stable;*

(ii) *(Barbosa and do Carmo* [1]*.) If the area of the Gaussian image satisfies* $A(N(\Sigma)) < 2\pi$*, then* S *is stable. (Here* $A(N(\Sigma))$ *is the area of the Gaussian image* disregarding multiplicities*);*

(iii) *(Fischer-Colbrie and Schoen* [23]*, do Carmo and Peng* [18]*, Pogorelov* [61]*.) If* S *is complete,* S *is stable if and only if* S *is a plane;*

(iv) *(Fischer-Colbrie [22], Gulliver [25], Gulliver and Lawson [26].)* Index$(\Sigma) < \infty$ *if and only if* $\int_\Sigma |K| dA < \infty$.

Without going too much into detail, we will give some of the background to the last two parts of the theorem above. The Gauss curvature K can be expressed as $-|\nabla N|^2/2$ and therefore we may write $L = -(\Delta + |\nabla N|^2)$. Observe that $Q_D(f,f) = \int_D (|\nabla f|^2 - |\nabla N|^2 f^2) dA$ is invariant under conformal change of metric and therefore $\operatorname{ind}(D)$ depends only on N.

We may therefore define, for a meromorphic mapping $\Phi: \Sigma \to S^2$, the index $\operatorname{ind}_\Phi(D)$ of a compact subdomain $D \subset \Sigma$ to be the number of negative eigenvalues of $L_\phi = -(\Delta + |\nabla\phi|^2)$, with Dirichlet boundary conditions on D. By definition, $\operatorname{ind}(D) = \operatorname{ind}_N(D)$. In particular if $\overline{\Sigma}$ is a compact Riemann surface and Φ a meromorphic function on $\overline{\Sigma}$,

$$\operatorname{index}(\Phi) := \operatorname{ind}_\Phi(\overline{\Sigma}) \tag{7.3}$$

is an integer invariant of $\Phi: \overline{\Sigma} \to S^2$.

For a complete immersed minimal surface $S \subset \mathbf{R}^3$, the condition of finite total curvature is equivalent to the conditions that the underlying Riemann surface Σ is a compact Riemann surface $\overline{\Sigma}$ with a finite number of points removed, and that the Gauss map N extends to a meromorphic function on $\overline{\Sigma}$ (See Theorem 2.2). Thus for any compact subdomain $D \subset \Sigma$,

$$\operatorname{ind}(D) = \operatorname{ind}_N(D) \le \operatorname{index}(N)\,,$$

In particular the condition of finite total curvature implies that

$$\operatorname{index}(\Sigma) = \sup\{\operatorname{ind}(D) \mid D \subset \Sigma, D \,\text{compact}\}$$

is finite. This is the easier implication of part (iii) of Theorem 7.1. In fact these two indices are equal:

Proposition 7.1 ([22]). *Let $X: \Sigma \to \mathbf{R}^3$ be a complete immersed minimal surface of finite total curvature. Let $N: \Sigma \to S^2$ be its Gauss map. Then* $\operatorname{ind}(\Sigma) = \operatorname{index}(N)$.

As a consequence of this Proposition, statement (iii) of Theorem 7.1 follows from the fact that the index of any nonconstant meromorphic function $\Phi: \overline{\Sigma} \to S^2$ is nonzero.

Remark 7.1. *Let ds_N^2 be the metric produced by pulling back the standard metric on S^2 by the Gauss map $N: \Sigma \to S^2$. If ds^2 denotes the metric induced by $X: \Sigma \to \mathbf{R}^3$, we have $ds_N^2 = \frac{1}{2}|\nabla N|^2 ds^2 = -K ds^2$. This metric is regular, away from the isolated zeros of K, and has constant curvature* 1. *The volume of $\overline{\Sigma}$ in this metric is therefore* 4π degree (Φ). *Because the Dirichlet integrand $|\nabla f|^2 dA$ is invariant under conformal change of metric*

$$Q(f,f) = \int (|\nabla f|^2 + 2Kf^2)dA = \int (|\widetilde{\nabla} f|^2 - 2f^2)d\widetilde{A},$$

where $\widetilde{\nabla} f$ and $d\widetilde{A}$ are computed in the metric ds_N^2.

The associated differential operator is $\widetilde{L} = -(\widetilde{\Delta}+2)$. Eigenfunctions of $-\widetilde{\Delta}$ can be defined by the standard variational procedure [54, 72]. *The spectrum of $-\widetilde{\Delta}$ is discrete, infinite and nonnegative and we may think of index (Φ) as the number of eigenvalues of $-\widetilde{\Delta}$ that are strictly less than two.*

It is natural to try to compute the index of stability of the minimal surfaces mentioned in this survey. We have seen that this is equivalent to computing the index of their extended Gauss maps. For the simplest case of the identity map from S^2 to itself, the index is just the number of eigenvalues of $\widetilde{\Delta} = \Delta$ (the spherical Laplacian) less than two. Since the first nonzero eigenvalue is equal to 2 and $\lambda_0 = 0$ is a simple eigenvalue, it follows from Remark 7.1 above that index$(Id) = 1$.

Proposition 7.2. *The index of stability of Enneper's surface and the catenoid is equal to one.*

This follows immediately from the fact that the extended Gauss map of these genus-zero surfaces is a conformal diffeomorphism to the sphere. Osserman [58, 59] proved that the catenoid and Enneper's surface were the only complete minimal surfaces with this property. It is natural to ask whether or not index one characterizes these surfaces among all complete minimal surfaces of finite total curvature. The following result of Montiel-Ros ([54]) can be used to resolve this question.

Theorem 7.2. *Suppose $\Phi: \overline{\Sigma} \to S^2$ is a nonconstant holomorphic map of* degree d *defined on a compact Riemann surface of genus k.*

(i) *If* $k = 0$, index$(\Phi) = 1 \iff d = 1$.
(ii) *If* $k = 1$, index$(\Phi) \geq 2$.
(iii) *If $k \geq 2$ then, for a generic conformal structure,* index$(\Phi) \geq 2$.
(iv) *If* index$(\Phi) = 1$, $d \leq 1 + \left[\frac{1+k}{2}\right]$.

Statements (i) and (ii) are straightforward. Statements (iii) and (iv) follow from more general results in [54] (see Corollary 8 and Theorem 6) and the Brill-Noether Theorem.

Now suppose that S is a complete minimal surface whose index of stability is equal to 1. From (2.9), we have for the Gauss map N of a complete minimal surface $X: \Sigma \to \mathbf{R}^3$ of genus k

$$d = \text{degree}(N) \geq k + 1\,.$$

Combined with statement (iv) of Theorem 7.2, this implies $k \leq 1$. Then statements (i) and (ii) imply that $k = 0$ and $d = 1$. As noted above, the catenoid and Enneper's surface are the only possibilities. Thus we have

Corollary 7.1 (Corollary 9 of [54, 50]. *A complete minimal surface with stability index equal to one must be the catenoid or Enneper's surface.*

Remark 7.2. *Montiel and Ros conjectured that it was possible to have a hyperelliptic Riemann surface whose two-fold covering map to S^2 had index equal to one, provided the branch values were sufficiently well distributed. This was proved by Souam [69].*

Theorem 7.3. *Given $\varepsilon > 0$, choose $\mathcal{P} = \{p_1 \dots p_{n_\varepsilon}\} \subset S^2$ such that the open ε-disks about the p_i are disjoint, but maximal in the sense that their complement does not contain an ε-disk. Suppose Σ is a hyperelliptic Riemann surface and $\phi\colon \Sigma \to S^2$ a two-fold covering of S^2 whose branch values are precisely $\mathcal{P}$. Then if $\varepsilon > 0$ is small enough,* $\mathrm{index}(\Phi) = 1$.

The following results allow us to compute the index of stability of the Meeks-Jorge n-noids, the Chen-Gackstatter surface and its generalizations to higher symmetry (see Remark 4.2).

Proposition 7.3 (Montiel and Ros [54]). *Suppose $\phi\colon \overline{\Sigma} \to S^2$ is a degree $d \geq 1$ holomorphic map. If all the branch values of ϕ lie on a great circle*

$$\mathrm{index}(\Phi) = 2d - 1\,. \tag{7.4}$$

Choe [12], proved a weaker version of this proposition.

Theorem 7.4 ([54]). *The index of stability of:*

(i) *the n-noid is $2n - 3$;*
(ii) *the Chen-Gackstatter surface is* 3.

Proof. From Section 2.2 we know that the Gauss map of the Chen-Gackstatter surface has degree $d = 2$, and all of its branch values are real. The n-noid $n \geq 2$ can be produced with the Weierstrass data $\overline{\Sigma} = S^2$,

$$g = z^{n-1}$$
$$dh = (z^n + z^{-n} + 2)z^{-1}dz\,.$$

(When $n = 2$ we have the catenoid.) In both cases we can apply Proposition 7.3. □

The equivalence of finite total curvature and finite index suggests that there should be an explicit relationship between d, the degree of the Gauss map, and the index of stability. Tysk [72] was the first to show that the index could be estimated by the degree:

$$\frac{\mathrm{Index}(N)}{\mathrm{degree}(N)} < c \text{ where } c \sim 7.68\,.$$

Choe [12] conjectured that a weakened version of (7.4) was true in general for all complete minimal surfaces M of finite total curvature $-4\pi d$:

Conjecture 7.1.

$$\text{index}(M) \le 2d - 1 \tag{7.5}$$

The conjectured inequality (7.5) cannot be replaced in general by an equality. Nayatani (private communication) has pointed out to us that an example written down by Kusner [44], Rosenberg-Toubiana [64, 65], R. Bryant [6, 7] with 4 flat ends, genus 0, and Gauss map of degree $d = 3$, gives strict inequality. On any surface the support function $u = X \cdot N$ satisfies $Lu = 0$, where L is defined in (7.2), and because the ends are flat, u is bounded. Using this fact, it can be shown that the index of this surface is 4. Since $d = 3$, index $= 4 < 5 = 2d - 1$.

Nayatani [55] computed the index of the Costa surface, M_2 of Theorem 3.2. This was done by computing the index of its Gauss map, which is the derivative of the Weierstrass $\wp$–function, geometrically normalized, on the square torus. The geometric normalization does not change the index.

He subsequently improved that result:

Theorem 7.5 (Nayatani [56]). *Let M_k be the Hoffman-Meeks surface of genus $k-1$, described in Theorem* 3.2. *For* $2 \le k \le 38$,

$$\text{index}(M_k) = 2k + 1\,.$$

The Gauss map of M_k has degree $k+1$, so equality holds in (7.5) for these surfaces. For genus 0 surfaces Montiel and Ros [54] and Ejiri and Kotani [20] independently proved the following result.

Theorem 7.6. *Formula* (7.4) *holds for a generic complete, genus zero, minimal surface of total curvature $-4\pi d$. In general,* index$(M) \le 2d - 1$ *for such a surface.*

This result holds even for branched minimal surfaces and involves an analysis of bounded Jacobi fields.

We end this chapter with a discussion of the work of Choe [12] about the *vision number* of a minimal surface with respect to a vector field V on $\mathbf{R}^3$. For a vector field V on $S \subset \mathbf{R}^3$, Choe defines the *horizon* of V to be

$$H(S, V) = \{s \in S \mid V \in T_s S\}\,,$$

and the vision number, to be

$$v(S, V) = \#\text{ components of } S - H(S, V)\,.$$

Suppose V is a variation vector field associated with a family of parallel translations, or rotations about a line. (Both are Killing vector fields on $\mathbf{R}^3$). Consider the restriction of such a vector field to S, and let $\widehat{V}$ be its projection onto the normal bundle of S. Then $\widehat{V}$ is a Jacobi field. That is, $\widehat{V} \cdot N$ is in the kernel of $L = -\Delta + 2K$. The same holds for the position vector field $V_0(p) = p - p_0$, for any fixed origin $p_0 \in \mathbf{R}^3$. This can be used to show that any connected subset of $S - H(S, V)$ is stable, for any one of these vector fields V. Choe uses this to prove

Theorem 7.7. *Let $S \subset \mathbf{R}^3$ be a complete minimal surface of finite total curvature $-4\pi d$. Let V_T, V_R denote the Killing vector fields generated by translation and rotation, respectively. Then*

(i) $\mathrm{Index}(S) \geq v(S, V_T) - 1$;
(ii) $d \geq v(S, V_T)/2$;

If all the ends are parallel

(iii) $\mathrm{Index}(S) \geq v(S, V_R) - 1$;

If all the ends are embedded and parallel
(iv) $d \geq v(S, V_R)/2$.

From statement (iii) of the above theorem we can get a lower bound for the index of the surfaces M_k of Theorem 3.2. Observe that the same is true for the $M_{k,\alpha}$ of Section 4. These surfaces have genus $k - 1$, parallel embedded ends and k vertical planes of reflective symmetry, which intersect in a vertical line, $\mathcal{L}$, orthogonal to the ends, and which divide the surface into $2k$ regions. Their Gauss maps have degree $d = k + 1$. If V_R is the Killing field generated by rotation about $\mathcal{L}$, then it is clear that $H(M_{k,\alpha}, V_R)$ contains the intersection of $M_{k,\alpha}$ with each plane of symmetry. Hence $v(M_{k,\alpha}, V_R) \geq 2k$. By Theorem 7.7(ii) we have

Corollary 7.2. *The index of stability of the surfaces $M_k = M_{k,0}$ and $M_{k,\alpha}$ satisfies*

$$\mathrm{Index}(M_{k,\alpha}) \geq 2k - 1 = 2(\mathrm{genus}\, M_{k,\alpha}) + 1\,. \tag{7.6}$$

Remark 7.3. *The symmetry of the surfaces $M_{k,\alpha}$ imply that $v(M_{k,\alpha}, V_R)$ is an integer multiple of $2k$. But by statement* (iv) *of Theorem* 7.7, $2(k+1) = 2d \geq v(M_{k,\alpha}, V_R)$. *Hence $v(M_{k,\alpha}, V_R) = 2k$. Thus this particular argument cannot be improved to give a better lower bound. As we know from Theorem* 7.5, $\mathrm{index}(M_k) = 2k + 1, 2 \leq k \leq 38$, *so this estimate is not sharp in this range.*

References *

1. Barbosa, J.L., and do Carmo, M.: On the size of a stable minimal surface in $\mathbf{R}^3$. Am. J. Math. **98** (2), 515–528 (1976), Zbl. 332.53006
2. Barbosa, L., and Colares, A.G.: Minimal surfaces in $\mathbf{R}^3$. Lecture Notes Math. 1195. Springer-Verlag 1986, Zbl. 609.53001

*For the convenience of the reader, references to reviews in Zentralblatt für Mathematik (Zbl.), compiled by means of the MATH database, and Jahrbuch über die Fortschritte der Mathematik (Jbuch) have, as far as possible, been included in this bibliography.

3. Bloß, D.: Elliptische Funktionen und vollständige Minimalflächen. PhD thesis, FU Berlin, 1989, Zbl. 708.53007
4. Boix, E., and Capdequi, L.: Deformation families of complete minimal surfaces with four catenoid ends. (Video tape) GANG, August 1991
5. Boix, E., and Wohlgemuth, M.: Numerical results on embedded minimal surfaces of finite total curvature. In preparation
6. Bryant, R.: A duality theorem for Willmore surfaces. J. Differ. Geom. **20**, 23–53 (1984), Zbl. 555.53002
7. Bryant, R.: Surfaces in conformal geometry. In: The mathematical heritage of Hermann Weyl, Proc. Symp. Durham/NC 1987, Proc. Symp. Pure Math. **48**, 227–240 (1988), Zbl. 654.53010
8. Callahan, M., Hoffman, D., and Hoffman, J.: Computer graphics tools for the study of minimal surfaces. Commun. ACM **31** (6), 648–661 (1988)
9. Callahan, M., Hoffman, D., and Meeks III, W.H.: Embedded minimal surfaces with an infinite number of ends. Invent. Math. **96**, 459–505 (1989), Zbl. 676.53004
10. Chen, C.C., and Gackstatter, F.: Elliptic and hyperelliptic functions and complete minimal surfaces with handles. Instituto de Matemática e Estatistica-Universidade de São Paulo 27, 1981
11. Chen, C.C., and Gackstatter, F.: Elliptische und hyperelliptische Funktionen und vollständige Minimalflächen vom Enneperschen Typ. Math. Ann. **259**, 359–369 (1982), Zbl. 468.53008
12. Choe, J.: Index, vision number, and stability of complete minimal surfaces. Arch. Ration. Mech. Anal. **109** (3), 195–212 (1990), Zbl. 695.53045
13. Choi, T., Meeks III, W.H., and White, B.: A rigidity theorem for properly embedded minimal surfaces in $\mathbf{R}^3$. J. Differ. Geom. **32**, No. *1*, 65–76 (1990), Zbl. 704.53008
14. Costa, C.: Imersöes minimas en $\mathbf{R}^3$ de gênero un e curvatura total finita. PhD thesis, IMPA, Rio de Janeiro, Brasil 1982
15. Costa, C.: Example of a complete minimal immersion in $\mathbf{R}^3$ of genus one and three embedded ends. Bol. Soc. Bras. Mat. **15**, 47–54 (1984), Zbl. 613.53002
16. Costa, C.: Uniqueness of minimal surfaces embedded in $\mathbf{R}^3$ with total curvature 12π. J. Differ. Geom. **30** (3), 597–618 (1989), Zbl. 696.53001
17. Dierkes, U., Hildebrandt, S., Küster, A., and Wohlrab, O.: Minimal Surfaces I. Grundl. math. Wiss. 295. Springer-Verlag 1992, Zbl. 777.53012
18. do Carmo, M., and Peng, C.K.: Stable minimal surfaces in $\mathbf{R}^3$ are planes. Bull. Am. Math. Soc., New Ser. **1**, 903–906 (1979), Zbl. 442.53013
19. do Espírito-Santo, N.: Superfícies mínimas completas em $\mathbf{R}^3$ com fim de tipo Enneper. PhD thesis, University of Niteroi, Brazil 1992. [See also Ann. Inst. Fourier **44**, No. *2*, 525–577 (1994), Zbl. 803.53006]
20. Ejiri, N., and Kotani, M.: Minimal surfaces in $S^{2m}(1)$ with extra eigenfunctions. Q. J. Math. Oxf., II. Ser. **172** 421–440 (1992), Zbl. 778.53047
21. Fang, Y., and Meeks III, W.H.: Some global properties of complete minimal surfaces of finite topology in $\mathbf{R}^3$. Topology **30** (1), 9–20 (1991), Zbl. 737.53010
22. Fischer-Colbrie, D.: On complete minimal surfaces with finite Morse index in 3-manifolds. Invent. Math. **82**, 121–132 (1985), Zbl. 573.53038
23. Fischer-Colbrie, D., and Schoen, R.: The structure of complete stable minimal surfaces in 3-manifolds of non-negative scalar curvature. Comm. Pure Appl. Math. **33**, 199–211 (1980), Zbl. 439.53060
24. Gackstatter, F.: Über die Dimension einer Minimalfläche und zur Ungleichung von St. Cohn-Vossen. Arch. Ration. Mech. Anal. **61** (2), 141–152 (1976), Zbl. 328.53002

25. Gulliver, R.: Index and total curvature of complete minimal surfaces. Proc. Symp. Pure Math. **44** 207–212 (1986), Zbl. 592.53006
26. Gulliver, R., and Lawson, H.B.: The structure of minimal hypersurfaces near a singularity. Proc. Symp. Pure Math. **44** 213–237 (1986), Zbl. 592.53005
27. Hadamard, J.-J.: Les surfaces à courbures opposées et leurs lignes géodésiques. J. Math. Pures Appl. **4**, 27–73 (1898)
28. Hoffman, D.: Embedded minimal surfaces computer graphics and elliptic functions. In: Proceedings of the Berlin Conference on Global Differential Geometry, Lecture Notes Math. **1156**, 204–215. Springer-Verlag 1985, Zbl. 566.53016
29. Hoffman, D.: The computer-aided discovery of new embedded minimal surfaces. Math. Intell. **9** (3), 8–21 (1987), Zbl. 616.53007
30. Hoffman, D., and Meeks III, W.H.: One-parameter families of embedded complete minimal surfaces of finite topology. In preparation
31. Hoffman, D., Karcher, H., and Rosenberg, H.: Embedded minimal annuli in $\mathbf{R}^3$ bounded by a pair of straight lines. Comment. Math. Helvet. **66**, 599–617 (1991), Zbl. 765.53004
32. Hoffman, D., Karcher, H., and Wei, F.: Adding handles to the helicoid. Bull. Am. Math. Soc., New Ser. **29** (1), 77–84 (1993), Zbl. 787.53003
33. Hoffman, D., Karcher, H., and Wei, F.: The genus one helicoid and the minimal surfaces that led to its discovery. In: Global Analysis and Modern Mathematics. Publish or Perish Press 1993. K. Uhlenbeck (ed.), 119–170
34. Hoffman, D., and Meeks III, W.H.: A variational approach to the existence of complete embedded minimal surfaces. Duke Math. J. **57** (3), 877–893 (1988), Zbl. 676.53006
35. Hoffman, D., and Meeks III, W.H.: The asymptotic behavior of properly embedded minimal surfaces of finite topology. J. Am. Math. **2** (4), 667–682 (1989), Zbl. 683.53005
36. Hoffman, D., and Meeks III, W.H.: Embedded minimal surfaces of finite topology. Ann. Math., II. Ser. **131**, 1–34 (1990), Zbl. 695.53004
37. Hoffman, D., and Meeks III, W.H.: Les surfaces minimales: la caténoïde par les deux bouts. Quadrature **5**, 31–47 (1990)
38. Hoffman, D., and Meeks III, W.H.: Limits of minimal surfaces and Scherk's Fifth Surface. Arch. Ration. Mech. Anal. **111** (2), 181–195 (1990), Zbl. 709.53006
39. Hoffman, D., and Meeks III, W.H.: Minimal surfaces based on the catenoid. Am. Math. Mon. **97** (8), 702–730 (1990), Zbl. 737.53006
40. Hoffman, D., and Meeks III, W.H.: The strong halfspace theorem for minimal surfaces. Invent. Math. **101**, 373–377 (1990), Zbl. 719.53037
41. Hoffman, J.T.: MESH manual. GANG preprint series II, #35
42. Jorge, L., and Meeks III, W.H.: The topology of complete minimal surfaces of finite total Gaussian curvature. Topology **22** (2), 203–221 (1983), Zbl. 517.53008
43. Karcher, H.: Construction of minimal surfaces. Surveys in Geometry, 1–96, 1989. University of Tokyo, 1989, and Lecture Notes No. *12*, SFB 256, Bonn, 1989
44. Kusner, R.: Conformal geometry and complete minimal surfaces. Bull. Am. Math. Soc. **17** (2), 291–295 (1987), Zbl. 634.53004
45. Kusner, R.: Global geometry of extremal surfaces in three-space. PhD thesis, University of California, Berkeley, 1988
46. Kusner, R.: Bubbles, conservation laws and balanced diagrams. In: Geometric Analysis and Computer Graphics, volume 17 of MSRI Publications. Springer-Verlag, Berlin Heidelberg New York 1990. P. Concus and R. Finn and D. Hoffman (eds.), Zbl. 776.68009
47. Lawson, H.B.: Complete minimal surfaces in S^3. Ann. Math., II. Ser. **92**, 335–374 (1970)

48. Lawson, Jr. H.B.: Lectures on Minimal Submanifolds. Publish or Perish Press, Berkeley, 1971, Zbl. 434.53006
49. Lopez, F.J.: The classification of complete minimal surfaces with total curvature greater than -12π. Trans. Am. Math. Soc. **334** (1), 49–74 (1992), Zbl. 771.53005
50. Lopez, F.J., and Ros, A.: Complete minimal surfaces with index one and stable constant mean curvature surfaces. Comment. Math. Helvet. **64**, 34–43 (1989), Zbl. 679.53047
51. Lopez, F.J., and Ros, A.: On embedded complete minimal surfaces of genus zero. J. Differ. Geom. **33** (1), 293–300 (1991), Zbl. 719.53004
52. Meeks III, W.H., and Rosenberg, H.: The maximum principle at infinity for minimal surfaces in flat three-manifolds. Comment. Math. Helvet. **65**, 255–270 (1990), Zbl. 713.53008
53. Meeks III, W.H., and Rosenberg, H.: The geometry and conformal structure of properly embedded minimal surfaces of finite topology in $\mathbf{R}^3$. Invent. Math. **114**, 625–639 (1993), Zbl. 803.53007
54. Montiel, S., and Ros, A.: Schrödinger operators associated to a holomorphic map. In: Global Differential Geometry and Global Analysis (Berlin, 1990), Proc. Conf. Berlin/Ger. 1990, Lecture Note Math. **1481**, 147–174. Springer-Verlag, Berlin Heidelberg New York 1990, Zbl. 744.58007
55. Nayatani, S.: On the Morse index of complete minimal surfaces in Euclidean space. Osaka J. Math. **27**, 441–451 (1990), Zbl. 704.53007
56. Nayatani, S.: Morse index of complete minimal surfaces. In: The Problem of Plateau, 181–189. World Scientific Press, 1992. Th. M. Rassias (ed.), Zbl. 794.58011
57. Nitsche, J.C.C.: A characterization of the catenoid. J. Math. Mech. **11**, 293–301 (1962), Zbl. 106, 146
58. Osserman, R.: Global properties of minimal surfaces in E^3 and E^n. Ann. Math., II. Ser. **80** (2), 340–364 (1964), Zbl. 134, 385
59. Osserman, R.: A Survey of Minimal Surfaces. Dover Publications, New York, 2nd edition, 1986. (1st ed. Van Nostrand, New York 1969, Zbl. 209, 529)
60. Pérez, J., and Ros, A.: Some uniqueness and nonexistence theorems for embedded minimal surfaces. Math. Ann. **295** (3), 513–525 (1993), Zbl. 789.53004
61. Pogorelov, A.V.: On the stability of minimal surfaces. Dokl. Akad. Nauk SSSR **260**, 293–295 (1981, Zbl. 495.53005) English transl.: Soviet Math. Dokl. **24**, 274–276 (1981)
62. Riemann, B.: Über die Fläche vom kleinsten Inhalt bei gegebener Begrenzung. Abh. Königl, d. Wiss. Göttingen, Mathem. Cl. **13**, 3–52 (1867)
63. Riemann, B.: Ouevres Mathématiques de Riemann. Gauthiers-Villars, Paris 1898, Jbuch 29, 9
64. Rosenberg, H.: Deformations of minimal surfaces. Trans. Am. Math. Soc. **295**, 475–290 (1986)
65. Rosenberg, H., and Toubiana, E.: Some remarks on deformations of minimal surfaces. Trans. Am. Math. Soc. **295**, 491–499 (1986), Zbl. 598.53005
66. Rosenberg, H., and Toubiana, E.: A cylindrical type complete minimal surface in a slab of $\mathbf{R}^3$. Bull. Sci. Math., II. Ser. **111**, 241–245 (1987), Zbl. 631.53012
67. Schoen, R.: Uniqueness, symmetry, and embeddedness of minimal surfaces. J. Differ. Geom. **18**, 791–809 (1983), Zbl. 575.53037
68. Schwarz, H.A.: Gesammelte Mathematische Abhandlungen, volume 1. Springer-Verlag, Berlin 1890, Jbuch 22, 31
69. Souam, R.: Exemples d'applications holomorphes d'indice un. Ann. Inst. Fourier **43** (2), 369–381 (1993), Zbl. 788.30037

70. Thayer, E.: Higher-genus Chen-Gackstatter minimal surfaces, Scherk's singly-periodic minimal surface and Riemann surfaces of infinite genus. Experimental Math. **4**, No. 1, 19–39 (1995)
71. Thayer, E.: Complete Minimal Surfaces in Euclidean 3-Space. PhD thesis, University of Massachusetts at Amherst, 1994
72. Tysk, J.: Eigenvalue estimates with applications to minimal surfaces. Pac. J. Math. **128**, 361–366 (1987), Zbl. 632.58016
73. Wohlgemuth, M.: Higher genus minimal surfaces by growing handles out of a catenoid. Manuscr. Math. **70**, 397–428 (1991), Zbl. 752.53008
74. Wohlgemuth, M.: Vollständige Minimalflächen höheren Geschlechts und endlicher Totalkrümmung. PhD thesis, Bonner Mathematische Schriften **262**, University of Bonn, April 1993
75. Wohlgemuth, M.: Higher genus minimal surfaces of finite total curvature. Arch. Ration. Mech. Anal. (to appear)
76. Yang, K.: Complete minimal surfaces of finite total curvature. Kluwer Acad. Pub., Dordrecht, Netherlands 1994

II. Nevanlinna Theory and Minimal Surfaces

Hirotaka Fujimoto

Contents

Introduction

In 1915, S. Bernstein proved that there is no nonflat minimal surface in $\mathbf{R}^3$ which is described as the graph of a C^2-function on the total plane $\mathbf{R}^2$ ([9]). This result was improved by many researchers in the field of differential geometry. E. Heinz studied nonflat minimal surfaces in $\mathbf{R}^3$ which are the graph of functions on discs $\Delta_R := \{(x, y); x^2+y^2 < R^2\}$ and showed that there exists a constant $C > 0$ such that $|K(0)| \leq C/R^2$ for the Gaussian curvature $K(0)$ at the origin ([43]). After some related results were given by E. Hopf ([45]), J. C. C. Nitsche ([52]) and so on, in 1961 R. Osserman proved that the Gauss map of a nonflat complete regular minimal surface in $\mathbf{R}^3$ cannot omit a set of positive logarithmic capacity in the Riemann sphere ([56]) and, in 1981 F. Xavier showed that the Gauss map of such a surface can omit at most six values. Moreover, in 1988 the author gave the best possible version of this, which asserts that the number of exceptional values of the Gauss map of a complete nonflat regular minimal surface is at most four. Recently, several related results were given by X. Mo and R. Osserman ([49]), S. J. Kao ([48]) and M. Ru ([60]).

These properties of the Gauss map of minimal surfaces are closely related to value-distribution-theoretic properties of meromorphic functions on the complex plane $\mathbf{C}$. Bernstein's theorem is related to Liouville's theorem for entire functions, Heinz's theorem is something like the classical Landau's theorem, Osserman's theorem resembles the Casorati-Weierstrass theorem and the author's result is very similar to the classical Picard theorem for meromorphic functions on $\mathbf{C}$. These classical theorems for meromorphic functions on $\mathbf{C}$ were systematically investigated in more precise form by R. Nevanlinna, who gave the first and second main theorems and the defect relation. Recently, the author obtained analogous results for the Gauss map of complete minimal surfaces, which we call the modified defect relation ([33], [34], [35]).

In this exposition, we first give a survey of Nevanlinna theory for holomorphic curves in $P^n(\mathbf{C})$ defined on an open Riemann surface with parabolic exhaustion, which was given by L. Ahlfors ([2]) and H. and J. Weyl ([66]), and secondly present some recent results concerning the analogous value-distribution-theoretic properties of the Gauss map of complete minimal surfaces in $\mathbf{R}^m$ defined on a Riemann surface with parabolic exhaustion, mainly with finite total curvature. We describe also Nevanlinna theory for minimal surfaces itself, which was developed by E. F. Beckenbach and his collaborators in [6], [5] and [4]. In the last chapter, we survey recent results on global properties of the Gauss map of complete minimal surfaces in $\mathbf{R}^m$ with infinite total curvature.

Chapter 1
Nevanlinna Theory for Holomorphic Curves

§1. Some Basic Formulas on a Riemann Surface

We first recall some basic properties of currents on a Riemann surface M. We denote by $\mathcal{D}_r$ the space of all C^∞ differentiable forms φ of degree r on M whose supports $\mathrm{Supp}(\varphi) := \overline{\{a \in M; \varphi(a) \neq 0\}}$ are compact and by $\mathcal{D}_{p,q}$ the space of all $\varphi \in \mathcal{D}_{p+q}$ of type (p,q) for nonnegative integers r, p, q, where we set $\mathcal{D}_r := \{0\}$ and $\mathcal{D}_{p,q} := \{0\}$ if $r < 0$, $p < 0$ or $q < 0$. We say that a sequence $\{\varphi_n\}$ in $\mathcal{D}_r$ converges to φ in $\mathcal{D}_r$ if there exists a compact set including all $\mathrm{Supp}(\varphi_n)$ and, for each local expression

$$\varphi_n = \sum_{p+q=r} \varphi_n^{p,q} \omega_{p,q}\,, \qquad \varphi = \sum_{p+q=r} \varphi^{p,q} \omega_{p,q}$$

in terms of nowhere zero (p,q)-forms $\omega_{p,q}$, all sequences of partial derivatives of $\varphi_n^{p,q}$ converge uniformly to those of $\varphi^{p,q}$. By definition, a current T of degree r is a $\mathbf{C}$-linear map of $\mathcal{D}_{2-r}$ into $\mathbf{C}$ such that $\lim_{n\to\infty} T(\varphi_n) = T(\varphi)$ for every sequence $\{\varphi_n\}$ converging to φ in $\mathcal{D}_r$.

A current T is called a (p,q)-current if $T(\varphi) = 0$ for all $\varphi \in \mathcal{D}_{r,s}$ with $(r,s) \neq (1-p, 1-q)$. Every current T of degree r is uniquely represented as the sum of (p,q)-currents with $p, q \geq 0$ and $r = p+q$. For a C^∞ (p,q)-form Ω on M we can define a (p,q)-current $[\Omega]$ by $[\Omega](\varphi) = \int_M \Omega \wedge \varphi$ $(\varphi \in \mathcal{D}_{1-p,1-q})$. Similarly, associated with each locally integrable function u on M, the $(0,0)$-current $[u]$ is defined by $[u](\varphi) = \int_M u\varphi$ $(\varphi \in \mathcal{D}_{1,1})$. They are simply denoted by Ω and u respectively if we have no confusion.

In this exposition, we mean by a divisor on M a map $\nu : M \to \mathbf{R}$ whose support has no accumulation points in M. For a point $a_0 \in M$ we define the divisor ν_{a_0} as $\nu_{a_0}(a_0) = 1$ and $\nu_{a_0}(a) = 0$ for any other point $a \in M$. For the case where M is compact, every divisor ν on M is represented as $\nu = \sum_i \nu(a_i)\nu_{a_i}$ with finitely many points a_i's. Classically, a divisor on a compact Riemann surface is defined as a $\mathbf{Z}$-linear combination of finitely many points in M, which may be rephrased as an integer-valued divisor in this exposition. Associated with each divisor ν the $(1,1)$-current $[\nu]$ is defined by $[\nu](\varphi) = \sum_{a\in M} \nu(a)\varphi(a)$ $(\varphi \in \mathcal{D}_{0,0})$.

We say that a $(1,1)$-current T is real if $T(\varphi) = \overline{T(\bar{\varphi})}$ $(\varphi \in \mathcal{D}_{0,0})$ and that a real current T is nonnegative if $T(\varphi) \geq 0$ for all nonnegative $\varphi \in \mathcal{D}_{0,0}$. For a nonnegative $(1,1)$-current T, we define the integral of T over a domain D by

$$\int_D T := \sup\{T(\varphi); \varphi \in \mathcal{D} \text{ such that } 0 \leq \varphi \leq 1 \text{ and } \mathrm{Supp}(\varphi) \subset D\} \ (\leq +\infty).$$

We can also define the integral of the difference $T := T_1 - T_2$ of two nonnegative $(1,1)$-currents and that of the sum $T := T_1 + iT_2$ of two real $(1,1)$-currents by $\int_D T := \int_D T_1 - \int_D T_2$ and $\int_D T := \int_D T_1 + i\int_D T_2$ respectively in the case where they have meanings. We have $\int_D [\Omega] = \int_D \Omega$ for every integrable C^∞ $(1,1)$-form Ω. We define the order of a divisor ν by

$$\mathrm{ord}(\nu) := \int_M [\nu] \equiv \sum_{z\in M} \nu(z) \ (\leq +\infty)$$

in the case $\nu \geq 0$ and by $\mathrm{ord}(\nu) := \mathrm{ord}(\nu^+) - \mathrm{ord}(\nu^-)$ in the case $\mathrm{ord}\ (\nu^+) < +\infty$ or $\mathrm{ord}\ (\nu^-) < +\infty$, where $\nu^+ := \max(\nu, 0)$ and $\nu^- := (-\nu)^+$.

Let T be an r-current on M. The differential of T is defined by setting $dT(\varphi) = (-1)^{r-1}T(d\varphi)$ for every $\varphi \in \mathcal{D}_{1-r}$. For a (p,q)-current T the $(p+1,q)$-current ∂T and the $(p,q+1)$-current $\bar{\partial} T$ are defined by

$$\partial T(\varphi) = (-1)^{p+q-1}T(\partial\varphi), \quad \bar{\partial} T(\varphi) = (-1)^{p+q-1}T(\bar{\partial}\varphi) \quad (\varphi \in \mathcal{D}_{1-p-q}),$$

respectively. For an r-current $T = \bigoplus_{r=p+q} T_{p,q}$ we define $\partial T := \bigoplus_{r=p+q} \partial T_{p,q}$, $\bar{\partial} T := \bigoplus_{r=p+q} \bar{\partial} T_{p,q}$. Then, $dT = (\partial + \bar{\partial})T$. We also define

$$d^c T := \frac{\sqrt{-1}}{4\pi}(\bar{\partial} - \partial)T .$$

As is easily shown by Stokes theorem, we have $d[\Omega] = [d\Omega], \partial[\Omega] = [\partial\Omega], \bar{\partial}[\Omega] = [\bar{\partial}\Omega]$ for every differential form Ω.

Take a 1-current T on a Riemann surface M and a relatively compact subdomain D of M with smooth boundary. If T is equal to a C^∞ one form on some neighborhood of ∂D, we have the following Stokes' theorem for currents:

$$\int_D dT = \int_{\partial D} T . \tag{1.1}$$

To see this, we write $T = [\eta] + T'$ with a C^∞ one form η and a current T' which vanishes on a neighborhood of ∂D. We know that (1.1) holds for $T = [\eta]$, and we have

$$\int_D dT' \equiv \int_D \varphi dT' = T'(d\varphi) = 0 = \int_{\partial D} T'$$

for every $\varphi \in \mathcal{D}_0$ with $\varphi = 1$ on a neighborhood of $\mathrm{Supp}(T')$.

Consider a nonzero complex-valued function u on a domain D possibly with isolated singularities. We call u a function with admissible singularities on D if u is of class C^∞ outside a discrete set E such that, for each $a \in D$, we can write

$$|u(z)| = |z-a|^{\sigma_a} u^*(z) \tag{1.2}$$

on a neighborhood U of a in terms of a holomorphic local coordinate $z = x+iy$, where $\sigma_a \in \mathbf{R}$ and u^* is a nonnegative continuous function such that, for $v := \log u^*$,

$$|v| = o\left(\frac{1}{|z-a|}\right), \quad \left|\frac{\partial v}{\partial x}\right| = o\left(\frac{1}{|z-a|}\right), \quad \left|\frac{\partial v}{\partial y}\right| = o\left(\frac{1}{|z-a|}\right) \tag{1.3}$$

$dd^c v$ is locally integrable.

The number σ_a in (1.2) is obviously unique. We define the divisor ν_u of such a function u by $\nu_u(a) := \sigma_a$ $(a \in D)$ for the number σ_a appearing in (1.2).

For example, if a function u has a local expression

$$|u(z)| = |z-a|^{\sigma_a} u^*(z) \prod_{j=1}^{q} |\log(|g_j(z)|v_j(z))|^{\tau_j}$$

in a neighborhood U of each $a \in M$, where $\tau_j \leq 0 (1 \leq j \leq q < +\infty)$, $\sigma_a \in \mathbf{R}$, g_j are nonzero holomorphic functions and u^* and v_j are some positive C^∞ functions on U, then u has admissible singularities. One of the simplest examples of such functions is a nonzero meromorphic function f on M, for which $\nu_f(a)$ is nothing but the order of f at a.

By a meromorphic s-form, where $s \in \mathbf{Z}$, we mean a form ω which is locally represented as $\omega = \omega_z dz^s$ with a meromorphic function ω_z such that $\omega_z = \omega_\zeta (d\zeta/dz)^s$ on the common domains of all holomorphic local coordinates z and ζ. We simply call a meromorphic 1-form a meromorphic form. For a nonzero meromorphic s-form ω we define the divisor ν_ω of ω by $\nu_\omega(a) := \nu_{\omega_z}(a)$ $(a \in M)$ for a local expression $\omega = \omega_z dz^s$. By a pseudo-metric on M we mean a conformal metric ds^2 possibly with isolated singularities which has a local expression $ds^2 = \lambda_z^2|dz|^2$ in terms of a nonnegative function λ_z with admissible singularities, where $\lambda_z = \lambda_\zeta |d\zeta/dz|$ for two holomorphic local coordinates z and ζ. We define the divisor of a pseudo-metric ds^2 by $\nu_{ds} := \nu_{\lambda_z}$ for local expressions $ds^2 = \lambda_z^2|dz|^2$. We say that a pseudo-metric ds^2 is continuous if $\nu_{ds} \geq 0$ everywhere.

For a function with admissible singularities, we can prove the following:

Proposition 1.4. $dd^c[\log|u|^2] = [\nu_u] + [dd^c \log|u|^2]$.

Proof. For each $\varphi \in \mathcal{D}_0$, let $\{a_1, \ldots, a_k\} := \operatorname{Supp}(\varphi) \cap \{a; \nu_u(a) \neq 0 \text{ or } u(a) = 0\}$. We write $|u| = |z_i|^{\nu_u(a_i)} u_i^*(z_i)$ with a function u_i^* satisfying the condition (1.3) on a neighborhood U_i of each a_i, where z_i is a holomorphic local coordinate on U_i with $z_i(a_i) = 0$. Choose some $\varepsilon > 0$ such that $D_i(\varepsilon) := \{a; |z_i(a)| < \varepsilon\} \subset \overline{D_i(\varepsilon)} \subset U_i$ $(1 \leq i \leq k)$, and $\bar{D}_i(\varepsilon) \cap \bar{D}_j(\varepsilon) = \emptyset$ $(i \neq j)$. Since $d\varphi \wedge d^c\psi = d\psi \wedge d^c\varphi$ $(\varphi, \psi \in \mathcal{D}_0)$, we have

$$\log|u|^2 dd^c\varphi = d(\log|u|^2 d^c\varphi - \varphi d^c \log|u|^2) + \varphi dd^c \log|u|^2$$

outside $\{a_1, \ldots, a_k\}$. By using Stokes theorem, we get

$$\begin{aligned}
\int_M \log|u|^2 dd^c\varphi &= \lim_{\varepsilon \to 0} \int_{M - \cup_i D_i(\varepsilon)} \log|u|^2 dd^c\varphi \\
&= \lim_{\varepsilon \to 0} \int_{\partial(M - (\cup_i D_i(\varepsilon)))} (\log|u|^2 d^c\varphi - \varphi d^c \log|u|^2) + \int_M \varphi dd^c \log|u|^2 \\
&= -\lim_{\varepsilon \to 0} \sum_{i=1}^{k} \int_{\partial D_i(\varepsilon)} (\log|u|^2 d^c\varphi - \varphi d^c \log|u|^2) + [dd^c \log|u|^2](\varphi).
\end{aligned}$$

and

$$\int_{\partial D_i(\varepsilon)} \log|u|^2 d^c\varphi - \varphi d^c \log|u|^2 = \nu_u(a_i)\left(\int_{\partial D_i(\varepsilon)} \log|z_i|^2 d^c\varphi - \varphi d^c \log|z_i|^2\right)$$
$$+ \int_{D_i(\varepsilon)} \log u_i^* dd^c\varphi - \varphi dd^c \log|u_i^*|^2 .$$

Here, the last term of the right hand side of this identity tends to zero as $\varepsilon \to 0$. On the other hand, the first term equals

$$\nu_u(a_i)\left(2\log\varepsilon \int_{\partial D_i(\varepsilon)} d^c\varphi - \frac{1}{2\pi}\int_0^{2\pi} \varepsilon\varphi(\varepsilon e^{i\theta})\frac{d\log r}{dr}\bigg|_{r=\varepsilon} d\theta\right)$$
$$= \nu_u(a_i)\left(2\log\varepsilon \int_{D_i(\varepsilon)} dd^c\varphi - \frac{1}{2\pi}\int_0^{2\pi} \varphi(\varepsilon e^{i\theta})d\theta\right),$$

which tends to $-\nu_u(a_i)\varphi(a_i)$ as $\varepsilon \to 0$. This gives the desired identity

$$\int_M \log|u|^2 dd^c\varphi = \sum_{a\in M} \nu_u(a)\varphi(a) + \int_M \varphi dd^c[\log|u|^2] \quad (\varphi \in \mathcal{D}_0).$$

Proposition 1.4 gives the following Poincaré-Lelong formula.

Corollary 1.5. *$dd^c[\log|g|^2] = [\nu_g]$ for all meromorphic functions $g(\not\equiv 0)$.*

In fact, we have $[dd^c \log u^2] = 0$ for $u := |g|$.

Corollary 1.6. *If M is compact,* $\mathrm{ord}(\nu_g) = 0$ *for every nonzero meromorphic function g on M.*

In fact, as a result of Corollary 1.5 and (1.1), we have

$$\mathrm{ord}(\nu_g) = \int_M [\nu_g] = \int_M dd^c[\log|g|^2] = \int_{\partial M} d^c[\log|g|^2] = 0 .$$

Corollary 1.7. *If M is compact, then* $\mathrm{ord}(\nu_{\omega_1}) = \mathrm{ord}\,(\nu_{\omega_2})$ *for all nonzero meromorphic forms ω_1 and ω_2 on M.*

In fact, we have $\mathrm{ord}(\nu_{\omega_1}) - \mathrm{ord}(\nu_{\omega_2}) = \mathrm{ord}(\nu_g) = 0$ for a well-defined meromorphic function $g := \omega_1/\omega_2$ on M.

Now, we recall the following two basic facts:

Theorem 1.8. *For an arbitrary Riemann surface M there exists at least one nonconstant meromorphic function on M. Moreover, if M is not compact, for every integer-valued divisor ν on M there exists a nonzero meromorphic function g on M such that $\nu_g = \nu$.*

For the proof, e.g., refer to [21, §16 and §26].

Theorem 1.9 (Riemann-Hurwitz formula). *Let $g : \tilde{M} \to M$ be a finite covering surface of a compact Riemann surface M and let d and V be the number of sheets and the total number of ramification indices respectively. Then, $\chi(\tilde{M}) = d\,\chi(M) - V$, where $\chi(M)$ denotes the Euler characteristic of M.*

For the proof, e.g., refer to [21, p. 140].

Let D be a relatively compact subdomain of a Riemann surface M with real analytic smooth boundary and let ω be a nonzero meromorphic one-form on D. We say that ω is purely imaginary on ∂D if it has a continuous extension to $\bar{D}$ satisfying the conditions $\mathrm{Re}\,\omega = 0$ and $\nu_\omega = 0$ on ∂D. If M is compact, we may say that every nonzero meromorphic one-form on M is purely imaginary on $\partial M (= \emptyset)$. If ∂D has at least two connected components, then we can construct a holomorphic form which is purely imaginary on ∂D. In fact, as a solution of Dirichlet problem (e.g., [21, Theorem 22.17]) there is a nonconstant continuous function h on $\bar{D}$ which is harmonic and a constant on each connected component of ∂D. As is easily seen, $\omega := \partial h$ is holomorphic and purely imaginary on ∂D.

Theorem 1.10. *Let D be a relatively compact domain in a Riemann surface M with real analytic smooth boundary and ω a nonzero meromorphic form on D which is purely imaginary on ∂D. Then,* $\mathrm{ord}\,(\nu_\omega|D) = -\chi(\bar{D})$. *In particular, if M is compact,* $\mathrm{ord}(\nu_\omega) = -\chi(M)$ *for every nonzero meromorphic form ω on M.*

To prove this, we consider first the case where M is compact and $D = M$. Take a nonconstant meromorphic function $g : M \to P^1(\mathbf{C})$ by the help of Theorem 1.8. Obviously, the branching order of g at a equals $\nu_{dg}(a)$ if $a \notin E_\infty := g^{-1}(\infty)$ and $\nu_{dg}(a) - 2\nu_g(a)$ if $a \in E_\infty$. Therefore, we get

$$V = \mathrm{ord}(\nu_{dg}) - 2 \sum_{a \in E_\infty} \nu_g(a) = \mathrm{ord}(\nu_{dg}) + 2d\,,$$

where d is the number of sheets of $g : M \to P^1(\mathbf{C})$. Corollary 1.7 and Theorem 1.9 yield that $\mathrm{ord}\,(\nu_\omega) = \mathrm{ord}\,(\nu_{dg}) = V - 2d = -\chi(M)$ because $\chi(P^1(\mathbf{C})) = 2$. For a general D we construct the double $\hat{D}$ of D, namely, the welding of the domain D and the conjugate of the surface D along ∂D(cf., [3, p. 119]). Then $\hat{D}$ is a compact Riemann surface with $\chi(\hat{D}) = 2\,\chi(\bar{D})$ and the meromorphic form ω extends to a meromorphic form $\hat{\omega}$ on $\hat{D}$. Since $2\,\mathrm{ord}(\nu_\omega) = \mathrm{ord}(\nu_{\hat{\omega}})$, we have $\mathrm{ord}(\nu_\omega) = -\chi(\bar{D})$.

Proposition 1.11. *Let D be a relatively compact domain in M with real analytic smooth boundary, ω a meromorphic form which is purely imaginary on ∂D and ds^2 a pseudo-metric on a neighborhood of $\bar{D}$ which has no singularities on ∂D. Then, for a function λ with $ds^2 = \lambda^2|\omega|^2$,*

$$\chi(\bar{D}) - \int_{\partial D} d^c \log \lambda^2 = -\int_D [dd^c \log \lambda^2] - \mathrm{ord}(\nu_{ds|D})\,.$$

For, since $[\nu_{ds}] = [\nu_\lambda] + [\nu_\omega]$, Proposition 1.4, (1.1) and Theorem 1.10 yield

$$\int_D [dd^c \log \lambda^2] = \int_D dd^c[\log \lambda^2] - \int_D [\nu_\lambda] = \int_{\partial D} d^c \log \lambda^2 - \mathrm{ord}(\nu_{ds|D}) - \chi(\bar{D}).$$

Here, we recall the classical Gauss-Bonnet formula. Let D be a relatively compact domain on a Riemann surface M with smooth boundary and ds^2 a metric on an open neighborhood of $\bar{D}$. By κ, K and Ω_{ds^2} we denote the geodesic curvature form of the curve $\gamma := \partial D$, the Gaussian curvature and the area form of ds^2 respectively. The classical Gauss-Bonnet formula asserts that

$$\chi(\bar{D}) - \frac{1}{2\pi} \int_{\partial D} \kappa = \frac{1}{2\pi} \int_D K \Omega_{ds^2}.$$

By the use of Proposition 1.11, we can prove the following version of the above Gauss-Bonnet theorem.

Theorem 1.12. *Let D be a relatively compact domain in M with real analytic smooth boundary and ds^2 a pseudo-metric on a neighborhood of $\bar{D}$ which has no singularities on ∂D. Then,*

$$\chi(\bar{D}) - \frac{1}{2\pi} \int_{\partial D} \kappa = \frac{1}{2\pi} \int_D K \Omega_{ds^2} - \mathrm{ord}\,(\nu_{ds|D}). \tag{1.13}$$

Proof. We write $ds^2 = \lambda_z^2 |dz|^2$ in terms of a holomorphic local coordinate z. Then, the Gaussian curvature is locally given by $K := -(1/\lambda_z^2) \Delta \log \lambda_z$ outside the singularities of ds^2, so that $K\Omega_{ds^2} = -2\pi dd^c \log \lambda_z^2$. On the other hand, on a sufficiently small neighborhood of each $a \in \partial D$, if we write $ds^2 = \lambda_\zeta^2 |d\zeta|^2$ with a holomorphic local coordinate ζ with $\mathrm{Re}(\zeta)$ =const on ∂D, the geodesic curvature form is given by $\kappa = 2\pi d^c \log \lambda_\zeta^2$ (cf., [68, pp. 27–28]).

Consider first the case that D is an open disk in $\mathbf{C}$ with center a and $\nu_{ds} \equiv 0$. As a holomorphic local coordinate ζ with $\mathrm{Re}\,\zeta$ = const on ∂D we take a branch of $\log(z - a)$ locally, where a is the center of D. Since $\lambda_\zeta = \lambda_z |dz/d\zeta| = \lambda_z |z - a|$, we have

$$\int_{\partial D} d^c \log \lambda_\zeta^2 = \int_{\partial D} d^c \log \lambda_z^2 + \int_{\partial D} d^c \log |z - a|^2 = \int_D dd^c \log \lambda_z^2 + 1.$$

This gives (1.13) for this particular case, because $\chi(\bar{D}) = 1$. For the general domain D we may assume that ∂D is not connected, because, otherwise, consider the domain D' obtained by removing one or two sufficiently small closed disks from D and add the formulas (1.13) applied for D' and for the removed disks. We choose a meromorphic form ω on D which is purely imaginary on ∂D. For each $a \in \partial D$ take a holomorphic local coordinate ζ around a such that $\mathrm{Re}\,\zeta$ =const on ∂D and $|d\zeta|^2 = |\omega|^2$. We have $\kappa = 2\pi d^c \log \lambda^2$ for a function λ with $ds^2 = \lambda^2 |\omega|^2$. The formula (1.13) is a direct result of Proposition 1.11.

§2. The First Main Theorem for Holomorphic Curves

Let M be an open Riemann surface. By an exhaustion τ of M we mean that τ is a continuous map of M into $[0, +\infty)$ which is proper, namely, whose inverse image of every compact set is compact. For an exhaustion τ of M we set $M_r := \{x; \tau(x) < r\}$ and $M_{s,r} := M_r - M_s$. We call τ a parabolic exhaustion of M if it is an exhaustion of M satisfying the condition that τ^2 is of class C^∞ on M and that $dd^c \log \tau \equiv 0$ on an open neighborhood of $M_{s,+\infty}$ for some fixed $s \geq 0$ and $dd^c\tau \not\equiv 0$. For example, the function $\tau(z) := |z|$ $(z \in \mathbf{C})$ is a parabolic exhaustion of $\mathbf{C}$, where we may take $s = 0$. If an open Riemann surface M is of finite type, namely, biholomorphic with a compact Riemann surface $\bar{M}$ with finitely many points $a_1, \ldots, a_k$ removed, then M has a parabolic exhaustion. In fact, as a parabolic exhaustion we can take a C^∞ function τ on M which equals $1/|z_i|$ on some neighborhood of each a_i, where z_i is a holomorphic local coordinate with $z_i(a_i) = 0$. It is known that an open Riemann surface M has a parabolic exhaustion if and only if M is of type O_G, namely, there is no nonconstant negative subharmonic function on M (cf., [64, Theorem 10.12]).

Let M be an open Riemann surface on which a fixed parabolic exhaustion τ is defined. We note that on $M_{s,+\infty}$

$$\tau dd^c\tau = \tau^2 dd^c \log \tau + d\tau \wedge d^c\tau = d\tau \wedge d^c\tau = |\partial\tau/\partial z|^2(\sqrt{-1}\,/\,2\pi)dz \wedge d\bar{z} \geq 0\,.$$

Moreover, since $\log \tau$ is harmonic, $d\tau$ vanishes only on a discrete subset of $M_{s,+\infty}$. For each $r > s$ set $\mathcal{C} = \int_{\partial M_r} d^c \log \tau^2$. By Stokes' theorem, $\mathcal{C}$ is independent of r and $\mathcal{C} = (2/r) \int_{\partial M_r} d^c\tau$, which is positive because $d^c\tau$ means the outwards normal derivative of τ and so $d^c\tau \geq 0$ on ∂M_r. We assume that $\mathcal{C} = 1$ in the following after replacing τ by τ^a for a suitable $a > 0$.

Let ν be a divisor on M. We choose and fix some $s > 0$ such that ∂M_s is smooth. We define the counting function of ν by

$$N(r, \nu) := \int_s^r \sum_{z \in M_t} \nu(z) \frac{dt}{t}\,.$$

Proposition 2.1. *Let u be a function with admissible singularities on M. Then, for $r > s > 0$,*

$$\begin{aligned}\int_s^r \frac{dt}{t} \int_{M_t} dd^c[\log |u|^2] &= \int_s^r \frac{dt}{t} \int_{M_t} [dd^c \log |u|^2] + N(r, \nu_u) \\ &= \frac{2}{r} \int_{\partial M_r} \log |u| d^c\tau - \frac{2}{s} \int_{\partial M_s} \log |u| d^c\tau\,. \qquad (2.2)\end{aligned}$$

The first identity is due to Proposition 1.4. To see the second, we first use (1.1) to see $\int_{M_t} dd^c[\log |u|^2] = \int_{\partial M_t} d^c \log |u|^2$ for the case where ∂M_t is smooth and u has no singularity on ∂M_t. Since $d \log \tau \wedge d^c \log |u|^2 = d \log |u|^2 \wedge$

$d^c \log\tau = d(\log|u|^2 \wedge d^c\log\tau)$, if u has no singularities on M_{r_1,r_2} $(s \le r_1 < r_2)$, we have

$$\begin{aligned}\int_{r_1}^{r_2}\frac{dt}{t}\int_{M_t} dd^c[\log|u|^2] &= \int_{r_1}^{r_2}\frac{dt}{t}\left(\int_{\partial M_t} d^c\log|u|^2\right)\\ &= \int_{M_{r_1,r_2}} d\log\tau\wedge d^c\log|u|^2 = \int_{\partial M_{r_1,r_2}}\log|u|^2 d^c\log\tau\\ &= \frac{2}{r_2}\int_{\partial M_{r_2}}\log|u|d^c\tau - \frac{2}{r_1}\int_{\partial M_{r_1}}\log|u|d^c\tau\,.\end{aligned}$$

This remains valid for all $r_1 = s$ and $r_2 = r$ because both sides are continuous in r_1, r_2 and $M_{r,s}$ can be represented as the finite union of sets M_{r_1,r_2} on the interior of which u has no singularities.

Remark 2.3. The first term of the right hand side of (2.2) is the mean value of $\log|u|$ on the set ∂M_r with respect to the density $d^c\log\tau^2$. For the particular case $M = \mathbf{C}$ and $\tau(z) = |z|$ we have $(2/\tau)d^c\tau = (1/2\pi)d\theta$ on ∂M_r if we use the polar coordinate $z = re^{i\theta}$.

As a direct result of Proposition 2.1, we have the following Jensen's formula:

Corollary 2.4. *For every nonzero meromorphic function g on M we have*

$$N(r,\nu_g) = \frac{2}{r}\int_{\partial M_r}\log|g|d^c\tau - \frac{2}{s}\int_{\partial M_s}\log|g|d^c\tau\,.$$

Now, we consider a holomorphic curve $f : M \to P^n(\mathbf{C})$, namely, a non-constant holomorphic map of M into $P^n(\mathbf{C})$. We denote by the notation $\langle W\rangle$, or $(w_0 : w_1 : \cdots : w_n)$ the point in $P^n(\mathbf{C})$ associated with a vector $W := (w_0, \ldots, w_n)$ in $\mathbf{C}^{n+1} - \{0\}$. For a domain $D \subset M$, we call $F := (f_0, f_1, \cdots, f_n)$ a reduced representation of f on D if $f_0, f_1, \ldots, f_n$ are holomorphic functions on D without common zeros such that $f(z) = \langle F(z)\rangle$ for every $z \in D$. Consider a hyperplane $H := \{\langle W\rangle; \langle A, W\rangle = 0\}$ in $P^n(\mathbf{C})$ with $f(M) \not\subseteq H$, where $A \in \mathbf{C}^{n+1} - \{0\}$ and $\langle A, W\rangle$ denotes the hermitian product of vectors A and W. For each $a \in M$, f has a reduced representation F on some neighborhood of a. Since another reduced representation of f is represented as a scalar product of F by a nowhere vanishing holomorphic function on the common domain of definition, the divisor $\nu_{<A,F>}$ is well-defined on the totality of M; it is denoted by $\nu(f, H)$, or $f^*([H])$, and called the pull-back of the hyperplane H considered as a divisor. We note here the following:

(2.5) *Every holomorphic curve f in $P^n(\mathbf{C})$ defined on an open (i.e., non-compact) Riemann surface M has a reduced representation on M.*

This is proved by the use of Theorem 1.8(cf., [39, Proposition 2.1.5]).

A holomorphic curve $f : M \to P^n(\mathbf{C})$ is said to be nondegenerate if $f(M) \not\subseteq H$ for any hyperplane H in $P^n(\mathbf{C})$. If we take a reduced representation $F = (f_0, \ldots, f_n)$ on some domain, f is nondegenerate if and only if $f_0, \ldots, f_n$ are linearly independent over $\mathbf{C}$. This is equivalent to the condition that the Wronskian $W(f_0, \ldots, f_n)$ of $f_0, \ldots, f_n$ does not vanish identically on the domain of a holomorphic local coordinate.

We now recall some basic facts concerning the Grassmannian manifold $\mathbf{C}^{n+1}$ for each $0 \leq k \leq n$. A vector A in $\bigwedge^{k+1} \mathbf{C}^{n+1} - \{0\}$ is called a decomposable k-vector if it is written as $A = A_0 \wedge \ldots \wedge A_k$ with $k+1$ vectors $A_i \in \mathbf{C}^{n+1}$. To each $P \in G(n,k)$, take a basis $\{A_0, \ldots, A_k\}$ of P and consider a decomposable vector $A := A_0 \wedge \ldots \wedge A_k$. If we choose another basis $\{\tilde{A}_0, \ldots, \tilde{A}_k\}$ of P, then $\tilde{A}_0 \wedge \ldots \wedge \tilde{A}_k$ is a nonzero scalar multiple of A. Therefore, $\Phi(P) := \pi_k(A) \in P^{N_k}(\mathbf{C})$ depends only on P, where $N_k = \binom{n+1}{k+1} - 1$ and π_k is the canonical projection of $\mathbf{C}^{N_k+1} - \{0\}$ onto $P^{N_k}(\mathbf{C})$. As is easily seen, the map $\Phi : G(n,k) \to P^{N_k}(\mathbf{C})$ is injective. This is called the Plücker imbedding of $G(n,k)$ into $P^{N_k}(\mathbf{C})$. In the following, $G(n,k)$ is identified with $\Phi(G(n,k)) \subset P^{N_k}(\mathbf{C})$.

Take a nondegenerate holomorphic curve $f : M \to P^n(\mathbf{C})$ with a reduced representation $F = (f_0, \ldots, f_n)$ on the domain D of a holomorphic local coordinate z. For $0 \leq k \leq n-1$ we consider the holomorphic map

$$F_k \equiv (F_k)_z := F^{(0)} \wedge F^{(1)} \wedge \ldots \wedge F^{(k)} : D \to \bigwedge^{k+1} \mathbf{C}^{n+1},$$

where $F^{(\ell)} := (f_0^{(\ell)}, \ldots, f_n^{(\ell)})$. For another reduced representation G of f, we can write $G = hF$ with a holomorphic function h without zeros, so that we have $G_k = h^{k+1} F_k$. On the other hand, for another holomorphic local coordinate ζ, we have $(F_k)_\zeta = (F_k)_z (dz/d\zeta)^{k(k+1)/2}$ (cf., [39, Proposition 2.1.6]). Take a nonzero holomorphic function g with $\nu_g = \nu_{|F_k|}$ and define the map $f^k := \pi_k \cdot (F_k/g)$ into $G(n,k)$. The map f^k is well-defined on M because it does not depend on a choice of a reduced representation, a holomorphic local coordinate ζ or a holomorphic function g. We call the map $f^k : M \to G(n,k)$ the k-th derived curve of f; the f^k are called the associated curves of f in [66]. The 0-th derived curve of f is nothing but the original map f itself.

The (normalized) Fubini-Study metric form on $P^{N_k}(\mathbf{C})$ is defined as the form ω_k such that $\pi_k^*(\omega_k) = dd^c \log |W|^2$ for homogeneous coordinates $\langle W \rangle$ on $P^{N_k}(\mathbf{C})$. The pull-backs of ω_k are given by $\Omega_{f^k} \equiv (f^k)^*(\omega_k) := [dd^c \log |F_k|^2]$. For convenience' sake, we set $|F_{-1}| \equiv 1$, and so $\Omega_{f^k} \equiv 0$ for $k = -1$. We can prove the following:

Proposition 2.6. $\Omega_{f^k} = \dfrac{|F_{k-1}|^2 |F_{k+1}|^2}{|F_k|^4} dd^c |z|^2 \ (0 \leq k < n).$

For the proof, refer to e.g. [39, Proposition 2.2.11].

Now, we define the order function of the k-th derived curve f^k by

$$T_f^k(r) := \int_s^r \frac{dt}{t} \int_{M_t} \Omega_{f^k}.$$

and set $T_f(r) := T_f^0(r)$. Let $H := \{\langle W\rangle; \langle A, W\rangle = 0\}$ be a hyperplane in $P^{N_k}(\mathbf{C})$, where A is a unit vector in $\mathbf{C}^{N_k+1}$. Taking a reduced representation $F = (f_0, \ldots, f_n)$ and setting $F_k := F^{(0)} \wedge \ldots \wedge F^{(k)}$ locally, we define the divisor $\nu(f^k, H) := \nu_{<A,F_k>} - \nu_{|F_k|}$, which we may call the pull-back of H via f^k considered as a divisor. We define the (m-truncated) counting function of H for f^k by $N_f^k(r, H)^{[m]} := N(r, \min(\nu(f^k, H), m))$, where m is a positive integer or $+\infty$. For particular cases, we set $N_f^k(r, H) := N_f^k(r, H)^{[+\infty]}$, $N_f(r, H)^{[m]} := N_f^0(r, H)^{[m]}$ and $N_f(r, H) := N_f(r, H)^{[+\infty]}$. We define also the proximity function of H for f^k by

$$m_f^k(r, H) := \frac{2}{r} \int_{\partial M_r} \log \psi(f^k, H) d^c\tau - \frac{2}{s} \int_{\partial M_s} \log \psi(f^k, H) d^c\tau,$$

where $\psi(f^k, H) := |F_k| / |\langle A, F_k\rangle|$. For brevity, we set $m_f(r, H) := m_f^0(r, H)$.

The first main theorem in Nevanlinna theory is stated as follows:

Theorem 2.7. *Let $f : M \to P^n(\mathbf{C})$ be a nondegenerate holomorphic map and $f^k : M \to P^{N_k}(\mathbf{C})$ $(0 \leq k < n)$ its derived curves. Then, for every hyperplane H in $P^{N_k}(\mathbf{C})$,*

$$T_f^k(r) = N_f^k(r, H) + m_f^k(r, H).$$

To see this, take a nonzero holomorphic function h with $\nu_h = \nu_{|F_k|}$. Then, f^k has a reduced representation F_k/h locally and $\nu(f^k, H) = \nu_{<A,F_k>/h}$. By applying Proposition 2.1 to $u := \psi(f^k, H)$ we have the desired conclusion

$$\begin{aligned} T_f^k(r) &= \int_s^r \frac{dt}{t} \int_{M_t} dd^c \log \frac{|F_k|^2}{|h|^2} \\ &= \int_s^r \frac{dt}{t} \int_{M_t} dd^c \left[\log \left|\frac{\langle A, F_k\rangle}{h}\right|^2\right] + \int_s^r \frac{dt}{t} \int_{M_t} dd^c [\log |\psi(f^k, H)|^2] \\ &= N_f^k(r, H) + m_f^k(r, H). \end{aligned}$$

We explain here a type of Crofton's formula for holomorphic curves. We first consider a holomorphic curve f in $P^n(\mathbf{C})$ defined on a compact Riemann surface M. For an arbitrarily chosen hyperplane $H_A := \{\langle W\rangle; \langle A, W\rangle = 0\}$ with $f(M) \not\subseteq H_A$, we define the degree of f by deg $(f) :=$ ord $\nu(f, H_A)$. It does not depend on the choice of a hyperplane H_A. For, if $f(M) \not\subseteq H_A \cup H_B$ for nonzero vectors A and B, then $\mathrm{ord}(\nu(f, H_A)) - \mathrm{ord}(\nu(f, H_B)) = \mathrm{ord}(\nu_\varphi) = 0$ for $\varphi := \langle A, F\rangle / \langle B, F\rangle$. We have

$$\deg(f) = \int_M \Omega_f \tag{2.8}$$

(cf., [15, p. 143]), since $\deg(f) - \int_M \Omega_f = -\int_M dd^c[\log \psi(f, H)^2] = 0$ by (1.1).

Go back to the case of a holomorphic curve defined on an open Riemann surface with a parabolic exhaustion τ. The set of all hyperplanes in $P^n(\mathbf{C})$ is considered as the dual of $P^n(\mathbf{C})$, which may be identified with $P^n(\mathbf{C})$ itself. We denote by dV the volume form associated with the normalized Fubini-Study metric form on $P^n(\mathbf{C})$. We can show the following:

Theorem 2.9. *For every nondegenerate holomorphic map $f : M \to P^n(\mathbf{C})$,*

$$T_f(r) = \int_{H \in P^n(C)} N(r, H) dV .$$

To see this, set $H_A := \{W; \langle A, W\rangle = 0\}$ and consider the function $u_{\langle A\rangle}(z) := |\langle A, F(z)\rangle|/|F(z)|$ for each unit vector $A \in \mathbf{C}^{n+1} - \{0\}$, where F is a reduced representation of f. By Proposition 2.1 we have

$$\begin{aligned} N_f(r, H_A) - T_f(r) &= \int_s^r \frac{dt}{t} \int_{M_r} dd^c[\log |u_{\langle A\rangle}|^2] \\ &= \frac{2}{r} \int_{z \in \partial M_r} \log |u_{\langle A\rangle}(z)| d^c\tau - \frac{2}{s} \int_{z \in \partial M_s} \log |u_{\langle A\rangle}(z)| d^c\tau . \end{aligned}$$

Consider each term of these identities as a function in $\langle A\rangle$ and integrate it with respect to dV. Since the volume of the total space $P^n(\mathbf{C})$ with respect to dV is 1, the left hand side becomes $\int_{H \in P^n(C)} N(r, H) dV - T_f(r)$. On the other hand, the right hand side vanishes. In fact, the function $G(z) = \int_{\langle A\rangle \in P^n(\mathbf{C})} u_{\langle A\rangle}(z) dV$ in z is a constant because $\int_{\langle A\rangle \in P^n(\mathbf{C})} \log(|\langle A, X\rangle|/|X|) dV$ does not depend on the choice of a vector $X \in P^n(\mathbf{C})$, and therefore

$$\int_{\langle A\rangle > \in \mathbf{P}^n(\mathbf{C})} \frac{2}{r} \int_{z \in \partial M_r} \log |u_{\langle A\rangle}(z)| d^c\tau dV = \frac{2}{r} \int_{z \in \partial M_r} G(z) d^c\tau$$

by Fubini's theorem, which does not depend on r.

§3. The Second Main Theorem for Holomorphic Curves

We recall first the notion of the stationary index for a nonconstant holomorphic map of a Riemann surface M into an n-dimensional complex manifold $\tilde{M}$. For a point $a \in M$, taking holomorphic local coordinates $u_1, \ldots, u_n$ around $f(a)$ and writing $f = (f_1, \ldots, f_n)$ with holomorphic functions f_i's around a, we define the stationary index of f at a by $s_f(a) = \min(\nu_{df_1}(a), \ldots, \nu_{df_n}(a))$, where we set $\nu_{df_i}(a) := +\infty$ if f_i is a constant. For a holomorphic curve f in $P^n(\mathbf{C})$ defined on a Riemann surface M, if we write $\Omega_f = u^2 dd^c|z|^2$ with a function u on the domain of each holomorphic local coordinate z, we have $s_f(a) = \nu_u(a)$ by Proposition 2.6, and for the derived curves $f^k : M \to P^{N_k}(\mathbf{C})$ of f we have $s_{f^k}(a) = \nu_{u_k}(a)$, where u_k is the function with $\Omega_{f^k} = u_k^2 dd^c|z|^2$.

Let D be a subdomain of M with real analytic smooth boundary and ω a meromorphic form on D which is purely imaginary on ∂D. Take the function h_k such that $\Omega_{f^k} = h_k^2\sqrt{-1}\omega\wedge\bar{\omega}$ and assume that $\nu_{h_k} = 0$ on ∂D. According to Propositions 1.11 and 2.6, we have the following:

$$\chi(\bar{D}) - \int_{\partial D} d^c \log h_k^2 + \operatorname{ord}(s_{f^k|D}) = -\int_D [\Omega_{f^{k-1}}] + 2\int_D [\Omega_{f^k}] - \int_D [\Omega_{f^{k+1}}] \tag{3.1}$$

By the help of (2.8), this yields the following Plücker's formula:

Theorem 3.2. *If M is compact, then*

$$\operatorname{ord}(s_{f^k}) + \deg(f^{k-1}) - 2\deg(f^k) + \deg(f^{k+1}) + \chi(M) = 0\,.$$

We can give also an analogous Plücker's formula for holomorphic curves in a hermitian manifold of constant holomorphic sectional curvature. For the details, refer to [67].

Now, we consider a holomorphic curve f in $P^n(\mathbf{C})$ defined on an open Riemann surface M with a parabolic exhaustion τ. We define the functions

$$E(r) := \int_s^r \chi(\bar{M}_t)\frac{dt}{t}\,, \qquad S_k(r) := N(r, s_{f^k})$$

and, taking a C^∞ form ω on M such that $\omega = \partial\log\tau$ on $M - M_s$, we set $\upsilon := \sqrt{-1}\omega\wedge\bar{\omega}$. Then, ω is a holomorphic form on $M - \bar{M}_s$ which is purely imaginary on each $\partial M_{s,r}$ except possibly for countably many r, because $\bar{\partial}(\partial\log\tau) = 0$ and $\operatorname{Re}\omega = \tau^{-1}\operatorname{Re}\partial\tau = (1/\tau)d\tau = 0$ on $\partial M_{s,r}$.

Theorem 3.3. *Take the function h_k with $\Omega_{f^k} = h_k^2\upsilon$. Then, for $r > s$,*

$$\begin{aligned} &E(r) + S_k(r) + T_f^{k-1}(r) - 2T_f^k(r) + T_f^{k+1}(r) \\ &= \frac{2}{r}\int_{\partial M_r} \log h_k d^c\tau - \frac{2}{s}\int_{\partial M_s} \log h_k d^c\tau\,. \end{aligned}$$

To see this, we add the identities obtained by applying Theorem 1.12 to the pseudo-metric associated with Ω_f on M_s and on $M_{r,s}$, and integrate each term of the obtained identity multiplied by $1/t$ from s to r to see

$$E(r) + S_k(r) - \int_s^r \frac{dt}{t}\int_{\partial M_t} d^c \log h_k^2 = -(T_{k-1}(r) - 2T_k(r) + T_{k+1}(r))\,.$$

Then, we obtain Theorem 3.3 because Proposition 2.1 gives

$$\begin{aligned} \int_s^r \frac{dt}{t}\int_{\partial M_t} d^c \log h_k^2 &= \int_s^r \frac{dt}{t}\int_{M_t} dd^c[\log h_k^2] \\ &= \frac{2}{r}\int_{\partial M_r} \log h_k d^c\tau - \frac{2}{s}\int_{\partial M_s} \log h_k d^c\tau\,. \end{aligned}$$

For each locally summable nonnegative function h on M we define

$$T^h(r) := \int_s^r \frac{dt}{t} \int_{M_t} hv\,.$$

We have the following:

Proposition 3.4. *For a locally summable nonnegative function h on M there are positive constants C_0 and C_1 such that*

$$\frac{2}{r}\int_{\partial M_r} \log h d^c\tau \leq C_0 \log T^h(r) + C_1$$

for each r except in a set with finite logarithmic measure.

The proof is similar to that of [39, Proposition 3.2.4].

Set $T(r) := \max_{0\leq k\leq n-1} T_f^k(r)$ and let $\alpha_T(t)$ denote a function in t such that $\alpha_T(r) \leq C\log T(r)$ $(r \notin E)$ for some $C > 0$ and a set E with finite logarithmic measure. By Theorem 3.3 and Proposition 3.4, we have

$$E(r) + S_k(r) + T_f^{k-1}(r) - 2T_f^k(r) + T_f^{k+1}(r) = \alpha_T(r)\,. \tag{3.5}$$

Similarly to [39, Corollary 3.2.11], we can show the following:

Proposition 3.6. *For a set E with finite logarithmic measure,*

$$T_f^k(r) \leq O(-E(r)) + O(T_f^\ell(r)) \qquad (r \notin E, 0 \leq k,\ell < n)\,.$$

To state the second main theorem, we review some results in linear algebra which were given by E. I. Nochka in [53](cf., [13]). Let $N \geq n$ and $q \geq N+1$. We consider q hyperplanes $H_j := \{\langle W\rangle; \langle A_j, W\rangle = 0\}$ $(1 \leq j \leq q)$ in $P^n(\mathbf{C})$, where $A_j \in \mathbf{C}^{n+1} - \{0\}$. Following W. Chen ([13]), we say that $H_1, \ldots, H_q$ (or $A_1, \ldots, A_q$) are in N-subgeneral position if $H_{i_0} \cap H_{i_1} \cap \cdots \cap H_{i_N} = \emptyset$ for every $1 \leq i_0 < \cdots < i_N \leq q$, where $q > 2N - n + 1$. By definition, $\{H_j; 1 \leq j \leq q\}$ are in general position if and only if they are in n-subgeneral position. E. I. Nochka showed the following:

Theorem 3.7. *Let $H_j (1 \leq j \leq q)$ be hyperplanes in $P^n(\mathbf{C})$ located in N-subgeneral position which are associated with vectors $A_j \in \mathbf{C}^{n+1} - \{0\}$. Then there exist constants $\omega(1), \ldots, \omega(q)$ and θ such that*

(i) $0 < \omega(j) \leq \theta \leq 1 \qquad (1 \leq j \leq q)$,

(ii) $\displaystyle\sum_{j=1}^{q} \omega(j) = \theta(q - 2N + n - 1) + n + 1$,

(iii) $\displaystyle\frac{n+1}{2N-n+1} \leq \theta \leq \frac{n+1}{N+1}$,

(iv) *if $R \subset Q := \{1, 2, \ldots, q\}$ and $0 < \#R \leq N+1$, then $\sum_{j\in R} \omega(j)$ is not larger than the dimension of the vector subspace of $\mathbf{C}^{n+1}$ generated by $\{A_j; j \in R\}$.*

In the following, we call constants $\omega(j)(1 \leq j \leq q)$ and θ with the properties (i) $\sim$ (iv) Nochka weights and a Nochka constant for the H_j's respectively. For $H_1, \dots, H_q$ in general position, we can take $\omega(1) = \cdots = \omega(q) = \theta = 1$.

Now, we shall state the second main theorem for holomorphic curves in $P^n(\mathbf{C})$, which was first proved by H. Cartan ([11]) for the case $M = \mathbf{C}$ and $N = n$ and generalized by H. and J. Weyl ([66]) and L. V. Ahlfors ([2]) to an arbitrary M with a parabolic exhaustion and by E. I. Nochka ([53]) to arbitrary N $(\geq n)$.

Theorem 3.8. *Let f be a nondegenerate holomorphic curve in $P^n(\mathbf{C})$ defined on an open Riemann surface with a parabolic exhaustion τ. Take hyperplanes $H_1, \dots, H_q$ in $P^n(\mathbf{C})$ located in N-subgeneral position and let $\omega(j)(1 \leq j \leq q)$ and θ be Nochka weights and a Nochka constant for them, where $q \geq 2N - n + 1$. Then, for every $\varepsilon > 0$ there is a set E with finite logarithmic measure such that, for all $r \notin E$,*

$$\begin{aligned} &\theta(q - 2N + n - 1)T_f(r) \\ &\qquad \leq \sum_{j=1}^{q} \omega(j) N_f(r, H_j)^{[n]} + O(-E(r)) + O(\log r) + O(\log T_f(r)). \end{aligned}$$

We give a sketch of the outline of the proof of Theorem 3.8 in a weaker form where $O(\log T_f(r))$ is replaced by $o(T_f(r))$. We first give the following:

Definition 3.9. Let $f : M \to P^n(\mathbf{C})$ be a holomorphic curve and H a hyperplane with $f(M) \not\subseteq H$. For each $a \in M$ choosing a reduced representation $F = (f_0, \cdots, f_n)$ on a neighborhood of a and a unit vector A with $H = \{\langle W\rangle; \langle A, W\rangle = 0\}$, we define the k-th contact function of f for H by

$$\varphi_k(H)(z) := \frac{|F_k(z) \vee A|^2}{|F_k(z)|^2},$$

where $F_k \vee A$ denotes the interior product of F_k and A as in [39]. It is easily seen that $\varphi_k(H)$ is globally well-defined on the totality of M.

For these contact functions, we can prove the following so-called "Sum to Product estimate" (cf., [39, Theorem 2.5.3]):

Theorem 3.10. *Let $H_1, \dots, H_q$ be hyperplanes in $P^n(\mathbf{C})$ located in N-subgeneral position and let $\omega(j)$ $(1 \leq j \leq q)$ and θ be Nochka weights and a Nochka constant for these hyperplanes. For every $\varepsilon > 0$ there exist constants $a > 1$ and $C > 0$ such that*

$$\begin{aligned} &dd^c \log \frac{\prod_{k=0}^{n-1} |F_k|^{2\varepsilon}}{\prod_{1 \leq j \leq q, 0 \leq k \leq n-1} \log^{2\omega(j)}(a/\varphi_k(H_j))} \\ &\qquad \geq C \left(\frac{|F_0|^{2\theta(q-2N+n-1)} |F_n|^2}{\prod_{j=1}^{q} (|F(H_j)|^2 \prod_{k=0}^{n-1} \log^2(a/\varphi_k(H_j)))^{\omega(j)}} \right)^{\frac{2}{n(n+1)}} dd^c|z|^2 \end{aligned}$$

on the domain of each holomorphic local coordinate z.

For the proof of Theorem 3.8, we use the form $\omega := \partial \log \tau$ and $\upsilon := \sqrt{-1}\omega \wedge \bar{\omega}$. Taking a nonzero function g_k with $\nu_{g_k} = \nu_{|F_k|}$ locally, we consider the function ψ_k such that $|F_k/g_k|\,|dz|^{k(k+1)/2} = \psi_k|\omega|^{k(k+1)/2}$. For an arbitrary given $\varepsilon > 0$, set

$$\hat{h} := \frac{\prod_{k=0}^{n-1}\psi_k^{\varepsilon}}{\prod_{1\le j\le q, 0\le k\le n-1}\log^{\omega(j)}(a/\varphi_k(H_j))}, \quad \tilde{h} := \frac{|F_0|^{\theta(q-2N+n-1)}|\psi_n||g_n|\hat{h}}{\prod_{k=0}^{n-1}\psi_k^{\varepsilon}\prod_{j=1}^{q}|F(H_j)|^{\omega(j)}}$$

and define the function h^* by $dd^c \log \hat{h}^2 = h^*\upsilon$. Then, by Theorem 3.10

$$h^* \ge C\tilde{h}^{4/n(n+1)} \tag{3.11}$$

for some $C > 0$. We apply the same argument as in the proof of Theorem 3.3 to the (1, 1)-form $\tilde{h}^{4/n(n+1)}\upsilon$ to see

$$\begin{aligned}\frac{n(n+1)}{2}E(r) + N(r, \nu_{\tilde{h}}) + \int_s^r \frac{dt}{t}\int_{M_t}[dd^c \log \tilde{h}^2] \\ = \frac{2}{r}\int_{\partial M_r}\log \tilde{h} d^c\tau - \frac{2}{s}\int_{\partial M_s}\log \tilde{h} d^c\tau\,.\end{aligned} \tag{3.12}$$

On the other hand, by the use of [39, Lemma 3.2.13] we can prove

$$N(r, \nu_{\tilde{h}}) = N(r, \nu_{|F_n|}) - \sum_{j=1}^{q}\omega(j)\nu(f, H_j) \ge -\sum_{j=1}^{q}N_f(r, H_j)^{[n]}\,.$$

Using the identity $[dd^c \log|\psi_n|^2] = [dd^c \log|F(H_j)|^2] = 0$, we have

$$\begin{aligned}&\int_s^r \frac{dt}{t}\int_{M_{s,t}}[dd^c \log \tilde{h}^2] \\ &= \theta(q-2N+n-1)T_f(r) + T^{h^*}(r) - \varepsilon\left(\sum_{k=0}^{n-1}\int_s^r \frac{dt}{t}\int_{M_t}[dd^c \log|\psi_k|^2]\right).\end{aligned}$$

Here, we may replace the last term by $-\varepsilon(\sum_{k=0}^{n-1}T_{f^k}(r))$ and the right hand side of (3.12) can be estimated from above by

$$\frac{n(n+1)}{4}\frac{2}{r}\int_{\partial M_r}\log h^* d^c\tau + O(1) \le O\left(\log T^{h^*}(r)\right) + O(1)$$

except on a set with finite logarithmic measure by the use of (3.11) and Proposition 3.4. Since $O(\log T^{h^*}(r)) - T^{h^*}(r)$ is bounded from above, we easily obtain

$$\theta(q-2N+n-1)T_f(r) \le \sum_{j=1}^{q}\omega(j)N_f(r, H_j)^{[n]} + O(-E(r)) + \varepsilon T_f(r) \tag{3.13}$$

with the help of Proposition 3.6.

§4. The Defect Relation and Its Applications

In this section, we present some applications of the main theorems for holomorphic curves in $P^n(\mathbf{C})$ which are related to the subjects presented in the following chapters. Let f be a holomorphic curve in $P^n(\mathbf{C})$ defined on an open Riemann surface with a parabolic exhaustion τ. We note first the following:

Proposition 4.1. *The order function $T_f(r)$ is increasing and convex with respect to $\log r$ on $[s, +\infty)$, and tends to ∞ as $r \to +\infty$.*

To see the first assertion, consider the function h with $\Omega_f = hd\log\tau \wedge d^c\log\tau$ and observe the identities

$$\frac{dT_f(r)}{d\log r} = \int_{M_r} \Omega_f \geq 0, \quad \frac{d^2T_f(r)}{(d\log r)^2} = \frac{1}{r}\int_{\partial M_t} hd^c\tau \geq 0.$$

To see the latter, take some $t_0 > s$ with $K := \int_{M_{t_0}} \Omega_f > 0$. We then have

$$T_f(r) \geq \int_{t_0}^{r} \frac{dt}{t} \int_{M_{t_0}} \Omega_f \geq K \log \frac{r}{t_0} \to \infty \quad (\text{as } r \to \infty).$$

We say that f is transcendental if $\lim_{r\to+\infty} T_f(r)/\log r = +\infty$. As in the proof of [39, Proposition 3.3.3], we can prove the following:

Proposition 4.2. *If M is of finite type, then the following four conditions are equivalent.*

(i) *f cannot be extended to a holomorphic curve defined on a compactification $\bar{M}$.*

(ii) *There is a hyperplane H in $P^n(\mathbf{C})$ with $f(M) \not\subseteq H$ such that $f^{-1}(H)$ contains infinitely many points.*

(iii) *There is some hyperplane H in $P^n(\mathbf{C})$ with $f(M) \not\subseteq H$ such that $\lim_{r\to+\infty} N_f(r, H)/\log r = +\infty$.*

(iv) *f is transcendental.*

Let H be a hyperplane in $P^n(\mathbf{C})$ and m a positive integer or $+\infty$. We define the defect (truncated by m) of H for f by

$$\delta_f(H)^{[m]} := 1 - \limsup_{r\to+\infty} \frac{N_f(r, H)^{[m]}}{T_f(r)}. \tag{4.3}$$

For convenience' sake, we set $\delta_f(H)^{[m]} = 0$ for a hyperplane H with $f(M) \subseteq H$ and $\delta_f(H) := \delta_f(H)^{[\infty]}$.

The following proposition gives some geometric meanings of the defect.

Proposition 4.4. (i) $0 \leq \delta_f(H)^{[m]} \leq 1$ *for every m.*

(ii) *If $f(M) \cap H = \emptyset$, or if $\#f^{-1}(H) < +\infty$ for the particular case where f is transcendental, then $\delta_f(H) = 1$.*

(iii) *If $\nu(f, H) \geq m$ on $f^{-1}(H)$ for some $m > n$, then $\delta_f(H) \geq 1 - n/m$.*

The assertions (i) and (ii) are easy results of Theorem 2.7 and (4.3). On the other hand, if $\nu(f,H) \geq m$ on $f^{-1}(H)$, we have

$$\min(\nu_f(r,H),n) \leq n\min(\nu_f(r,H),1) \leq \frac{n}{m}\nu_f(r,H),$$

and so $1 - N_f(r,H)^{[n]}/T_f(r) \geq 1-(n/m)N_f(r,H)/T_f(r)$, which gives (iii).

Proposition 4.5. *Let f be a nondegenerate holomorphic curve in $P^n(\mathbf{C})$ defined on a Riemann surface M with a parabolic exhaustion. Then, for almost all hyperplanes H in $P^n(\mathbf{C})$, $\delta_f(H)=0$ and so $f(M)\cap H \neq \emptyset$.*

For, by Theorem 2.9, we have

$$0 \leq \int_{H\in P^n(C)} \delta_f(H)dV \leq \liminf_{r\to+\infty}\int_{H\in P^n(C)}(1-N_f(r,H)/T_f(r))\,dV = 0.$$

Now, we give the defect relation for a holomorphic curve in $P^n(\mathbf{C})$.

Theorem 4.6. *Let M be a Riemann surface of finite type, $f: M \to P^n(\mathbf{C})$ a nondegenerate transcendental holomorphic curve and $H_j (1 \leq j \leq q)$ hyperplanes located in N-subgeneral position with Nochka weights $\omega(j)$. Then,*

$$\sum_{j=1}^{q}\omega(j)\delta_f(H_j)^{[n]} \leq n+1.$$

Proof. For every $\varepsilon > 0$, by (3.13) and Theorems 3.7, (ii) we have

$$\begin{aligned}
&\sum_{j=1}^{q}\omega(j)\left(1-\frac{N_f(r,H_j)^{[n]}}{T_f(r)}\right)\\
&\qquad = \theta(q-2N+n-1)+n+1-\sum_{j=1}^{q}\omega(j)\frac{N_f(r,H_j)^{[n]}}{T_f(r)}\\
&\qquad \leq n+1+\varepsilon+\frac{O(-E(r))+O(\log(rT_f(r))))}{T_f(r)}
\end{aligned}$$

for all r except on a set with finite logarithmic measure. Taking lower limits of both sides of this inequality and letting ε tend to zero, we easily conclude Theorem 4.6 because $-\chi(M_r)$ is bounded and so $\liminf_{r\to+\infty} -E(r)/T_f(r) = O(\log r/T_f(r)) = 0$.

In the same situation as in Theorem 4.6, we have always

$$\sum_{j=1}^{q}\delta_f(H_j)^{[n]} \leq 2N-n+1 \tag{4.7}$$

(cf., [39, Corollary 3.3.9]).

Corollary 4.8. *Let f be a transcendental holomorphic curve in $P^N(\mathbf{C})$ defined on an open Riemann surface of finite type such that the image of f is not included in any projective linear subspace of dimension less than n. If there are hyperplanes $H_1,\ldots,H_q$ located in general position and positive integers*

$m_1, \dots, m_q$ larger than n such that $\nu(f, H_j) \geq m_j$ $(1 \leq j \leq q)$ on $f^{-1}(H_j)$ excluding a compact set, then

$$\sum_{j=1}^{q} \left(1 - \frac{n}{m_j}\right) \leq 2N - n + 1 (\leq 2N) .$$

To see this, consider the smallest projective linear subspace of $P^N(\mathbf{C})$ including $f(H)$, which we denote by $P^m(\mathbf{C})$. By assumption, $m \geq n$ and f is considered as a nondegenerate holomorphic map into $P^m(\mathbf{C})$ and, moreover, the intersections of $P^m(\mathbf{C})$ with H_j are in N-subgeneral position in $P^m(\mathbf{C})$. These show that Corollary 4.8 is an immediate consequence of (4.7).

In Corollary 4.8, if $f(M) \cap H_j = \emptyset$, we may take $m_j := +\infty$. Therefore, a transcendental holomorphic curve in $P^N(\mathbf{C})$ defined on a Riemann surface of finite type can omit at most $2N$ hyperplanes in general position.

For the particular case where $M = \mathbf{C}$, we can show the following generalization of Picard's theorem.

Theorem 4.9. *Every holomorphic curve f in $P^n(\mathbf{C})$ defined on $\mathbf{C}$ can omit at most $2n$ hyperplanes in $P^n(\mathbf{C})$ located in general position. For the particular case where f is nondegenerate, it can omit at most $n+1$ hyperplanes in general position.*

We need only prove this for the case f is not transcendental because of Corollary 4.8. In this case, f can be extended to a holomorphic map into $P^n(\mathbf{C})$ defined on $P^1(\mathbf{C}) = \mathbf{C} \cup \{\infty\}$ by Proposition 4.2. This means that we can take a reduced representation $F = (f_0, \dots, f_n)$ on $\mathbf{C}$ with polynomials f_i's. If $f(\mathbf{C}) \cap H = \emptyset$ for a hyperplane $H = \{\langle W \rangle; \langle A, W \rangle = 0\}$, then $\varphi := \langle A, F \rangle$ is a holomorphic function which has no zero on $\mathbf{C}$, which is necessarily a constant. Therefore, the proof of Theorem 4.9 is reduced to some easy algebraic considerations.

From Theorem 4.9 we can conclude the following classical Borel's theorem ([10]):

Theorem 4.10 ([22]). *Let $f_1, f_2, \dots, f_p$ be nowhere vanishing holomorphic functions on a domain in $\mathbf{C}$ and assume that $c_1 f_1 + \dots + c_p f_p \equiv 0$ for some nonzero constants c_j. Consider the partition*

$$\{1, 2, \dots, p\} = I_1 \cup \dots \cup I_k$$

such that i and j are in the same class I_ℓ if and only if f_i/f_j is a constant. Then, $\sum_{i \in I_\ell} c_i f_i \equiv 0$ for any ℓ.

For the proof, refer to [39, pp. 115–116].

For a nondegenerate holomorphic curve f in $P^n(\mathbf{C})$, the derived curves f^k are also holomorphic curves in the higher dimensional projective space $P^{N_k}(\mathbf{C})$. Although they are not necessarily nondegenerate, we can prove the following analogue of Theorem 4.6([66], [17] and [39, §3.6]).

Theorem 4.11. *Let f be a nondegenerate transcendental holomorphic curve in $P^n(\mathbf{C})$ defined on an open Riemann surface of finite type and f^k $(1 \leq k \leq n-1)$ its derived curves in $P^{N_k}(\mathbf{C})$. Take q decomposable vectors $A_1, \ldots, A_q$ in $\mathbf{C}^{N_k+1}$ and consider the hyperplanes $H_j := \{\langle W\rangle; \langle A_j, W\rangle = 0\}$. If they are located in general position, then*

$$\sum_{j=1}^{q} \delta_{f^k}(H_j)^{[k(n-k)]} \leq \binom{n+1}{k+1}.$$

We present here another application of Nevanlinna theory. In [50], R. Nevanlinna showed that two nonconstant meromorphic functions φ and ψ on $\mathbf{C}$ coincide with each other if $\varphi^{-1}(\alpha_j) = \psi^{-1}(\alpha_j)$ for five distinct values $\alpha_1, \ldots, \alpha_5$. Related to this result is the following uniqueness theorem that can be proved by the use of Theorem 4.10.

Theorem 4.12. *Let f and g be holomorphic curves in $P^n(\mathbf{C})$ defined on $\mathbf{C}$ at least one of which is nondegenerate. If there are $3n+2$ hyperplanes $H_1, \ldots, H_{3n+2}$ located in general position such that $f(\mathbf{C}) \not\subseteq H_j$, $g(\mathbf{C}) \not\subseteq H_j$ and $\nu(f, H_j) = \nu(g, H_j)$ for $j = 1, \ldots, 3n+2$, then $f \equiv g$.*

For the proof, see the original paper [23].

We can also prove the following:

Theorem 4.13. *Let f and g be holomorphic curves in $P^n(\mathbf{C})$ defined on $\mathbf{C}$. Assume that at least one of them is algebraically nondegenerate, namely, the image is not contained in any proper algebraic subset of $P^n(\mathbf{C})$. If there are $2n+3$ hyperplanes $H_1, \ldots, H_{2n+3}$ located in general position such that $f(\mathbf{C}) \not\subseteq H_j$, $g(\mathbf{C}) \not\subseteq H_j$ and $\nu(f, H_j) = \nu(g, H_j)$ $(1 \leq j \leq 2n+3)$, then $f \equiv g$.*

For the proof, see the original papers [24] and [25].

For other results relating to the uniqueness problems of holomorphic curves, refer to [26] $\sim$ [28].

Chapter 2
Minimal Surfaces of Parabolic Type

§5. Minimal Surfaces and Their Gauss Maps

We begin with recalling some basic notions concerning a surface $x : M \to \mathbf{R}^m$, where M is a connected, oriented real 2-dimensional differentiable manifold and x is a differentiable map of M into $\mathbf{R}^m$. We say that a surface $x : M \to \mathbf{R}^m$ is regular if x is an immersion, namely, of rank two everywhere. For each $a \in M$ we denote by $T_a(M)$ and $N_a(M)$ the tangent plane and the normal space of M at a respectively. The first fundamental form is given by $ds^2 = \sum_{1\le i,j\le 2} g_{ij}du_idu_j$ with $g_{ij} = (\partial x/\partial u_i, \partial x/\partial u_j)$ in terms of local coordinates u_1, u_2, and for each $N \in N_a(M)$ the second fundamental form in the normal direction N is given by $d\sigma^2 = \sum_{1\le i,j\le 2} b_{ij}du_idu_j$ with $b_{ij} = (N, \partial x^2/\partial u_i\partial u_j)$. Take a vector $N \in N_a(M)$. For each nonzero vector $T \in T_a(M)$ we choose a regular curve $\gamma : x = \gamma(s)$ in M with arc-length parameter s such that $\gamma(0) = a$ and $(d\gamma/ds)(0) = T$, and define the normal curvature of M in the direction T with respect to the normal vector N by $k_N(T) := (d^2\gamma/ds^2, N)$, which depends only on N and T. By an elementary calculation, we can show that $|k_N(T)|$ is the reciprocal of the radius of curvature for the curve which is the intersection of the surface M and the plane containing N and T. By definition, the mean curvature $H_a(M)$ of M for the direction N at a is the mean of the maximum and the minimum of the quantities $k_N(T)$ for all unit vectors $T \in T_a(M)$. We can show that

$$H_a(N) = \frac{g_{11}b_{22} - 2g_{12}b_{12} + g_{22}b_{11}}{2(g_{11}g_{22} - g_{12}^2)} \tag{5.1}$$

(cf, [58, p. 13]).

A regular surface M in $\mathbf{R}^m$ is called a minimal surface if $H_a(N) = 0$ for all $a \in M$ and $N \in N_a(M)$. To explain a geometric meaning of minimal surfaces, take a subdomain D of M with smooth boundary and a normal vector field N on a neighborhood of $\bar{D}$ which vanishes on ∂D. Consider the surfaces

$$D^t : x = x^t := x(u_1, u_2) + tN(u_1, u_2) \qquad ((u_1, u_2) \in D)$$

for a sufficiently small t and denote by $A(D_t)$ the area of D_t. Then

$$\frac{d}{dt}A(D_t)\Big|_{t=0} = -2\int\int_{a\in D} H_a(N)\Omega_{ds^2}\,, \tag{5.2}$$

where $\Omega_{ds^2} := \sqrt{g_{11}g_{22} - g_{12}^2}du_1du_2$ is the area form of M(cf., [58, p. 22]). It follows that a regular surface M is minimal if and only if $(d/dt)A(D_t)\big|_{t=0} = 0$ for all subdomains D with smooth boundary and normal vector fields N vanishing on ∂D. Therefore, we may say that a surface M is minimal if and only if for all local perturbations of M to the normal directions the area function takes a critical value at M.

A system of local coordinates u, v on an open set U in M is called isothermal coordinates on U if we can write $ds^2 = \lambda^2(du^2 + dv^2)$ with some positive C^∞ function λ on U. Local coordinates u, v on a surface $x : M \to \mathbf{R}^m$ are isothermal coordinates if and only if $g_{11} = g_{22}, g_{12} = g_{21} = 0$, or

$$\left(\frac{\partial x}{\partial u}, \frac{\partial x}{\partial u}\right) = \left(\frac{\partial x}{\partial v}, \frac{\partial x}{\partial v}\right), \quad \left(\frac{\partial x}{\partial u}, \frac{\partial x}{\partial v}\right) = 0\,. \tag{5.3}$$

Theorem 5.4. *Every regular surface M is covered by the domains of isothermal local coordinates.*

For the proof, refer to e.g. [14]. A simple proof of Theorem 5.4 is known for the particular case where M is a minimal surface in $\mathbf{R}^m$. For the details, refer to [58, pp. 31–32].

Consider two systems of positively oriented isothermal local coordinates (x, y) and (u, v) on an oriented surface M with a metric ds^2. Then, we can easily show that $w := u + \sqrt{-1}v$ is a holomorphic function in $z := x + \sqrt{-1}y$ on the common domain of definition. According to this fact and Theorem 5.4, every regular surface $x = (x_1, \ldots, x_m) : M \to \mathbf{R}^m$ can be considered as a Riemann surface by associating a holomorphic local coordinate $z := u + \sqrt{-1}v$ with each system of positively oriented isothermal local coordinates (u, v). Set $\omega_i := \partial x_i \equiv (\partial/\partial z)x_i dz$ $(1 \leq i \leq m)$. Then the metric ds^2 is given by

$$ds^2 = 2(|\omega_1|^2 + \cdots + |\omega_m|^2).$$

Proposition 5.5. *For a regular surface $x : M \to \mathbf{R}^m$, take a system of isothermal local coordinates (u, v) such that the induced metric is represented as $ds^2 = \lambda^2(du^2 + dv^2)$. Then,*

(i) $(\Delta x, X) = 0$ *for each* $X \in T_a(M)$,

(ii) $(\Delta x, N) = 2\lambda^2 H_a(N)$ *for each* $N \in N_a(M)$,

where Δ denotes the Laplacian $\Delta := \dfrac{\partial^2}{\partial u^2} + \dfrac{\partial^2}{\partial v^2}$.

In fact, by (5.3) we have

$$\left(\Delta x, \frac{\partial x}{\partial u}\right) = \frac{1}{2}\frac{\partial}{\partial u}\left\{\left(\frac{\partial x}{\partial u}, \frac{\partial x}{\partial u}\right) - \left(\frac{\partial x}{\partial v}, \frac{\partial x}{\partial v}\right)\right\} + \frac{\partial}{\partial v}\left(\frac{\partial x}{\partial u}, \frac{\partial x}{\partial v}\right) = 0$$

and, similarly, $(\Delta x, \partial x/\partial v) = 0$, which gives (i). On the other hand, by (5.1)

$$H_a(N) = \frac{b_{11}(N) + b_{22}(N)}{2\lambda^2} = \frac{(\Delta x, N)}{2\lambda^2} \qquad (N \in N_a(M)),$$

because $g_{11} = g_{22} = \lambda^2$ and $g_{12} = 0$, which gives (ii).

According to Proposition 5.5, we can easily show the following:

Theorem 5.6. *Let $x := (x_1, \ldots, x_m) : M \to \mathbf{R}^m$ be a regular surface in $\mathbf{R}^m$, which is considered as a Riemann surface with a conformal metric. Then, M is minimal if and only if each x_i is a harmonic function on M.*

Corollary 5.7. *There is no compact regular minimal surface in $\mathbf{R}^m$.*

In fact, for a compact regular minimal surface $x = (x_1, \ldots, x_m) : M \to \mathbf{R}^m$ each component x_i takes its maximum value at a point in M. By the maximum principle, x_i is a constant and so x can not be an immersion.

We next explain the Gauss map of a regular surface $x : M \to \mathbf{R}^m$. Consider the set Π of all oriented 2-planes in $\mathbf{R}^m$ containing the origin. For each $P \in \Pi$, taking a positively oriented basis $\{X, Y\}$ of P such that $|X| = |Y|, (X, Y) = 0$ and setting $\Phi(P) := \langle X - \sqrt{-1}Y\rangle \in P^{m-1}(\mathbf{C})$ we define the map $\Phi : \Pi \to P^{m-1}(\mathbf{C})$. This is well-defined, because an arbitrary positive basis $\{\tilde{X}, \tilde{Y}\}$ of

P with $|\tilde{X}| = |\tilde{Y}|, (\tilde{X}, \tilde{Y}) = 0$ satisfies the identity $\tilde{X} - \sqrt{-1}\tilde{Y} = re^{i\theta}(X - \sqrt{-1}Y)$ for some $r > 0$ and $\theta \in \mathbf{R}$. Since

$$(X - \sqrt{-1}Y,\ X - \sqrt{-1}Y) = (X,X) - 2\sqrt{-1}(X,Y) - (Y,Y) = 0$$

for the above (X,Y), $\Phi(\Pi)$ is included in the quadric

$$Q_{m-2}(\mathbf{C}) := \{(w_1 : \cdots : w_m); w_1^2 + \cdots + w_m^2 = 0\}(\subset P^{m-1}(\mathbf{C}))\,.$$

Moreover, as is easily shown, the map $\Phi : \Pi \to Q_{m-2}$ is bijective and therefore we may identify Π with $Q_{m-2}(\mathbf{C})$.

Let $x = (x_1, x_2, \cdots, x_m) : M \to \mathbf{R}^m$ be a regular surface in $\mathbf{R}^m$. The Gauss map of M is defined as the map of M into $Q_{m-2}(\mathbf{C})$ which maps each $a \in M$ to $\Phi(T_a(M))$, where each oriented tangent plane $T_a(M)$ is regarded as an element in Π after the parallel translation which maps a to the origin. The Gauss map defined above is equal to the conjugate of the usually defined Gauss map. The reason why we adopt the above definition is to simplify the description of function-theoretic properties of minimal surfaces.

For a system of positively oriented isothermal local coordinates (u,v), if we take the vectors $X = \partial x/\partial u$ and $Y = \partial x/\partial v$, X and Y satisfy the condition $|X| = |Y|, (X,Y) = 0$ by (5.3) and so the Gauss map of M is locally given by

$$G = \langle X - \sqrt{-1}Y \rangle = \left(\frac{\partial x_1}{\partial z} : \frac{\partial x_2}{\partial z} : \cdots : \frac{\partial x_m}{\partial z}\right),$$

where $z = u + \sqrt{-1}v$. We may write $G = (\omega_1 : \cdots : \omega_m)$ with globally defined holomorphic forms ω_i $(1 \leq i \leq m)$.

We have the following criterion for minimal surfaces.

Proposition 5.8. *A surface $x : M \to \mathbf{R}^m$ is minimal if and only if the Gauss map $G : M \to P^{m-1}(\mathbf{C})$ is holomorphic.*

In fact, if M is minimal, then $\bar{\partial}\omega_i = (1/4)\Delta x_i d\bar{z} \wedge dz = 0$ and so each ω_i is holomorphic, whence G is holomorphic. For the proof of the converse, refer to [44, Theorem 1.1].

We say that a holomorphic form ω on a Riemann surface M has no real period if $\mathrm{Re}\int_\gamma \omega = 0$ for every closed piecewise smooth continuous curve γ in M. If ω has no real period, then $x(z) = \mathrm{Re}\int_{\gamma_{z_0}^z} \omega$ depends only on z_0 and z for each piecewise smooth continuous curve $\gamma_{z_0}^z$ in M joining z_0 and z and so $x(z)$ is a well-defined function on M. For new constructions of minimal surfaces, the following theorem is useful.

Theorem 5.9. *Let M be an open Riemann surface and ω_i $(1 \leq i \leq m)$ holomorphic forms on M which have no common zero and no real periods and satisfy the identity $\sum_{i=1}^m f_i^2 = 0$ for local expressions $\omega_i = f_i(z)dz$. Then, for an arbitrary fixed $z_0 \in M$ the surface $x = (x_1, \ldots, x_m) : M \to \mathbf{R}^m$ defined by the functions $x_i = 2\mathrm{Re}\int_{z_0}^z \omega_i$ is a regular minimal surface in $\mathbf{R}^m$ whose Gauss map is given by $G = (\omega_1 : \cdots : \omega_m) : M \to Q_{m-2}(\mathbf{C})$.*

In fact, for the local expressions $\omega_i = f_i dz$ with a holomorphic local coordinate $z = u + \sqrt{-1}v$, we have $(\partial/\partial z)x_i = f_i$ and the condition $\sum_{i=1}^m f_i^2 = 0$ is equivalent to (5.3). Therefore, the metric is conformal. Since ω_i are holomorphic and have no common zero, M is a regular minimal surface.

Let M be a Riemann surface with a conformal metric ds^2 with local expressions $ds^2 = \lambda_z^2|dz|^2$. By definition, the Gaussian curvature of M is given by $K \equiv K_{ds^2} := -(1/\lambda_z^2)\Delta_z \log \lambda_z$. For a regular minimal surface $x : M \to \mathbf{R}^m$ we have $\lambda_z^2 = 2\sum_{j=1}^m |f_j|^2$, where $f_j := \partial x_j/\partial z$. Therefore, the Gaussian curvature is locally given by

$$K_{ds^2} = \frac{-2}{|F|^2}\frac{\partial^2}{\partial \bar{z}\partial z}\log|F| = \frac{-1}{|F|^6}\left(\sum_{i<j}|f_i f_j' - f_j f_i'|^2\right) (\leq 0),$$

where $|F| = (|f_1|^2 + \cdots + |f_m|^2)^{1/2}$. By definition, a surface with a metric is flat if and only if the Gaussian curvature vanishes identically. A regular minimal surface M in $\mathbf{R}^m$ is flat if and only if G is a constant, or M lies in a plane.

We now consider the particular case $m = 3$. In $\mathbf{R}^3$, to each oriented plane there corresponds a unique positive unit normal vector $N \in S^2$, where S^2 denotes the unit sphere in $\mathbf{R}^3$. On the other hand, S^2 is bijectively mapped onto the extended complex plane $\bar{\mathbf{C}} = \mathbf{C} \cup \{\infty\}$ by the stereographic projection ϖ. Instead of the above-mentioned Gauss map, we may study the map G whose value at a is the positively oriented unit normal vector $N_p \in S^2$ of M at a. We call the map $G : M \to S^2$, or occasionally $g := \varpi \cdot G : M \to \bar{\mathbf{C}}$, the classical Gauss map of M in the following. It can be shown that the classical Gauss map g is represented as $g = f_3/(f_1 - \sqrt{-1}f_2)$ with the functions $f_i := \partial x_i/\partial z$([58]). As a result of Proposition 5.8, a regular surface M in $\mathbf{R}^3$ is minimal if and only if the classical Gauss map g is a meromorphic function on M.

We explain here the Enneper-Weierstrass representation theorem.

Theorem 5.10. *Let $x = (x_1, x_2, x_3) : M \to \mathbf{R}^3$ be a nonflat regular minimal surface in $\mathbf{R}^3$ and g the classical Gauss map of M. Then, in terms of the holomorphic form $\omega := \partial x_1 - \sqrt{-1}\partial x_2$,*

$$\begin{aligned} x_1 &= \operatorname{Re}\int_{z_0}^{z}(1-g^2)\omega + \text{const.} \\ x_2 &= \operatorname{Re}\sqrt{-1}\int_{z_0}^{z}(1+g^2)\omega + \text{const.} \\ x_3 &= \operatorname{Re} 2\int_{z_0}^{z} g\omega + \text{const.} \end{aligned} \tag{5.11}$$

and the metric is given by

$$ds^2 = (1+|g|^2)^2|\omega|^2. \tag{5.12}$$

Moreover, ω has a zero of order $2k$ when and only when g has a pole of order k.

Conversely, we have the following:

Theorem 5.13. *Let M be an open Riemann surface, ω a nonzero holomorphic form and g a nonconstant meromorphic function on M. Assume that ω has a zero of order $2k$ when and only when g has a pole of order k and that the holomorphic forms*

$$\omega_1 := \frac{1-g^2}{2}\omega, \quad \omega_2 := \frac{\sqrt{-1}(1+g^2)}{2}\omega, \quad \omega_3 := g\omega$$

have no real periods. Then, for the functions

$$x_i := \mathrm{Re} \int^z \omega_i \qquad (1 \leq i \leq 3)$$

the surface $x = (x_1, x_2, x_3) : M \to \mathbf{R}^3$ is a regular minimal surface in $\mathbf{R}^3$ whose classical Gauss map is given by the map g.

For the proofs of Theorems 5.10 and 5.13, refer to [58, p. 64].

§6. Minimal Surfaces with Finite Total Curvature

Let $x = (x_1, \ldots, x_m) : M \to \mathbf{R}^m$ be a regular minimal surface in $\mathbf{R}^m$ and ds^2 the metric on M induced from $\mathbf{R}^m$. By definition, the Ricci curvature of M is given by $K_{ds^2}\Omega_{ds^2}$, where K_{ds^2} and Ω_{ds^2} denote the Gaussian curvature and the area form of M respectively. For each local expression $ds^2 = \lambda^2|dz|^2$ in terms of a holomorphic local coordinate z, the Ricci curvature equals $-2\pi dd^c \log \lambda^2$. For convenience' sake of description, we call $-dd^c \log \lambda^2$ itself the Ricci curvature of M and denote it by $\mathrm{Ric}\,[ds^2]$. As mentioned previously, $\mathrm{Ric}\,[ds^2] \leq 0$ everywhere. We define the total curvature of M by $C(M) := \int_M \mathrm{Ric}[ds^2]\ (\geq -\infty)$.

Consider the forms $\omega_i := \partial x_i$ $(1 \leq i \leq m)$. As explained in §5, these are holomorphic and the Gauss map G of M is given by $G = (\omega_1 : \cdots : \omega_m) : M \to P^{m-1}(\mathbf{C})$. Minimal surfaces with finite total curvature are characterized as follows:

Theorem 6.1. *Let $x : M \to \mathbf{R}^m$ be a nonflat complete minimal surface in $\mathbf{R}^m$, which is considered as a Riemann surface with conformal metric. Then, the following conditions are equivalent.*

(i) *M has finite total curvature.*

(ii) *M is biholomorphic with a compact Riemann surface $\bar{M}$ with finitely many points removed and each ω_i $(1 \leq i \leq m)$ extends to $\bar{M}$ as a meromorphic form.*

(iii) *M is biholomorphic with an open subset of a compact Riemann surface $\bar{M}$ and the Gauss map $G : M \to P^{m-1}(\mathbf{C})$ extends to a holomorphic map of $\bar{M}$ into $P^{m-1}(\mathbf{C})$.*

For the proof, we need the following result which was given by A. Huber in [46] and afterwards improved by S. S. Chern and R. Osserman in [16].

Theorem 6.2. *If an open Riemann surface M has a complete metric $d\rho^2$ such that $\int_M \max(-K_{d\rho^2}, 0)\Omega_{d\rho^2} < +\infty$, then M is a Riemann surface of finite type.*

We use also the following:

Lemma 6.3. *Let h be a nowhere zero holomorphic function on $\Delta^* := \{z; 0 < |z| < 1\}$ and consider the metric $ds^2 := |h|^2|dz|^2$. If every piecewise smooth continuous curve in Δ^* converging to the origin has infinite length, then h has at most a pole at the origin.*

This is a restatement of [58, Lemma 9.6].

We now give the proof of Theorem 6.1. Assume that M has finite total curvature. By Theorem 6.2 M is biholomorphic with a compact Riemann surface $\bar{M}$ with points $a_1, \ldots, a_k$ removed. The induced metric ds^2 is written as $ds^2 = \sum_{i=1}^m |f_i|^2|\omega|^2$ for a nowhere vanishing holomorphic form ω and holomorphic functions f_i with $\partial x_i = f_i\omega$. This implies that $-\mathrm{Ric}[ds^2]$ coincides with the pull-back Ω_G of the Fubini-Study metric form on $P^{m-1}(\mathbf{C})$ via G. Take a parabolic exhaustion τ. We may rewrite the order function of G as

$$T_G(r) = \int_s^r \frac{dt}{t} \int_{M_t} -\mathrm{Ric}[ds^2]\,.$$

By the assumption, $\int_M \Omega_G < +\infty$, so that $\lim_{r\to\infty} T_G(r)/\log r < +\infty$. We can apply here Proposition 4.2 to show that G has a removable singularity at each a_ℓ. Now, for each a_ℓ we take a neighborhood $U_\ell := \{z_\ell; |z_\ell| < 1\}$ of a_ℓ, where z_ℓ is a holomorphic local coordinate with $z_\ell(a_\ell) = 0$, and consider holomorphic functions g_i $(1 \le i \le m)$ on $U_\ell - \{0\}$ such that $\partial x_i = g_i dz_\ell$. Here, by changing indices i's if necessary we assume that g_1 has no zero on $U_\ell - \{0\}$ and each g_i/g_1 has a holomorphic extension to U_ℓ, so that they are bounded near the origin. Therefore, we can find a positive constant C such that

$$ds^2 = 2|g_1|^2 \left(\sum_{i=1}^m \left|\frac{g_i}{g_1}\right|^2\right) |dz_\ell|^2 \le C|g_1|^2|dz_\ell|^2\,.$$

We now apply Lemma 6.3 to show that g_1 has at most a pole at the origin. This proves that each ∂x_i extends to $\bar{M}$ as a meromorphic form. Thus, we have (ii). It is obvious that (ii) implies (iii). On the other hand, under the assumption (iii), we can conclude (i) because

$$0 \ge C(M) = \int_M \mathrm{Ric}[ds^2] \ge -\int_{\bar{M}} \Omega_G > -\infty\,.$$

For a complete regular minimal surface with finite total curvature in $\mathbf{R}^m$, a compact Riemann surface $\bar{M}$ satisfying conditions (ii) of Theorem 6.1 is unique up to a biholomorphic isomorphism. We call the surface $\bar{M}$ the compactification of M and an open connected neighborhood of each point a_ℓ in $\bar{M} - M$, or occasionally a_ℓ itself, an end of M.

Let $x : M \to \mathbf{R}^m$ be a nonflat complete regular minimal surface with finite total curvature in $\mathbf{R}^m$ and let $\bar{M}$ be the compactification of M with $\bar{M} - M = \{a_1. \dots, a_k\}$. As a result of Theorem 6.1 the forms $\omega_i := (\partial/\partial z)x_i dz$ $(1 \leq i \leq m)$ and the metric ds^2 on M extends to meromorphic forms $\tilde{\omega}_i$ and a pseudo-metric on $\bar{M}$ respectively. We can also extend the divisor ν_{ds} of ds^2 to $\bar{M}$ by setting $\nu_{ds} := \min(\nu_{\tilde{\omega}_1}, \dots, \nu_{\tilde{\omega}_m})$.

Definition 6.4. For each end a_ℓ we define the multiplicity of M at a_ℓ by $I_\ell := -(\nu_{ds}(a_\ell) + 1)$.

Proposition 6.5 ([16, Lemma 2]). *Each multiplicity I_ℓ is a positive integer.*

Proof. For each fixed ℓ $(1 \leq \ell \leq k)$, set $\nu_i := \nu_{\tilde{\omega}_i}(a_\ell)$ $(1 \leq i \leq m)$. Taking a holomorphic local coordinate z with $z(a_\ell) = 0$, we write $\omega_i = |z|^{2\nu_i} h^2 |dz|^2$ with a positive C^∞ function h. Our purpose is to show that $\min_{1 \leq i \leq m} \nu_i \leq -2$. On the contrary, we assume that $\nu_i \geq -1$ for all i. Expand the functions $f_i := \partial x_i / \partial z$ $(1 \leq i \leq m)$ as $f_i(z) = \sum_{n \geq -1} c_n^i z^n$, we then have

$$x_i(z) - x_i(z_0) = 2\,\mathrm{Re}\left(\int_{z_0}^{z} f_i(z) dz\right) = 2\,\mathrm{Re}(c_{-1}^i \log z) + 2\,\mathrm{Re} \sum_{n=0}^{\infty} \frac{c_n^i}{n+1} z^{n+1}.$$

Here, c_{-1}^i is the residue of ω_i at a_ℓ. Since ω_i has no real period, c_{-1}^i are real. On the other hand,

$$0 = f_1^2 + \dots + f_m^2 = \frac{(c_{-1}^1)^2 + \dots + (c_{-1}^m)^2}{z^2} + \frac{d_{-1}}{z} + \sum_{n=0}^{\infty} d_n z^n$$

for some $d_n \in \mathbf{C} (n \geq -1)$. It follows that $\sum_{i=1}^m (c_{-1}^i)^2 = 0$, so that $c_{-1}^i = 0$ $(1 \leq i \leq m)$. This is impossible because M is complete. Thus Proposition 6.5 is proved.

For the multiplicities of ends, we can show the following:

Proposition 6.6. *Let $\{D_\nu; \nu = 1, 2, \dots\}$ be a sequence of simply connected open neighborhoods of an end a_ℓ such that they have real analytic smooth boundary and $\bigcap_{\nu=1}^\infty \bar{D}_\nu = \{a_\ell\}$. Then,*

$$I_\ell = -\lim_{\nu \to \infty} \frac{1}{2\pi} \int_{\partial D_\nu} \kappa_{ds},$$

where κ_{ds} denotes the geodesic curvature of ∂D_ν.

Proof. Taking a holomorphic local coordinate z on a neighborhood of a_ℓ with $z(a_\ell) = 0$, we consider discs $\Delta_\nu := \{z; |z| < \delta_\nu\}$ such that $D_\nu \supset \bar{\Delta}_\nu$ and $\lim_{\nu \to \infty} \delta_\nu = 0$. Then, by (1.13) we have

$$\chi(\bar{D}_\nu - \Delta_\nu) - \frac{1}{2\pi}\left(\int_{\partial D_\nu} \kappa_{ds} - \int_{\partial \Delta_\nu} \kappa_{ds}\right) = -\frac{1}{2\pi}\int_{D_\nu - \Delta_\nu} K\Omega_{ds^2},$$

the right hand side of which tends to zero as ν tends to ∞ because M has finite total curvature. Therefore, we have only to show that

$$I_\ell = -\lim_{\nu\to\infty} \frac{1}{2\pi}\int_{\partial\Delta_\nu} \kappa_{ds}.$$

By definition, we can write $ds^2 = |z|^{-2(I_\ell+1)}v(z)|dz|^2$ for some positive function $v(z)$. As in the proof of Theorem 1.12, we have

$$\frac{1}{2\pi}\int_{\partial\Delta_\nu} \kappa_{ds} = \chi(\bar{\Delta}_\nu) - (I_\ell + 1)\int_{\partial\Delta_\nu} d^c \log|z|^2 + \int_{\partial\Delta_\nu} d^c \log v(z),$$

which tends to $-I_\ell$.

Related to this is the following theorem given by L. P. Jorge and W. H. Meeks III in [47, Theorem 1].

Theorem 6.7. *Let $x : M \to \mathbf{R}^3$ be a complete regular surface in $\mathbf{R}^3$ which is diffeomorphic to a compact oriented surface $\bar{M}$ with finitely many points $a_1, a_2, \ldots, a_k$ removed. Suppose that the Gauss map of M is continuously extended to $\bar{M}$. For each $r > 0$, set $X_r = (1/r)M \cap S_r^2$, where S_r^2 denotes the sphere of radius r in $\mathbf{R}^3$. Then*

(i) *The map $x : M \to \mathbf{R}^3$ is proper.*

(ii) *For sufficiently large r, X_r consists of k immersed closed curves $\gamma_1^r, \ldots, \gamma_k^r$ on S^2 which are associated with ends of M.*

(iii) *Each $\{\gamma_\ell^r\}$ $(1 \le \ell \le k)$ converges to a geodesic with multiplicity $m_\ell(\ge 1)$ on S^2 as r tends to infinity.*

By Theorem 6.1, this result can be applied to complete regular minimal surfaces with finite total curvature in $\mathbf{R}^3$, and we have similar results for complete regular minimal surfaces in $\mathbf{R}^m$ ([47, Theorem 2]). For a complete regular minimal surface M in $\mathbf{R}^3$ with ends $a_1, \ldots, a_k$, it is easily shown that the multiplicities I_ℓ as in Definition 6.4 coincide with the number m_ℓ in Theorem 6.7, (iii). We state here the following geometric formula for the total curvature given in [47, §4].

Theorem 6.8. *If a complete regular minimal surface M in $\mathbf{R}^m$ with finite total curvature $C(M)$ has k ends with multiplicities $I_\ell (1 \le \ell \le k)$, then*

$$C(M) = \chi(M) - \sum_{\ell=1}^{k} I_\ell.$$

Proof. For each end a_ℓ choose a holomorphic local coordinate z_ℓ with $z_\ell(a_\ell) = 0$ and set $D_\delta := \bar{M} - \bigcup_{\ell=1}^k \bar{\Delta}_\delta^\ell$ for sufficiently small $\delta > 0$, where $\Delta_\delta^\ell := \{z_\ell; |z_\ell| < \delta\}$. By applying (1.13) to the domain D_δ, we obtain

$$\frac{1}{2\pi}\int_{D_\delta} K\Omega_{ds^2} = \chi(\bar{D}_\delta) - \frac{1}{2\pi}\int_{\partial D_\delta}\kappa = \chi(\bar{D}_\delta) + \sum_{\ell=1}^{k}\frac{1}{2\pi}\int_{\partial\Delta_\delta^\ell}\kappa$$

As $\delta \to 0$, we have the desired results by the use of Proposition 6.6.

Combining Theorem 6.8 with Proposition 6.5, we have the following:

Corollary 6.9 ([16, Theorem 2]). *For a complete regular minimal surface M in $\mathbf{R}^m$ with finite total curvature and its compactification $\bar{M}$, it holds that*

$$C(M) \leq \chi(M) - \#(\bar{M} - M).$$

Moreover, by the full use of these results we can prove the following theorems concerning the value distribution of the Gauss map of complete minimal surfaces in $\mathbf{R}^m$ which have finite total curvature.

Theorem 6.10. *The Gauss map of a nonflat complete regular minimal surface in $\mathbf{R}^m$ with finite total curvature cannot omit $m(m+1)/2$ hyperplanes H in $P^{m-1}(\mathbf{C})$ with $f(M) \not\subseteq H$ which are located in general position.*

This was proved for the case where the Gauss map G is nondegenerate by S. S. Chern and R. Osserman ([16]) and generalized to the above-mentioned case by M. Ru ([61]). For details, refer to the original papers or [39, Theorem 5.2.10].

For the particular case of $m = 3$, we have the more precise result for the classical Gauss map as follows:

Theorem 6.11. *Let $x : M \to \mathbf{R}^3$ be a nonflat complete minimal surface immersed in $\mathbf{R}^3$ with finite total curvature. Then, the classical Gauss map of M can omit at most three distinct points.*

For the proof, refer to the original paper [57] or [39, Theorem 5.2.1].

The catenoid provides an example of (normalized) total curvature -2 whose classical Gauss map omits two values. It is still an open problem to determine whether there exists a nonflat complete regular minimal surface in $\mathbf{R}^3$ with finite total curvature whose classical Gauss map omits three distinct values or not.

The followings are some necessary conditions for the classical Gauss map of complete minimal surfaces with finite total curvature in $\mathbf{R}^3$ to omit three distinct values. Let M be a nonflat complete minimal surface in $\mathbf{R}^3$ with finite total curvature whose Gauss map g omits three values. Let γ, k and $C(M)$ be the genus of the compactification, the number of the ends and the total curvature of M, respectively. In [57] R. Osserman showed that $\gamma \geq 1$ and $C(M) < -4$, and in [65] A. Weitsman and F. Xavier proved that $C(M) < -6$. Recently, Y. Fang improved this to $C(M) < -8$ ([19]).

Related to these subjects are some results for minimal surfaces defined on the complex plane $\mathbf{C}$ which are given by P. Hall. Consider a minimal surface $x : \mathbf{C} \to \mathbf{R}^m$ and let $F = (f_1, \ldots, f_m)$ be a reduced representation of the

Gauss map $G : \mathbf{C} \to P^{m-1}(\mathbf{C})$ of x. A direction $v = (v_1, \dots, v_m) \in \mathbf{R}^m$ is called a normal to M at $a \in M$ if it is orthogonal to $T_a(M)$, or equivalently $\sum_{i=1}^m v_i f_i(a) = 0$. P. Hall gave the following:

Theorem 6.12. *If the normals to a minimal surface $x : \mathbf{C} \to \mathbf{R}^m$ omit m directions in general position, then $x : \mathbf{C} \to \mathbf{R}^m$ has a holomorphic factor, namely, there is an orthogonal decomposition $\mathbf{R}^m = \mathbf{R}^2 \oplus \mathbf{R}^{m-2}$ such that the projection of x into the first factor is holomorphic or antiholomorphic with respect to an orthogonal complex structure on $\mathbf{R}^2$.*

For a particular case $m = 4$, Theorem 6.12 implies the following:

Corollary 6.13. *If the normals to a minimal surface $x : \mathbf{C} \to \mathbf{R}^4$ omit four directions in general position, then x is holomorphic in some orthogonal complex structure. Moreover, if they omit five directions in general position, then x is planar.*

Moreover, he proved the following:

Theorem 6.14. *Let $x : \mathbf{C} \to \mathbf{R}^m$ be a regular minimal surface such that the normals of x omit $n+k (k \geq 0)$ directions in general position. Denote by d and a the dimension of the linear subspace of $P^{m-1}(\mathbf{C})$ generated by the image of the Gauss map of x and the dimension of the affine subspace of $\mathbf{R}^m$ generated by the image of x respectively. Then,*

$$1 \leq d \leq \frac{m-3}{k+1}, \qquad d + 3 \leq a \leq \min(n - kd, 2d + 2).$$

The proofs of these results are given by the use of Theorem 4.10. For other results and the details, refer to the papers [41] and [42].

§7. Nevanlinna Theory for Minimal Surfaces of Parabolic Type

In this section, we explain some of the results given by E. F. Beckenbach and his collaborators concerning Nevanlinna theory for minimal surfaces. They developed their theory for 'meromorphic minimal surfaces'. Roughly speaking, these are minimal surfaces in $\mathbf{R}^m$ defined on the complex plane which have at worst pole-like singularities. In our context, we state another version of his theory for regular minimal surfaces in $\mathbf{R}^m$ defined on a Riemann surface of parabolic type.

Let $x = (x_1, \dots, x_m) : M \to \mathbf{R}^m$ be a regular minimal surface and regard M as a Riemann surface with conformal metric. We first note the following:

Proposition 7.1. *For each $c \in \mathbf{R}^m$, $h_c(z) := |x(z) - c|$ is a function with admissible singularities on M.*

Proof. If $x(a) \neq c$, $h_c(z)$ is obviously of class C^∞ near a. In a neighborhood of a point a with $x(a) = c$, choosing a holomorphic local coordinate $z = u + \sqrt{-1}v = a + re^{i\theta}$, we can expand as

$$x(u,v) = c + \sum_{k=1}^{\infty} r^k (d_k \cos k\theta + e_k \sin k\theta)\,,$$

where $d_k, e_k \in \mathbf{R}^m$, $d_1 \neq 0$ or $e_1 \neq 0$, because each component of x is harmonic. Since u, v are isothermal coordinates, we have (5.3) and so

$$r\left\|\frac{\partial x}{\partial r}\right\| = \left\|\frac{\partial x}{\partial \theta}\right\|, \quad \left(\frac{\partial x}{\partial r}, \frac{\partial x}{\partial \theta}\right) = 0\,.$$

Hence $|d_1| = |e_1| \neq 0$, $(d_1, e_1) = 0$. Then, we can write

$$|x(z) - c|^2 = |z - a|^2 u^*(z)$$

with some function $u^*(z) = |d_1|^2 + O(|z-a|)$. We can easily check that u^* satisfies the condition (1.3) and so u_c has an admissible singularity at a.

Take a holomorphic local coordinate $\zeta = u + \sqrt{-1}v$ and set $ds^2 = \lambda^2|d\zeta|^2$. Then, $\lambda = |x_u| = |x_v|$ and the vectors

$$e_1 := \frac{x_u}{\lambda}, \quad e_2 := \frac{x_v}{\lambda} \tag{7.2}$$

give an orthonormal basis of the tangent plane of M, where x_u and x_v denote the partial derivatives of x with respect to u and v respectively.

Proposition 7.3 ([6]). *For each* $c \in \mathbf{R}^m$,

$$\Delta \log |x - c|^2 = \frac{4\lambda^2(|x-c|^2 - (x-c, e_1)^2 - (x-c, e_2)^2)}{|x-c|^4} \geq 0\,.$$

In fact, by the use of harmonicity of each x_i we have

$$\begin{aligned}\Delta \log |x - c|^2 &= \frac{2((|x_u|^2 + |x_v|^2)|x-c|^2 - 2(x-c, x_u)^2 - 2(x-c, x_v)^2)}{|x-c|^4} \\ &= \frac{4|x_u|^2(|x-c|^2 - (x-c, e_1)^2 - (x-c, e_2)^2)}{|x-c|^4}\,.\end{aligned}$$

On the other hand, if we take vectors e_i $(3 \leq i \leq m)$ such that $e_1, e_2, \ldots, e_m$ give an orthonormal basis of $\mathbf{R}^m$, we have Proposition 7.3 because

$$|x-c|^2 - (x-c, e_1)^2 - (x-c, e_2)^2 = \sum_{i=3}^{m} (x-c, e_i)^2 \geq 0\,.$$

By the same calculations as the above, we obtain also

$$\Delta \log(1 + |x|^2) = \frac{4\lambda^2(1 + |x|^2 - (x, e_1)^2 - (x, e_2)^2)}{(1+|x|^2)^2} \geq \frac{4\lambda^2}{(1+|x|^2)^2}\,. \tag{7.4}$$

As a result of Proposition 7.3, the function $\log|x-c|$ is always subharmonic on M. E. F. Beckenbach and T. Radó proved the following related result:

Theorem 7.5. *Let $x : M \to \mathbf{R}^3$ be a regular surface in $\mathbf{R}^3$ defined by a conformal immersion of a domain in $\mathbf{C}$. It is a minimal surface if and only if $\log|x-c|$ is subharmonic for every point $c \in \mathbf{R}^3$.*

For the proof, see [7].

Now, we regard the space $\mathbf{R}^m$ as a subspace of $\mathbf{R}^{m+1}$ by identifying a point $(c_1, \ldots, c_m) \in \mathbf{R}^m$ with the point $(c_1, \ldots, c_m, 0) \in \mathbf{R}^{m+1}$. Consider the one point compactification $\overline{\mathbf{R}^m} := \mathbf{R}^m \cup \{\infty\}$ and the stereographic projection $\varpi : S^m \to \overline{\mathbf{R}^m}$, namely, the map which maps $N := (0, \ldots, 0, 1)$ to ∞ and each $C(\neq N) \in S^m$ to the point $c \in \mathbf{R}^m$ such that C, c and N are collinear. For each $c := (c_1, \ldots, c_m)$, $d := (d_1, \ldots, d_m) \in \mathbf{R}^m$ we denote a half of the distance between $\varpi^{-1}(c)$ and $\varpi^{-1}(d)$ by $|c, d|$. By elementary calculations, we have

$$|c, d| := \frac{|c-d|}{\sqrt{1+|c|^2}\sqrt{1+|d|^2}} (\leq 1)$$

and $|c, \infty| := 1/\sqrt{1+|c|^2}$.

We give the following analogue to [39, Lemma 1.4.8].

Proposition 7.6. *For each $\varepsilon > 0$ there exists a constant δ_0 depending only on ε such that, for every $c \in \mathbf{R}^m$ and $\delta \geq \delta_0$,*

$$\Delta \log \frac{1}{\log(\delta/|x,c|^2)} \geq \left(\frac{1}{1+|c|^2} \frac{1}{|x,c|^2 \log^2(\delta/|x,c|^2)} - \varepsilon \right) \Delta \log(1+|x|^2).$$

Proof. For some $\delta > e$ set $\varphi := -\log\log(\delta/|x,c|^2)$. We easily obtain

$$\Delta\varphi = \frac{\Delta \log|x-c|^2 - \Delta\log(1+|x|^2)}{\log(\delta/|x,c|^2)} + 4\frac{\left(\dfrac{(x,x_u)}{1+|x|^2} - \dfrac{(x-c,x_u)}{|x-c|^2}\right)^2 + \left(\dfrac{(x,x_v)}{1+|x|^2} - \dfrac{(x-c,x_v)}{|x-c|^2}\right)^2}{\log^2(\delta/|x,c|^2)}.$$

We can rewrite this as

$$\begin{aligned}
\Delta\varphi = &\left(\frac{1}{\log(\delta/|x,c|^2)} - \frac{1}{\log^2(\delta/|x,c|^2)}\right)\Delta\log|x-c|^2 \\
&- \left(\frac{1}{\log(\delta/|x,c|^2)} + \frac{1}{\log^2(\delta/|x,c|^2)}\right)\Delta\log(1+|x|^2) \\
&+ \frac{1}{\log^2(\delta/|x,c|^2)}\Bigg\{\Delta\log|x-c|^2 + \Delta\log(1+|x|^2) \\
&+ 4\left(\frac{(x,x_u)}{1+|x|^2} - \frac{(x-c,x_u)}{|x-c|^2}\right)^2 + 4\left(\frac{(x,x_v)}{1+|x|^2} - \frac{(x-c,x_v)}{|x-c|^2}\right)^2\Bigg\}.
\end{aligned}$$

Take δ_0 with $\delta_0 > e$ and $1/\log \delta_0 + 1/\log^2 \delta_0 < \varepsilon$. Then, for every $\delta \geq \delta_0$ the first term is nonnegative and the second term is not larger than $-\varepsilon \Delta \log(1+|x|^2)$. Substitute $x_i = (x, e_i)$ and $x_i - c_i = (x - c, e_i)$ into the third term. Using (7.2), Proposition 7.3 and (7.4), we obtain the desired conclusion

$$\begin{aligned}
&\Delta\varphi + \varepsilon \Delta \log(1+|x|^2) \\
&\quad \geq \frac{4|x_u|^2}{\log^2(\delta/|x,c|^2)} \frac{1+|x|^2+\sum_{i=1}^m (x_i - c_i)^2 - 2(x_1^2 + x_2^2 - c_1x_1 - c_2x_2)}{(1+|x|^2)|x-c|^2} \\
&\quad = \frac{4|x_u|^2}{\log^2(\delta/|x,c|^2)} \frac{1+|x|^2 - x_1^2 - x_2^2 + c_1^2 + c_2^2 + \sum_{i=3}^m (x_i - c_i)^2}{(1+|x|^2)|x-c|^2} \\
&\quad \geq \frac{1}{\log^2(\delta/|x,c|^2)} \frac{(1+|x|^2)\Delta \log(1+|x|^2)}{|x-c|^2} \\
&\quad = \frac{1}{1+|c|^2} \frac{1}{|x,c|^2 \log^2(\delta/|x,c|^2)} \Delta \log(1+|x|^2)\,.
\end{aligned}$$

Take mutually distinct points $c_1, \ldots, c_q \in \mathbf{R}^m$ and set $L := \min_{i<j} |c_i, c_j|$.

(7.7) *For every $w \in \mathbf{R}^m$, we have $|w, c_i| \geq \dfrac{L}{2}$ for all i but at most one.*

In fact, if there are two distinct i and j such that $|w, c_i| < L/2$, then we have an absurd conclusion $L \leq |c_i, c_j| \leq |c_i, w| + |w, c_j| < L$.

We can give also the following analogue to Theorem 3.10.

Proposition 7.8. *Let $c_1, \ldots, c_q$ be mutually distinct points in $\mathbf{R}^m$. For an arbitrary given $\varepsilon > 0$ take some δ_0 with $1/\log^2 \delta_0 + 1/\log \delta_0 < \varepsilon/q$. Then there is a positive constant $C > 0$ depending only on $c_1, \ldots, c_q$ such that*

$$\Delta \log \frac{(1+|x|^2)^{\varepsilon}}{\prod_{j=1}^q \log(\delta_0/|x,c_j|^2)} \geq C \Delta \log(1+|x|^2) \prod_{j=1}^q \frac{1}{|x,c_j|^2 \log^2(\delta_0/|x,c_j|^2)}\,.$$

Proof. By Proposition 7.6, we can find a positive constant C such that

$$\begin{aligned}
\Delta \log \frac{(1+|x|^2)^{\varepsilon}}{\prod_{j=1}^q \log(\delta_0/|x,c_j|^2)} &= \varepsilon \Delta \log(1+|x|^2) + \sum_{j=1}^q \Delta \log \frac{1}{\log(\delta_0/|x,c_j|^2)} \\
&\geq C \left(\sum_{j=1}^q \frac{1}{|x,c_j|^2 \log^2(\delta_0/|x,c_j|^2)} \right) \Delta \log(1+|x|^2)\,.
\end{aligned}$$

For an arbitrary fixed point $a_0 \in M$ we change the indices of c_j's so that

$$|x(a_0), c_1| \leq |x(a_0), c_2| \leq \cdots \leq |x(a_0), c_q|\,.$$

Then, for $j \geq 2$, we have $|x, c_j| \geq L/2$ by (7.7) and so $|x, c_j|^2 \log^2(\delta_0/|x,c_j|^2)$ is bounded from below by a positive constant depending only on L. Therefore, we can easily find a positive constant C' such that

$$\sum_{j=1}^{q} \frac{1}{|x, c_j|^2 \log^2(\delta_0/|x, c_j|^2)}$$
$$\geq \frac{1}{|x, c_1|^2 \log^2(\delta_0/|x, c_1|^2)} \geq C' \prod_{j=1}^{q} \frac{1}{|x, c_j|^2 \log^2(\delta_0/|x, c_j|^2)}.$$

This gives Proposition 7.8.

Corollary 7.9. *In the same situation as in Proposition* 7.8, *for each* $\varepsilon > 0$ *there exist positive constants* δ *and* C *such that*

$$\Delta \log \frac{(1+|x|^2)^{\varepsilon}}{\prod_{j=1}^{q} \log(\delta/|x, c_j|^2)} \geq C^2 \frac{|x_u|^2}{(1+|x|^2)^2} \prod_{j=1}^{q} \frac{1}{|x, c_j|^2 \log^2(\delta/|x, c_j|^2)}.$$

This is an immediate consequence of Proposition 7.8 and (7.4).

Now, we consider a regular minimal surface $x = (x_1, \dots, x_m) : M \to \mathbf{R}^m$ and assume that the surface M, considered as a Riemann surface with conformal metric, has a parabolic exhaustion τ. We regard M as a surface immersed in $\overline{\mathbf{R}^m}$. We define the hyperspherical proximity function for M by

$$m^0(r; c, M) := \frac{2}{r} \int_{\partial M_r} \log \frac{1}{|x, c|} d^c\tau - \frac{2}{s} \int_{\partial M_s} \log \frac{1}{|x, c|} d^c\tau$$

for $c \in \overline{\mathbf{R}^m}$. For the particular case $c = \infty$, we see

$$m^0(r; \infty, M) := \frac{2}{r} \int_{\partial M_r} \log(1+|x|^2)^{1/2} d^c\tau - \frac{2}{s} \int_{\partial M_s} \log(1+|x|^2)^{1/2} d^c\tau (\geq 0).$$

We also define the counting function and the visibility function by

$$N(r; c, M) := N(r, \nu_{|x,c|}),$$
$$H(r; c, M) := \int_s^r \frac{dt}{t} \int_{M_t} dd^c \log |x - c|^2$$

respectively and set $N(r; \infty, M) \equiv H(r; \infty, M) \equiv 0$.

Moreover, we define the hyperspherical affinity of c by

$$A^0(r; c, M) := m^0(r; c, M) + N(r; c, M) + H(r; c, M) \qquad (c \in \overline{\mathbf{R}^m})$$

and the order function of M by

$$T^0(r, M) := A(r; \infty, M) \left(= m^0(r; \infty, M) = \int_s^r \frac{dt}{t} \int_{M_t} dd^c \log(1+|x|^2) \right).$$

By the same argument as in the proof of Proposition 4.1, we can show that $T^0(r, M)$ is increasing and convex with respect to $\log r$ and tends to ∞ as $r \to \infty$.

The first main theorem for minimal surfaces given by E. F. Beckenbach and T. A. Cootz is stated as follows:

Theorem 7.10 ([5], [8]). *$T^0(r;M) = A(r;c,M)$ for all $c \in \overline{\mathbf{R}^m}$.*

Proof. Apply Proposition 2.1 to the function $u := 1/|x,c|$ to see

$$\begin{aligned}\int_s^r \frac{dt}{t}\int_{M_t} dd^c \log(1+|x|^2) &= \int_s^r \frac{dt}{t}\int_{M_t} dd^c[\log|x-c|^2] + \frac{1}{r}\int_{\partial M_r}\log|u| d^c\tau - \frac{1}{s}\int_{\partial M_s}\log|u| d^c\tau \\ &= H(r;c,M) + N(r;c,M) + m^0(r;c,M)\,.\end{aligned}$$

Thus we obtain Theorem 7.10.

As is stated earlier, we identify the extended euclidean space $\overline{\mathbf{R}^m}$ with the unit sphere S^m in $\mathbf{R}^{m+1}$ by stereographic projection and denote by dV the standard volume form on S^m normalized so that the volume of S^m is 1. By the same method as in the proof of Proposition 2.9 we can prove

$$T^0(r,M) = \int_{c\in S^m} N(r;c,M)dV + \int_{c\in S^m} H(r;c,M)dV\,.$$

Now, we give the second main theorem for minimal surfaces.

Theorem 7.11 ([5], [8], [40]). *Let M be a regular minimal surface in $\mathbf{R}^m$ which has a parabolic exhaustion τ. For mutually distinct points $c_1, \dots, c_q \in \overline{\mathbf{R}^m}$ and a positive number ε there is a set E_ε with $\int_{E_\varepsilon} d\rho/\rho < +\infty$ such that*

$$\sum_{j=1}^q m^0(r;c_j,M) + T_G(r) \le (2+\varepsilon)T^0(r,M) - E(r) + O(1) \tag{7.12}$$

for all $r \notin E_\varepsilon$.

Proof. As in §3, we consider the form $\omega := \partial \log\tau$ on M, which is holomorphic on $M - M_s$ and is purely imaginary on each smooth ∂M_r. Consider the functions g_i with $\partial x_i = g_i\omega$ $(1 \le i \le m)$, which are meromorphic on $M_{s,r}$. Set $|G| := (|g_1|^2 + \cdots + |g_m|^2)^{1/2}$. Then $ds^2 = 2|G|^2|\omega|^2$ and so $\nu_{|G|} + \nu_\omega = \nu_{ds} = 0$. For a given $\varepsilon > 0$ take some $\delta > 0$ such that the inequality in Corollary 7.9 holds, and set $h := (1+|x|^2)^{\varepsilon/2}/\prod_{j=1}^q \log(\delta/|c_j,x|^2)$. Consider the function h^* with admissible singularities such that $dd^c\log h = {h^*}^2\sqrt{-1}\omega\wedge\bar{\omega}$. As a consequence of Corollary 7.9 we get

$$h^* \ge C\frac{|G|h}{(1+|x|^2)^{(2+\varepsilon)/2}\prod_{j=1}^q |x,c_j|}$$

for some $C > 0$. By the monotonicity of the integral, we have

$$\begin{aligned}\frac{2}{r}\int_{\partial M_r}&\log h^* d^c\tau + O(1)\\ &\geq \frac{2}{r}\int_{\partial M_r}\log |G| d^c\tau + \frac{2}{r}\int_{\partial M_r}\log h d^c\tau\\ &\quad -(2+\varepsilon)\frac{2}{r}\int_{\partial M_r}\log(1+|x|^2)^{1/2}d^c\tau + \sum_{j=1}^{q}\frac{2}{r}\int_{\partial M_r}\log\frac{1}{|x,c_j|}d^c\tau\end{aligned}$$

for every $r > s$. On the other hand, as in the proof of Theorem 3.3 we obtain

$$\begin{aligned}T_G(r)+E(r) &= \int_s^r \frac{dt}{t}\int_{M_t}[dd^c \log|G|^2] + N(r,\nu_{|G|})\\ &= \frac{2}{r}\int_{\partial M_r}\log|G|d^c\tau - \frac{2}{s}\int_{\partial M_s}\log|G|d^c\tau\,.\end{aligned}$$

It follows that

$$\begin{aligned}T_G(r)+E(r)-(2+\varepsilon)T^0(r,M)+\sum_{j=1}^{q} m^0(r;c_j,M)+\frac{2}{r}\int_{\partial M_r}\log h d^c\tau\\ \leq \frac{2}{r}\int_{\partial M_r}\log h^* d^c\tau + O(1)\,.\end{aligned}$$

We apply Proposition 3.4 to find some constant $C > 0$ such that

$$\frac{2}{r}\int_{\partial M_r}\log h^* d^c\tau \leq C\log T^{h^*}(r) + O(1)$$

for all r except in a set E_ε with $\int_{E_\varepsilon}(1/\rho)d\rho < \infty$. Then, we have

$$\log T^{h^*}(r) = \log\int_s^r\frac{dt}{t}\int_{M_t}dd^c\log h \leq \log\left(\frac{2}{r}\int_{\partial M_r}\log h d^c\tau + \text{const}\right).$$

On the other hand, $O(\log((2/r)\int_{\partial M_r}\log h d^c\tau)) - (2/r)\int_{\partial M_r}\log h d^c\tau$ is bounded from above. Therefore,

$$\begin{aligned}T_G(r)&+E(r)+\sum_{j=1}^{q} m^0(r;c_j,M)\\ &\leq \frac{2}{r}\int_{\partial M_r}\log h^* d^c\tau - \frac{2}{r}\int_{\partial M_r}\log h d^c\tau + (2+\varepsilon)T^0(r,M)+O(1)\\ &\leq O\left(\log\left(\frac{2}{r}\int_{\partial M_r}\log h d^c\tau\right)\right) - \frac{2}{r}\int_{\partial M_r}\log h d^c\tau + (2+\varepsilon)T^0(r,M)+O(1)\\ &\leq (2+\varepsilon)T^0(r,M)+O(1)\end{aligned}$$

for all r except in a set with finite logarithmic measure. This gives Theorem 7.11.

Now, we define the defect of c for M by

$$\delta(c,M):=\liminf_{r\to\infty}\frac{m^0(r;c,M)}{T^0(r,M)}\ (\leq 1).$$

Moreover, we consider the quantities

$$\Psi(M):=\liminf_{r\to\infty}\frac{T_G(r)}{T^0(r,M)},\quad \mathcal{E}(M):=\liminf_{r\to\infty}\frac{-E(r)}{T^0(r,M)}.$$

We have the following defect relation defect relation for minimal surfaces:

Theorem 7.13. *Let M be a regular minimal surface which have a parabolic exhaustion. Then, for arbitrarily given distinct points $c_1,\dots,c_q$ in $\overline{\mathbf{R}^m}$,*

$$\sum_{j=1}^{q}\delta(c_j,M)+\Psi(M)\leq 2+\mathcal{E}(M).$$

Proof. Divide each term of (7.12) by $T^0(r,M)$ and take the limit as r and ε tend to $+\infty$ and 0 respectively. We then have the desired conclusion.

Chapter 3 Value Distribution of the Gauss Map of Minimal Surfaces

§8. Some Global Properties of Minimal Surfaces in $\mathbf{R}^3$

In 1915, S. Bernstein showed that a minimal surface in $\mathbf{R}^3$ which is described as the graph of a differentiable function on the total plane $\mathbf{R}^2$ is necessarily a plane ([9]). Thirty-seven years later, E. Heinz gave the following estimate of the Gaussian curvature of minimal surfaces ([43]):

Theorem 8.1. *Let M be a minimal surface in $\mathbf{R}^3$ which is described as the graph of a C^2 function $z:=z(x,y)$ on a disk $\Delta_R:=\{(x,y);x^2+y^2<R^2\}$ in the (x,y)-plane. Then, for the Gaussian curvature $K(0)$ of M at the origin,*

$$|K(0)|\leq\frac{4\pi^3}{3R^2}.$$

If a minimal surface M in $\mathbf{R}^3$ is described as the graph of a function on $\mathbf{R}^2$, we can replace Δ_R in Theorem 8.1 by an arbitrary disc in $\mathbf{R}^2$ and so the curvature vanishes identically, so that Theorem 8.1 is an improvement of the classical Bernstein theorem. Following this, E. Hopf showed that, in the same situation as in Theorem 8.1,

$$|K(0)| \leq C \frac{\zeta^2(1+\zeta^2)}{R^2}$$

for some constant $C > 0$, where $\zeta := 1/\sqrt{1 + z_x^2 + z_y^2}$ ([45]). Moreover, J. C. C. Nitsche obtained the better estimate $|K(0)| \leq 49/(4R^2)$ ([52]).

Later, R. Osserman gave some generalizations of Theorem 8.1 to minimal surfaces which need not be the graph of a function ([54]–[57]). To state some of them, we consider the distance $d(a)$ from each point a in a regular minimal surface M in $\mathbf{R}^3$ to its boundary, namely, the infimum of the lengths of differentiable curves $\gamma : z = \gamma(t)$ $(0 \leq t < 1)$ in M with $\gamma(0) = a$ such that, for every compact subset K of M, there is some t_0 with $\gamma(t) \notin K$ $(t_0 \leq t < 1)$. In the case where M is described as the graph of a C^2 function on Δ_R we see easily $d(0) \geq R$.

Theorem 8.2 ([54]). *Let M be a regular minimal surface in $\mathbf{R}^3$ such that, for some fixed nonzero vector v_0 and a number $\theta_0 > 0$, all normals to M make angles of at least θ_0 with v_0. Then,*

$$|K(p)|^{1/2} \leq \frac{1}{d(p)} \frac{2\cos(\theta_0/2)}{\sin^3(\theta_0/2)} \qquad (p \in M).$$

As is easily seen, this implies that the Gauss map of a nonflat complete regular minimal surface in $\mathbf{R}^3$ cannot omit a nonempty open set. The Gauss map of a minimal surface which is written as the graph of a function on $\mathbf{R}^2$ always omit the set which corresponds to the lower half of the unit sphere, which shows that Theorem 8.2 is an improvement of Theorem 8.1.

In relation to these, R. Osserman proved also the following fact in [56]:

Theorem 8.3. *Let $x : M \to \mathbf{R}^3$ be a nonflat complete regular minimal surface and $g : M \to P^1(\mathbf{C})$ the classical Gauss map of M. Then the complement of $g(M)$ in $P^1(\mathbf{C})$ is of logarithmic capacity zero, namely, locally included in a set E which is written as $E = \{z; v(z) = -\infty\}$ for a subharmonic function $v(\not\equiv -\infty)$.*

For minimal surfaces in $\mathbf{R}^m$ $(m \geq 3)$, he showed the following:

Theorem 8.4 ([57]). *Let M be a minimal surface in $\mathbf{R}^m$ and let $a \in M$. Suppose that all normals to points of M make angles of at least $\theta_0 > 0$ with some fixed direction. Then,*

$$|K(a)| \leq \frac{1}{d(a)^2} \cdot \frac{16(m-1)}{\sin^4 \theta_0}.$$

In 1981, F. Xavier gave the following improvement of Theorem 8.3 by the use of a result of [70], which is an analogue to Theorem 6.11 on the case of minimal surfaces of finite total curvature.

Theorem 8.5. *The Gauss map of a nonflat complete regular minimal surface in* $\mathbf{R}^3$ *can omit at most six points in* $P^1(\mathbf{C})$.

Afterwards, the author gave some generalizations of Theorem 8.5 to the Gauss map of complete minimal surfaces in $\mathbf{R}^m$ in [29], [30] and [31]. In 1988, R. S. Earp and H. Rosenberg proved that, for a complete minimal surface M, if M is of finite topological type and not a plane, then the classical Gauss map g of M assumes every value infinitely many often, with the possible exception of six values, or else M has finite total curvature and g omits at most three values. Moreover, in the same year, the author obtained the following precision of Theorem 8.5:

Theorem 8.6 ([32]). *The number of exceptional values of the Gauss map of a nonflat complete regular minimal surface in* $\mathbf{R}^3$ *is at most four.*

The number four in Theorem 8.6 is the best-possible. The most famous example of a complete minimal surfaces in $\mathbf{R}^3$ whose Gauss map omits exactly four values is the classical Scherk surface, namely, the surface defined as

$$M \ : \ z = \log\cos x - \log\cos y\,, \qquad (x,y) \in \bigcup_{k+\ell=even} D_{k,\ell}\,,$$

where $D_{k,\ell} := \{(x,y); |x-k\pi| < \pi/2, |y-\ell\pi| < \pi/2\}$ for integers k,ℓ. Furthermore, for arbitrarily given at most four points in $P^1(\mathbf{C})$ we can construct a nonflat complete regular minimal surface whose Gauss map omits exactly these values(cf., [58, Theorem 8.3]).

Here, we shall describe the outline of the proof of Theorem 8.6. We first recall the notation introduced in §7. For two distinct values α and $\beta \in \bar{\mathbf{C}}$ we set

$$|\alpha,\beta| := \frac{|\alpha-\beta|}{\sqrt{1+|\alpha|^2}\sqrt{1+|\beta|^2}}$$

if $\alpha \neq \infty$ and $\beta \neq \infty$, and $|\alpha,\beta| := |\beta,\alpha| = 1/\sqrt{1+|\alpha|^2}$ if $\beta = \infty$. As noted ealier, if we take $v_1, v_2 \in S^2$ with $\alpha = \varpi(v_1)$ and $\beta = \varpi(v_2)$, we have $|\alpha,\beta|$ is a half of the chordal distance between v_1 and v_2, where ϖ denotes the stereographic projection of S^2 onto $\bar{\mathbf{C}}$.

Lemma 8.7 (cf., [39, Lemma 1.4.8]). *Let g be a meromorphic function on $\Delta_R := \{z; |z| < R\}$ which omits a value $\alpha \in \bar{\mathbf{C}}$. For given $\rho > 0$ there is some δ_0 such that, for $\delta \geq \delta_0$,*

$$\Delta \log \frac{1}{\log(\delta/|g,\alpha|^2)} \geq \frac{4|g'|^2}{(1+|g|^2)^2}\left(\frac{1}{|g,\alpha|^2\log^2(\delta/|g,\alpha|^2)} - \rho\right). \tag{8.8}$$

Proof. Set $\varphi := |g,\alpha|^2$. By elementary calculations, we have

$$\left|\frac{\partial\varphi}{\partial z}\right|^2 = (\varphi-\varphi^2)\frac{|g'|^2}{(1+|g|^2)^2}, \quad \frac{\partial^2}{\partial z\partial\bar{z}}\log(1+|g|^2) = \frac{|g'|^2}{(1+|g|^2)^2}.$$

This implies that, for every $\delta > 0$,

$$\frac{1}{4}\Delta \log \frac{1}{\log(\delta/\varphi)} = -\frac{\dfrac{\partial^2}{\partial z \partial \bar{z}} \log(1+|g|^2)}{\log(\delta/\varphi)} + \frac{\left|\dfrac{\partial \varphi}{\partial z}\right|^2}{\varphi^2 \log^2(\delta/\varphi)}$$
$$= \frac{|g'|^2}{(1+|g|^2)^2}\left(\frac{1}{\varphi \log^2(\delta/\varphi)} - \left(\frac{1}{\log^2(\delta/\varphi)} + \frac{1}{\log(\delta/\varphi)}\right)\right).$$

If we take δ_0 with $\log^{-2}(\delta_0/\varphi) + \log^{-1}(\delta_0/\varphi) < \rho$, we obtain (8.8).

Proposition 8.9. *Let g be a nonzero meromorphic function on Δ_R which omit q distinct values $\alpha_1, \ldots, \alpha_q$. If $q > 2$, there exist some positive constants δ and C such that*

$$\frac{|g'|}{1+|g|^2} \prod_{j=1}^{q} \frac{1}{|g, \alpha_j| \log(\delta/|g, \alpha_j|^2)} \leq C \frac{R}{R^2 - |z|^2}. \tag{8.10}$$

Proof. We take some δ satisfying (8.8) for all given α_j's and $\rho := (q-2)/q$ and denote the left hand side of (8.10) by v. We may write

$$v^2 = \frac{(1+|g|^2)^{q-2}|g'|^2 \prod_{j=1}^{q}(1+|\alpha_j|^2)}{\prod_{j=1}^{q} |g-\alpha_j|^2 \log^2(\delta/|g, \alpha_j|^2)}.$$

According to Lemma 8.7, we have

$$\Delta \log v^2 \geq \frac{4|g'|^2}{(1+|g|^2)^2}\left(q-2+\sum_{j=1}^{q}\left(\frac{2}{|g,\alpha_j|^2 \log^2(\delta/|g,\alpha_j|^2)} - \frac{q-2}{q}\right)\right)$$
$$= \frac{8|g'|^2}{(1+|g|^2)^2}\sum_{j=1}^{q}\frac{1}{|g,\alpha_j|^2 \log^2(\delta/|g,\alpha_j|^2)}.$$

For each fixed $z \in \Delta_R$, we easily see $|g(z), \alpha_j| \geq \min_{k<l} |\alpha_k, \alpha_l|/2$ for all j but at most one j_0. Therefore, we can find a positive constant C such that

$$\Delta \log v \geq \frac{4|g'|^2}{(1+|g|^2)^2} \frac{1}{|g,\alpha_{j_0}|^2 \log^2(\delta/|g,\alpha_{j_0}|^2)}$$
$$\geq C \frac{|g'|^2}{(1+|g|^2)^2} \prod_{j=1}^{q} \frac{1}{|g,\alpha_j|^2 \log^2(\delta/|g,\alpha_j|^2)} = Cv^2.$$

Proposition 8.9 is a result of the following Ahlfors-Schwarz lemma.

Lemma 8.11 ([1]). *If a C^2 function v on Δ_R satisfies the condition $\Delta \log v \geq C^2 v^2$ for some positive constant C, then*

$$v(z) \leq \frac{2}{C} \frac{R}{R^2 - |z|^2}.$$

As a consequence of Proposition 8.9, we have the following:

(8.12) *Under the same assumption as in Proposition 8.9, for each positive* η *with* $\eta < (q-2)/q$ *there is some constant* $C > 0$ *such that*

$$\frac{|g'|}{1+|g|^2} \frac{1}{\prod_{j=1}^{q} |g, \alpha_j|^{1-\eta}} \leq C \frac{R}{R^2 - |z|^2}.$$

In fact, since the function $\log(\delta x^2)/x^\eta$ $(1 \leq x < +\infty)$ is bounded, each factor $\log(\delta/|g,\alpha_j|^2)$ in (8.10) can be replaced by $|g,\alpha_j|^{2\eta}$ after a suitable change of the constant C.

Theorem 8.6 is proved by contradiction. Suppose that the classical Gauss map g of a nonflat complete minimal surface M in $\mathbf{R}^3$ omits q distinct values $\alpha_1, \ldots, \alpha_q$, where $q = 5$. For our purpose, we may assume that $\alpha_q = \infty$ and, moreover, that M is biholomorphic with the unit disc because M can be replaced by its universal covering surface and Theorem 8.6 is obvious in the case $M = \mathbf{C}$ by the small Picard theorem. Taking some positive η with $q - 6 < q\eta < q - 4$ and setting $\tau := 2/(q - 2 - q\eta)$, we define a new metric

$$d\sigma^2 = |f_z|^{2/(1-\tau)} \left(\frac{1}{|g'_z|} \prod_{j=1}^{q-1} \left(\frac{|g-\alpha_j|}{(1+|\alpha_j|^2)^{1/2}} \right)^{1-\eta} \right)^{2\tau/(1-\tau)} |dz|^2 \tag{8.13}$$

on the set $M' := \{z; g'_z(z) \neq 0\}$, where $g'_z = (d/dz)g$ and f_z is the function with $ds^2 = |f_z|^2(1+|g|^2)^2|dz|^2$. Take a point $a \in M'$. Since $d\sigma^2$ is flat on M', there is an isometry Φ satisfying $\Phi(0) = a$ from a disc Δ_R with the standard metric onto an open neighborhood of a in M' with the metric $d\sigma^2$. We take the largest such R. For brevity, we denote the function $g \cdot \Phi$ on Δ_R by g in the followings. According to (8.12), we have

$$R \leq C \frac{1+|g(0)|^2}{|g'_z(0)|} \prod_{j=1}^{q} |g(0), \alpha_j|^{1-\eta} < +\infty. \tag{8.14}$$

Then, there is some point w_0 with $|w_0| = R$ such that, for the line segment $\Gamma : w = tw_0$ $(0 \leq t < 1)$, $\gamma := \Phi(\Gamma)$ tends to the boundary of M' as $t \to 1$. Suppose that γ tends to a point $a_0 \in M - M'$. Taking a holomorphic local coordinate ζ with $\zeta(a_0) = 0$ in a neighborhood U of a_0, we can write $d\sigma^2 = |\zeta|^{-\tau/(1-\tau)} w |d\zeta|^2$ with some positive C^∞ function w. Since $\tau/(1-\tau) > 1$, we have

$$R = \int_\Gamma d\sigma \geq C' \int_{\Gamma \cap U} \frac{1}{|\zeta|^{\tau/(1-\tau)}} |d\zeta| = +\infty$$

for some $C' > 0$, which contradicts (8.14). A similar argument shows that $\gamma(t_n) \to a_0$ is impossible for any sequence $t_n \to 1$. Therefore, γ diverges outside any compact subset of M as $t \to 1$.

On the other hand, since $d\sigma^2 = |dz|^2$, we obtain by (8.13)

$$|f_z| = \left(|g_z'| \prod_{j=1}^{q-1} \left(\frac{(1+|\alpha_j|^2)^{1/2}}{|g-\alpha_j|} \right)^{1-\eta} \right)^{\tau} . \tag{8.15}$$

By the use of (8.12) we have

$$\begin{aligned} d(p) &\leq \int_\gamma ds = \int_\Gamma \Phi^* ds = \int_\Gamma |f_z|(1+|g|^2)|dz| \\ &= \int_\Gamma \left(|g_z'|(1+|g|^2)^{1/\tau} \prod_{j=1}^{q-1} \frac{(1+|\alpha_j|^2)^{(1-\eta)/2}}{|g-\alpha_j|^{1-\eta}} \right)^{\tau} |dz| \\ &= \int_\Gamma \left(\frac{|g_z'|}{1+|g|^2} \frac{1}{\prod_{j=1}^{q} |g,\alpha_j|^{1-\eta}} \right)^{\tau} |dz| \\ &\leq C'' \int_\Gamma \left(\frac{2R}{R^2-|z|^2} \right)^{\tau} |dz| < +\infty \end{aligned}$$

for some $C'' > 0$, which contradicts the assumption that M is complete. We have Theorem 8.6.

In [49], X. Mo and R. Osserman improved Theorem 8.6 as follows:

Theorem 8.16. *Let M be a nonflat complete regular minimal surface with infinite total curvature in $\mathbf{R}^m$. Then, for the classical Gauss map g of M there are at most four distinct points $\alpha \in P^1(\mathbf{C})$ such that $g^{-1}(\alpha)$ is finite.*

Combining this with Theorem 6.11, they gave also the following:

Corollary 8.17. *Let M be a nonflat complete minimal surface. If the classical Gauss map of M omits four distinct values, then it takes all other values infinitely many times.*

In [48], S. J. Kao generalized Corollary 8.17 to the case where M is a complete minimal surface in $\mathbf{R}^3$ with annular ends. He showed that the Gauss map of an end of a complete minimal surface in $\mathbf{R}^3$ that is conformally an annulus $\{z | 1/r < |z| < r\} (r > 0)$ must assume every value infinitely often, with the possible exception of at most four values.

Recently, the author gave the following estimate for the Gaussian curvature of a minimal surface in $\mathbf{R}^3$([36]).

Theorem 8.18. *Let M be a regular minimal surface $x : M \to \mathbf{R}^3$ whose classical Gauss map omits q distinct values $\alpha_1, \ldots, \alpha_q$. Set $L := \min_{i<j} |\alpha_i, \alpha_j|$. If $q > 4$, then there exists a universal constant C such that*

$$|K(p)|^{1/2} \leq \frac{C}{d(p)} \frac{\log^2 \frac{1}{L}}{L^3} \qquad (p \in M) . \tag{8.19}$$

Theorem 8.6 is a particular case of Theorem 8.18. In fact, if M is complete, then we may set $d(p) \equiv \infty$ in Theorem 8.18, so that $K(p) \equiv 0$, or M is a plane. Theorem 8.18 can be proved in the following more precise form ([36]):

Theorem 8.20. *Let $x : M \to \mathbf{R}^3$ be a regular minimal surface and g its classical Gauss map. Suppose that there are distinct values $\alpha_1, \ldots, \alpha_q$ and positive integers $m_1, \ldots, m_q$ such that $\nu_{g-\alpha_j}(z) \geq m_j$ on $g^{-1}(\alpha_j)$ for each j and*

$$\gamma := \sum_{j=1}^{q} \left(1 - \frac{1}{m_j}\right) > 4 . \tag{8.21}$$

Then, (8.19) holds for some $C > 0$ not depending on any data M, α_j's or m_j's.

The proof of Theorem 8.20 is accomplished by more precise observations of constants appearing in the proof of Theorem 8.6. We omit the details.

In Theorem 8.20, if the Gauss map g omits a value α_j, we may set $m_j := +\infty$ in (8.21) and, if g omits all values $\alpha_j (1 \leq j \leq q)$, (8.21) means $q > 4$, which is the case of Theorem 8.18.

Here, we note that the exponent 3 of L in the denominator on the right hand side of (8.19) is in a sense the best-possible. In fact, for every $\varepsilon > 0$, we can construct a family of minimal surfaces in $\mathbf{R}^3$ such that there is no positive constant C satisfying the condition

$$|K(p)|^{1/2} \leq \frac{C}{d(p)} \frac{1}{L^{3-\varepsilon}} . \tag{8.22}$$

To see this, for each $R (\geq 1)$ we consider Enneper's surface M whose domain of definition is restricted to $\Delta_R := \{z; |z| < R\}$, namely, the surface defined as in Theorem 5.13 by using the function $g(z) = z$ and the form $\omega = dz$ on the disc Δ_R. Take five values $\alpha_1 := R$, $\alpha_2 := \sqrt{-1}R$, $\alpha_3 := -R$, $\alpha_4 := -\sqrt{-1}R$, $\alpha_5 := \infty$ in $\bar{\mathbf{C}}$, which are omitted by the classical Gauss map g of M and the quantities of (8.19) are given by $d(0) = R + R^3/3$, $|K(0)|^{1/2} = 2|g'(0)|/(|f(0)|(1 + |g(0)|^2)) = 2$ and $L = 1/\sqrt{1 + R^2}$. Therefore, we obtain

$$|K(0)|^{1/2} d(0) L^{3-\varepsilon} = \frac{2(R + \frac{1}{3}R^3)}{(1 + R^2)^{(3-\varepsilon)/2}} ,$$

which converges to $+\infty$ as $R \to +\infty$. This shows that there is no constant $C > 0$ not depending on each minimal surface satisfying (8.22).

Now, we give another type of value-distribution-theoretic property of the Gauss map of complete minimal surfaces in $\mathbf{R}^3$. As is stated in §4, two non-constant meromorphic functions on $\mathbf{C}$ coincide with each other if they have the same inverse images for five distinct values. The author gave the following analogue to this unicity theorem.

Theorem 8.23 ([37]). *Let M and $\tilde{M}$ be two complete nonflat regular minimal surfaces and assume that there is a sense preserving isometry Φ of M*

onto $\tilde{M}$. *If there are seven distinct values* $\alpha_1, \ldots, \alpha_7$ *such that* $g^{-1}(\alpha_j) = (\tilde{g} \cdot \Phi)^{-1}(\alpha_j)$ $(1 \leq j \leq 7)$ *for the classical Gauss map* g *of* M *and* $\tilde{g}$ *of* $\tilde{M}$, *then* $g \equiv \tilde{g} \cdot \Phi$.

Moreover, he showed that, under the same assumption as in Theorem 8.23, if both of M and $\tilde{M}$ have finite total curvature, then the number seven in Theorem 8.23 can be replaced by six. We have also an analogous unicity theorem for complete minimal surfaces in $\mathbf{R}^m$ $(m \geq 4)$. For the details, refer to [38].

The number seven in Theorem 8.23 is the best-possible. To see this, take a number α with $\alpha \neq 0, \pm 1$ and consider the function $g(z) = z$ and the form $\omega := dz/(z(z-\alpha)(\alpha z - 1))$ on the universal covering surface M of $\mathbf{C} - \{0, \alpha, 1/\alpha\}$. By using them, we construct a regular minimal surface M in $\mathbf{R}^3$ by the use of Theorem 5.13. Moreover, we construct another minimal surface $\tilde{x} : \tilde{M} \to \mathbf{R}^3$ in the same manner by the use of the meromorphic function $\tilde{g}(z) = 1/z$ and the above ω. Then, we can easily check that $\tilde{M}$ is isometric with M. For the classical Gauss maps g and $\tilde{g}$ we have $g \not\equiv \tilde{g}$ and $g^{-1}(\alpha_j) = \tilde{g}^{-1}(\alpha_j)$ for six values $\alpha_1 := 0, \alpha_2 := \infty, \alpha_3 := \alpha, \alpha_4 := 1/\alpha, \alpha_4 := 1,\ \alpha_5 := -1$. This shows that the number seven in Theorem 8.23 cannot be replaced by six.

§9. Modified Defect Relations for Holomorphic Curves in $P^n(\mathbf{C})$

Let M be an open Riemann surface with a continuous pseudo-metric ds^2 and let Ω_1, Ω_2 be two (1,1)-currents on an open set U in M. By $\Omega_1 \prec_c \Omega_2$, we mean that there are a divisor ν and a bounded continuous real-valued function k with admissible singularities on U such that $\nu(z) > c$ on $\mathrm{Supp}(\nu)$ and

$$\Omega_1 + [\nu] = \Omega_2 + dd^c[\log |k|^2]$$

on U. The notation $\Omega_1 \prec \Omega_2$, or $\Omega_2 \succ \Omega_1$, means $\Omega_1 \prec_c \Omega_2$ for some $c > 0$. We can easily show the following:

(9.1) (i) *If* $c \geq d$ *and* $\Omega_1 \prec_c \Omega_2$, *then* $\Omega_1 \prec_d \Omega_2$.
(ii) *If* $\Omega_1 \prec_c \Omega_2$ *and* $\Omega_2 \prec_c \Omega_3$ *for some* $c > 0$, *then* $\Omega_1 \prec_c \Omega_3$.
(iii) *If* $\Omega_1 \prec \Omega_2$, *then* $a\Omega_1 + \Omega \prec a\Omega_2 + \Omega$ *for every* $a > 0$ *and* Ω.

Let f be a holomorphic map of an open Riemann surface M into $P^n(\mathbf{C})$ and consider a hyperplane $H := \{\langle W \rangle; \langle W, A \rangle = 0\}$ in $P^n(\mathbf{C})$ with $f(M) \not\subseteq H$, where A is a unit vector in $\mathbf{C}^{n+1}$. As in §2, taking a reduced representation $f = (f_0 : \cdots : f_n)$ on M and setting $F := (f_0, \ldots, f_n)$, we consider the divisor $\nu(f, H) = \nu_{\langle A, F \rangle}$ and, for a positive integer m or $m = \infty$, the m-truncated pull-back $f^*(H)^{[m]} := [\min(\nu(f, H), m)]$ of H considered as a current.

Definition 9.2. We define the modified defect of H for f by

$$D_f(H) := 1 - \inf\{\eta \geq 0; f^*(H)^{[n]} \prec \eta[\Omega_f] \text{ outside a compact set in } M\},$$

where Ω_f is the pull-back of the normalized Fubini-Study metric on $P^n(\mathbf{C})$.

We note here that the constant $\eta := 1$ satisfies the condition $f^*(H)^{[n]} \prec \eta[\Omega_f]$ on M. For, if we take a nonzero holomorphic function g with $\nu_g = f^*(H)^{[n]}$ and set $u := |\langle A, F\rangle|/|g|$ and $k := |\langle A, F\rangle|/|F|(\leq 1)$, then we have

$$f^*(H)^{[n]} + dd^c[\log u^2] = dd^c \log |F|^2 + dd^c[\log k^2]$$

and $dd^c[\log u^2] = [\nu_u]$.

For convenience's sake, we set $D_f(H) = 0$ if $f(M) \subseteq H$.

The modified defect has the following properties (cf., Proposition 4.4):

Proposition 9.3. (i) $0 \leq D_f(H) \leq 1$.

(ii) *If there exists a compact set K and a bounded nonzero holomorphic function g on $M - K$ such that $\nu_g \geq \min(\nu(f, H), n)$ on $M - K$, or in particular, if $\# f^{-1}(H) < \infty$, then $D_f(H) = 1$.*

(iii) *If there are some $m > n$ and a compact set K such that $\nu(f, H)(a) \geq m$ for every $a \in f^{-1}(H) - K$, then $D_f(H) \geq 1 - n/m$.*

The assertion (i) is obvious. Take g as in (ii) and set $k := |g|$, which is a bounded continuous function with admissible singularities. For $\nu := \nu_g - \min(\nu(f, H), n)$ and $\eta := 0$, we have $\nu \geq 1$ on Supp (ν) and $f^*(H)^{[n]} + [\nu] = \eta\Omega_f + dd^c[\log k^2]$ on $M - K$, which gives (ii). Next, consider $k := (|F(H)|/|F|)^{n/m}$. We then have

$$f^*(H)^{[n]} + \left[\frac{n}{m}\nu(f, H) - \min(\nu(f, H), n)\right] = \frac{n}{m} dd^c \log |F|^2 + dd^c[\log k^2]$$

on $M - K$. Since $\nu(f, H) \geq m$ on $f^{-1}(H) - K$, we have $(n/m)\nu(f, H) \geq n \geq \min(\nu(f, H), n)$ on $f^{-1}(H) - K$ and therefore $f^*(H)^{[n]} \prec (n/m)\ \Omega_f$, which proves (iii).

We now discuss the relation between the classical defects and the modified defects for a particular case. Consider the case where f is nondegenerate and M contains an open set G such that there is a biholomorphic map Φ of a not relatively compact neighborhood D of $M - G$ onto a neighborhood of $\Delta_{s,+\infty} := \{z; s \leq |z| < +\infty\}$ with $\Phi(\partial G) = \{z; |z| = s\}$ $(s > 0)$. Consider a C^∞ map $\tilde{f} : \mathbf{C} \to P^n(\mathbf{C})$ such that $\tilde{f} \cdot \Phi = f$ on D. We can define the order function $T_{\tilde{f}}(r)$, the counting function $N_f(r, H)$ of a hyperplane H in the same manner as in §2, disregarding the quantities which arise from the integrals over Δ_s. Then, as in Proposition 4.2 we can show that $\tilde{f}$ has an essential singularity at ∞ if and only if $\lim_{r\to\infty} T_{\tilde{f}}(r)/\log r = +\infty$(cf., [39, Chapter 3]). In this case, the defect $\delta_f(H)$ defined by (4.3) depends only on the map f and H, and Theorem 4.6 remains true for these defects.

Theorem 9.4. *If $\tilde{f}$ has an essential singularity at ∞, then* $D_f(H) \leq \delta_{\tilde{f}}(H)^{[n]}$.

To see this, consider an arbitrary number η such that

$$f^*(H)^{[n]} + [\nu] = \eta \Omega_f + dd^c[\log k^2]$$

outside a compact set K for a divisor $\nu \geq 0$ and a bounded nonnegative continuous function k with admissible singularities on M. We may assume that $0 \leq k \leq 1$ and $\bar{\Delta}_{s,\infty} \cap K = \emptyset$. Since $f^*(H)^{[n]} \leq dd^c[\log(k|F|^\eta)^2]$, we get

$$\begin{aligned} N_{\tilde{f}}(r,H)^{[n]} &= \int_s^r \frac{dt}{t} \int_{\Delta_{s,t}} f^*(H)^{[n]} \\ &\leq \eta \int_s^r \frac{dt}{t} \int_{\Delta_{s,t}} dd^c[\log|F|^2] + \int_s^r \frac{dt}{t} \int_{\Delta_{s,t}} dd^c \log k^2 \\ &\leq \eta\, T_{\tilde{f}}(r) + O(\log r) \end{aligned}$$

by the use of Proposition 2.1. This implies that

$$\frac{N_{\tilde{f}}(r,H)^{[n]}}{T_{\tilde{f}}(r)} \leq \eta + O\left(\frac{\log r}{T_{\tilde{f}}(r)}\right).$$

Since $\lim_{r\to\infty} \log r / T_{\tilde{f}}(r) = 0$, as $r \to +\infty$ we have $1 - \eta \leq \delta_{\tilde{f}}(H)^{[n]}$. Therefore, we obtain Theorem 9.4, because η tends to $1 - D_{\tilde{f}}(H)$.

Let M be an open Riemann surface of finite type and $f : M \to P^n(\mathbf{C})$ a nondegenerate holomorphic map. Assume that f is transcendental, namely, f has no holomorphic extension to a compactification $\bar{M}$. By Theorems 4.6 and 9.4, for finitely many arbitrary hyperplanes H_j's in general position with Nochka weights $\omega(j)$'s, we see

$$\sum_j \omega(j) D_f(H_j) \leq n + 1. \tag{9.5}$$

Now, consider an open Riemann surface with a continuous pseudo-metric ds^2 and a holomorphic map $f = (f^1, \ldots, f^L)$ of M into the so-called Osgood space $P^{n_1 n_2 \ldots n_L}(\mathbf{C}) := P^{n_1}(\mathbf{C}) \times \cdots \times P^{n_L}(\mathbf{C})$. We say that f is nondegenerate if each component $f^\ell : M \to P^{n_\ell}(\mathbf{C})$ of f is nondegenerate. A nondegenerate holomorphic map $f = (f^1, \ldots, f^L) : M \to P^{n_1 n_2 \ldots n_L}(\mathbf{C})$ is said to satisfy the condition $(C)_{\rho_1 \cdots \rho_L}$ for nonnegative constants $\rho_1, \ldots, \rho_L$ if

$$-\operatorname{Ric}[ds^2] \prec \rho_1 \Omega_1 + \cdots + \rho_L \Omega_L$$

holds outside a compact set in M, where Ω_ℓ denotes the pull-back of the Fubini-Study metric form on $P^{n_\ell}(\mathbf{C})$ via f^ℓ.

We now give the following modified defect relation for holomorphic curves in $P^{n_1 \ldots n_L}(\mathbf{C})$.

Theorem 9.6. *Let M be an open Riemann surface with a complete continuous pseudo-metric ds^2 and $f = (f^1, \dots, f^L) : M \to P^{n_1 \dots n_L}(\mathbf{C})$ a nondegenerate holomorphic map satisfying the condition $(C)_{\rho_1 \dots \rho_L}$. Assume that M is not of finite type or else that some $f^\ell (1 \le \ell \le L)$ is transcendental. For each ℓ take hyperplanes $H_{\ell 1}, \dots, H_{\ell q_\ell}$ in $P^{n_\ell}(\mathbf{C})$ located in N_ℓ-subgeneral position and choose Nochka weights $\omega_\ell(1), \dots, \omega_\ell(q_\ell)$ for these hyperplanes, where $q_\ell > 2N_\ell - n_\ell + 1$. Set $D_\ell := \sum_{j=1}^{q_\ell} \omega_\ell(j) D_{f^\ell}(H_{\ell j})$. If $D_\ell > n_\ell + 1$ for all ℓ, then*

$$\sum_{\ell=1}^{L} \frac{n_\ell(n_\ell+1)}{2} \frac{\rho_\ell}{D_\ell - n_\ell - 1} \ge 1 .$$

The proof is given by the full use of the following theorem. For the details, see [39, §4.4].

Theorem 9.7 ([35, Theorem 4.6]). *Let M be an open Riemann surface with a complete continuous pseudo-metric ds^2, and $d\tau^2$ a continuous pseudo-metric on M whose curvature is strictly negative outside a compact set K. If there exists some p with $0 < p < 1$ such that*

$$-\mathrm{Ric}[ds^2] \prec_{1-p} p(-\mathrm{Ric}[d\tau^2])$$

on $M - K$, then M is of finite type.

We describe here some consequences of Theorem 9.6. Consider a nondegenerate holomorphic map f into $P^n(\mathbf{C})$ defined on an open Riemann surface M with a complete continuous pseudo-metric ds^2. We define the order of f by

$$\rho_f := \inf\{\rho > 0; -\mathrm{Ric}[ds^2] \prec \rho\, \Omega_f \text{ outside a compact set in } M\} .$$

As a particular case with $L = 1$ of Theorem 9.6, we have the following:

Corollary 9.8. *Assume that M is not of finite type or else that f is transcendental. Then, for arbitrary hyperplanes H_j in N-subgeneral position with Nochka weights $\omega(j)$ $(1 \le j \le q)$,*

$$\sum_{j=1}^{q} \omega(j) D_f(H_j) \le n + 1 + \frac{\rho_f n(n+1)}{2} .$$

Corollary 9.9. *Under the same assumption as in Corollary 9.8, we have*

$$\sum_{j=1}^{q} D_f(H_j) \le 2N - n + 1 + \frac{\rho_f n(2N - n + 1)}{2}$$

for arbitrary hyperplanes $H_1, \dots, H_q$ in $P^n(\mathbf{C})$ located in N-subgeneral position.

In particular, if the given hyperplanes H_j are in general position, then

$$\sum_{j=1}^{q} D_f(H_j) \leq n+1+\frac{\rho_f n(n+1)}{2}\,. \tag{9.10}$$

This is easily seen from Corollary 9.8 by using Theorem 3.7(cf., [39, p. 158]).

It is an interesting problem to get a good generalization of Corollary 9.9 to the case of holomorphic maps of a complete Kähler manifold M into $P^N(\mathbf{C})$. In the particular case where the universal covering of M is biholomorphic with a ball in $\mathbf{C}^n$, the author obtained a weaker type of estimate, say non-integrated defect relation, which is analogous to (9.10). For the details, see [31].

§10. The Gauss Map of Complete Minimal Surfaces

We can apply the modified defect relations given in the previous section to the study of the Gauss map of complete minimal surfaces in $\mathbf{R}^m$ possibly with branch points, where we mean by a minimal surface with branch points a nonconstant differentiable map $x=(x_1,\dots,x_m)$ of an open Riemann surface M into $\mathbf{R}^m$ such that each x_i is harmonic and (5.3) holds for each holomorphic local coordinate $z=u+\sqrt{-1}v$. We consider the holomorphic forms $\omega_i := \partial x_i$. Set $S := \{a; \nu_{\omega_i}(a) > 0$ for all $i\}$, which has no accumulation points in M. Points in S are called branch points of M. As is seen from Theorem 5.9, $x: M-S \to \mathbf{R}^m$ is a regular minimal surface in $\mathbf{R}^m$. In this section, minimal surfaces will mean minimal surfaces with branch points unless specified otherwise. We consider the continuous pseudo-metric

$$ds^2 := 2(|\omega_1|^2+\cdots+|\omega_m|^2) \tag{10.1}$$

and the divisor $\nu_{ds} = \min\{\nu_{\omega_i}; 1\leq i\leq m\}$. Take a nonzero holomorphic form ω_0 on M with $\nu_{\omega_0} = \nu_{ds}$ and the function g_i with $\omega_i = g_i\omega_0$. Then $g_i(1\leq i\leq m)$ are holomorphic and have no common zero. We can define the holomorphic map $\tilde{G} := (g_1:\cdots:g_m): M \to P^{m-1}(\mathbf{C})$, which gives the Gauss map of the regular minimal surface $M-S$. We define the Gauss map of M by the map $\tilde{G}$.

We define the branching order of M by

$$\rho_{ds} := \inf\{\rho\geq 0; [\nu_{ds}] \prec \rho\,\Omega_G \text{ outside a compact set in } M\},$$

where we set $\rho_{ds} = +\infty$ if there is no ρ with the above property. Obviously, if $x: M \to \mathbf{R}^m$ has only finitely many branch points, or in particular, if x is an immersion, then $\rho_{ds}=0$.

Theorem 10.2. *Let $x: M \to \mathbf{R}^m$ be a complete minimal surface and $G: M \to P^{m-1}(\mathbf{C})$ the Gauss map of M. Consider the smallest projective linear subspace $P^n(\mathbf{C})$ of $P^{m-1}(\mathbf{C})$ which includes $G(M)$. Take q hyperplanes*

$H_1, \dots, H_q$ in $P^{m-1}(\mathbf{C})$ located in general position. If M has infinite total curvature, then

$$\sum_{j=1}^{q} \omega(j) D_G(\tilde{H}_j) \leq n+1+\frac{(\rho_{ds}+1)n(n+1)}{2},$$

where $\omega(j)$ are Nochka weights for the hyperplanes $\tilde{H}_j := H_j \cap P^n(\mathbf{C})(1 \leq j \leq q)$ in $P^n(\mathbf{C})$ located in $(m-1)$-subgeneral position.

To see this, we represent the pseudo-metric ds^2 as $ds^2 = 2\sum_{i=1}^{m} |g_i|^2 |\omega_0|^2$ with the above-mentioned form ω_0 and holomorphic functions g_i. We then have

$$-\operatorname{Ric}[ds^2] = \Omega_G + [\nu_{ds}], \tag{10.3}$$

where $\Omega_G = dd^c \log(|g_1|^2 + \cdots + |g_m|^2)$ is the pull-back of the Fubini-Study metric form on $P^{m-1}(\mathbf{C})$ via the Gauss map G. After a suitable unitary transformation, G may be regarded as a nondegenerate holomorphic map of M into $P^n(\mathbf{C})$ and the pull-back of the Fubini-Study metric form on $P^{m-1}(\mathbf{C})$ coincides with that on $P^n(\mathbf{C})$. Take an arbitrary $\rho \geq 0$ such that $[\nu_{ds}] \prec \rho\, \Omega_G$ on M outside a compact set K. By (10.3) we have $-\operatorname{Ric}[ds^2] \prec (\rho+1)\Omega_G$ on $M-K$. Taking the infimum of the right hand side for various ρ, we get $\rho_G \leq \rho_{ds}+1$. Theorem 10.2 is now an immediate consequence of Corollary 9.8.

Corollary 10.4. *Let $x : M \to \mathbf{R}^m$ be a complete nonflat minimal surface with infinite total curvature and $G : M \to P^{m-1}(\mathbf{C})$ the Gauss map of M. Assume that $G(M)$ is not included in any projective linear subspace of dimension less than n. Then, for arbitrary hyperplanes $H_1, \dots, H_q$ in $P^{m-1}(\mathbf{C})$ located in general position,*

$$\sum_{j=1}^{q} D_G(H_j) \leq 2m-n-1+\frac{(\rho_{ds}+1)n(2m-n-1)}{2}. \tag{10.5}$$

This is a direct result of Corollary 9.9 because we have $\rho_G \leq \rho_{ds}+1$.

Corollary 10.6. *Let M be a nonflat complete regular minimal surface in $\mathbf{R}^m$ with infinite total curvature and let G be the Gauss map of M. Then, for arbitrary hyperplanes $H_1, \dots, H_q$ in $P^{m-1}(\mathbf{C})$ located in general position,*

$$\sum_{j=1}^{q} D_G(H_j) \leq \frac{m(m+1)}{2}.$$

In fact, we can apply Corollary 10.4 for some n with $1 \leq n \leq m-1$. Since $\rho_{ds} = 0$ in this case, we have

$$\sum_{j=1}^{q} D_G(H_j) \leq 2m-n-1+\frac{n(2m-n-1)}{2} \leq \frac{m(m+1)}{2}.$$

Combining this with Theorem 6.10, we have the following:

Corollary 10.7. *The Gauss map of a nonflat complete regular minimal surface in* $\mathbf{R}^m$ *can omit at most* $m(m+1)/2$ *hyperplanes in* $P^{m-1}(\mathbf{C})$ *located in general position.*

We can show that, for an arbitrary odd number m, the number $m(m+1)/2$ of Corollary 10.7 is the best-possible; namely, there exist complete minimal surfaces in $\mathbf{R}^m$ whose Gauss maps are non-degenerate and omit $m(m+1)/2$ hyperplanes in general position. Refer to [39, §5.5] for concrete examples of complete minimal surfaces with these properties.

Now, we consider a nonconstant holomorphic map $w = (w_1, w_2, \dots, w_m)$ of an open Riemann surface M into $\mathbf{C}^m$. By associating a point $(x_1, y_1, \dots, x_m, y_m)$ with $(x_1 + \sqrt{-1}y_1, \dots, x_m + \sqrt{-1}y_m) \in \mathbf{C}^m$, the holomorphic curve w is identified with the surface $w = (x_1, y_1, \dots, x_m, y_m) : M \to \mathbf{R}^{2m}$ in $\mathbf{R}^{2m}$, which is a minimal surface because x_i's and y_i's are harmonic and satisfies the condition (5.3). Since $f_i = (\partial/\partial z)x_i = \sqrt{-1}(\partial/\partial z)y_i$ by the Cauchy-Riemann equations, the Gauss map of M is given by $G = (f_1 : -\sqrt{-1}f_1 : \cdots : f_m : -\sqrt{-1}f_m)$ and so $G(M)$ is included in the $(m-1)$-dimensional projective linear subspace $\{(u_1 : v_1 : \cdots : u_m : v_m); u_i = -\sqrt{-1}v_i\ (1 \le i \le m)\}$, which we denote by $P^{m-1}(\mathbf{C})$. In particular, if M is not included in any proper affine subspace of $\mathbf{C}^m$, G is nondegenerate as a map into $P^{m-1}(\mathbf{C})$. As an easy consequence of Theorem 10.2, we get the following:

Corollary 10.8 ([34, Theorem 2.10]). *Let* $w : M \to \mathbf{C}^m$ *be a complete regular holomorphic curve in* $\mathbf{C}^m$ *not included in any affine hyperplane and let* $G : M \to P^{m-1}(\mathbf{C})$ *be the above-mentioned Gauss map of* M. *If* M *is not of finite type or else if* G *is transcendental, then*

$$\sum_{j=1}^{q} D_G(H_j) \le m + \frac{(\rho_{ds}+1)m(m-1)}{2}$$

for arbitrary hyperplanes $H_1, \dots, H_q$ *in* $P^{m-1}(\mathbf{C})$ *in general position.*

Now, we consider the classical Gauss map g of a nonflat minimal surface $x = (x_1, x_2, x_3) : M \to \mathbf{R}^3$ possibly with branch points. As in Theorem 5.10, setting $\omega_i := \partial x_i$ $(i = 1, 2, 3)$ and $\omega := \omega_1 - \sqrt{-1}\omega_2$, we have $ds^2 = (1+|g|^2)^2|\omega|^2$. Take a reduced representation (g_1, g_2) of g. Then, we can write $ds^2 = (|g_1|^2 + |g_2|^2)^2|\omega_0|^2$ with a meromorphic form $\omega_0 := (1/g_2^2)\omega$. Since ds^2 is continuous and $|g_1|^2 + |g_2|^2$ has no zero, ω_0 has no pole and $\nu_{\omega_0} = \nu_{ds}$, so that $-\operatorname{Ric}[ds^2] = [\nu_{ds}] + 2\Omega_g \prec (\rho+2)\Omega_g$ whenever $[\nu_{ds}] \prec \rho\Omega_g$. This yields $\rho_g \le \rho_{ds} + 2$.

The map g is nondegenerate if and only if M is nonflat, or equivalently M is not a plane. On the other hand, when the set of all hyperplanes in $P^1(\mathbf{C})$ is identified with $P^1(\mathbf{C})$ itself, with each $\alpha := (a_0 : a_1) \in P^1(\mathbf{C})$ corresponding to the hyperplane $H := \{(w_0 : w_1); a_1w_0 - a_0w_1 = 0\}$ in $P^1(\mathbf{C})$, mutually distinct points in $P^1(\mathbf{C})$ are associated with hyperplanes in general position. Therefore, Corollary 9.9 implies the following result.

Theorem 10.9. *Let $x : M \to \mathbf{R}^3$ be a complete minimal surface with infinite total curvature and let $g : M \to P^1(\mathbf{C})$ be the classical Gauss map. Then, for arbitrary distinct points $\alpha_1, \dots, \alpha_q$ in $P^1(\mathbf{C})$,*

$$\sum_{j=1}^{q} D_g(\alpha_j) \leq 4 + \rho_{ds}\,. \tag{10.10}$$

Here, we note that Theorem 10.9 can be applied to give another proof of Theorem 8.16. In relation to Theorem 10.9, we can give an example of a minimal surface with infinite total curvature, for which the branching order is equal to two and equality holds for the modified defect relation (10.10). For the details, see the original paper [35, §6].

We next consider a complete regular minimal surface $x = (x_1, x_2, x_3, x_4) : M \to \mathbf{R}^4$. The Gauss map G of M is a map into $Q_2(\mathbf{C})$. We shall inquire into the structure of $Q_2(\mathbf{C})$. We define the map $\psi_1 : Q_2(\mathbf{C}) - E \to P^1(\mathbf{C})$ by

$$\psi_1(w) := (w_1 - \sqrt{-1}w_2 : w_3 + \sqrt{-1}w_4)\,, \qquad w = (w_1 : \dots : w_4) \in Q_2(\mathbf{C}) - E\,,$$

where $E := \{(w_1 : \dots : w_4) \in Q_2(\mathbf{C}); w_1 - \sqrt{-1}w_2 = w_3 + \sqrt{-1}w_4 = 0\}$. For each $u := (u_1 : \dots : u_4) \in Q_2(\mathbf{C})$, we have

$$\begin{aligned}\psi_1(u) &= ((u_1 - \sqrt{-1}u_2)(u_3 - \sqrt{-1}u_4) : u_3^2 + u_4^2) \\ &= ((u_1 - \sqrt{-1}u_2)(u_3 - \sqrt{-1}u_4) : -(u_1^2 + u_2^2)) \\ &= (u_3 - \sqrt{-1}u_4 : -(u_1 + \sqrt{-1}u_2))\end{aligned}$$

except for the case where these have no meanings. For $w := (w_1 : w_2 : w_3 : w_4) \in E$, since $w_1 + \sqrt{-1}w_2 \neq 0$ or $w_3 - \sqrt{-1}w_4 \neq 0$, the value $\lim_{u \notin E, u \to w} \psi_1(u)$ exists in $P^1(\mathbf{C})$, by which we define $\psi_1(w)$. Similarly, for each $w := (w_1 : w_2 : w_3 : w_4) \in Q_2(\mathbf{C})$ we define

$$\psi_2(w) = \begin{cases} (w_1 - \sqrt{-1}w_2 : -w_3 + \sqrt{-1}w_4) & w \notin E' \\ \lim_{u \notin E', u \to w} \psi_2(u) & \text{otherwise}\,. \end{cases}$$

where $E' := \{w_1 - \sqrt{-1}w_2 = w_3 - \sqrt{-1}w_4 = 0\}$. By using these maps, we define the map $\Psi := (\psi_1, \psi_2) : Q_2(\mathbf{C}) \to P^1(\mathbf{C}) \times P^1(\mathbf{C})$. If we consider the map $\Psi^* : P^1(\mathbf{C}) \times P^1(\mathbf{C}) \to Q_2(\mathbf{C})$ defined by

$$\Psi^*((z : w), (u : v)) := (zu + wv : \sqrt{-1}(zu - wv) : wu - zv : -\sqrt{-1}(wu + zv))\,,$$

we can easily check that $\Psi^* \cdot \Psi$ and $\Psi \cdot \Psi^*$ are both identity maps. Therefore, Ψ is bijective and so the quadric $Q_2(\mathbf{C})$ is biholomorphic with $P^1(\mathbf{C}) \times P^1(\mathbf{C})$. For a more precise description, refer to [44, §2].

As in the case of surfaces in $\mathbf{R}^3$, regarding M as a Riemann surface with a conformal metric and taking a holomorphic local coordinate z, we set $\omega_i = \partial x_i$ $(i = 1, 2, 3, 4)$ and define the map

$$g = (g_1, g_2) := ((\omega_1 - \sqrt{-1}\omega_2 : \omega_3 + \sqrt{-1}\omega_4), (\omega_1 - \sqrt{-1}\omega_2 : -\omega_3 + \sqrt{-1}\omega_4)).$$

Instead of the Gauss map $G : M \to Q_2(\mathbf{C})$ we consider the map $g : M \to P^1(\mathbf{C}) \times P^1(\mathbf{C})$, which we call the classical Gauss map of M in the following.

Now, we describe some results concerning on the value distributions of the classical Gauss map g of a complete nonflat regular minimal surface M in $\mathbf{R}^4$. In the followings, g_1 and g_2 denote the first and second components of the map $g : M \to P^1(\mathbf{C}) \times P^1(\mathbf{C})$ respectively.

In [44] D. A. Hoffman and R. Osserman showed the following:

Theorem 10.11. *Assume that M has finite total curvature. If both g_1 and g_2 omit four values, or if one of them is a constant and the other omits three values, then M is a plane.*

The author gave the following analogous result without the assumption of finiteness of total curvature.

Theorem 10.12 ([32]). (i) *Assume that g_1 and g_2 are both nonconstant and omit q_1 and q_2 values respectively. If $q_1 > 2$, $q_2 > 2$, then*

$$\frac{1}{q_1 - 2} + \frac{1}{q_2 - 2} \geq 1.$$

(ii) *If either g_1 or g_2, say g_2, is a constant, then g_1 can omit at most three values.*

Afterwards, this was improved by X. Mo and R. Osserman as follows:

Theorem 10.13 ([49]). *If both g_1 and g_2 are nonconstant and omit four values, then both of them must take on every other value infinitely often. If one of them is a constant and the other is not a constant and omits three values, it takes on every other value infinitely often.*

They also showed by examples that both parts of this theorem are the best-possible. For details, refer to the original paper.

As an application of Theorem 9.6, we can give the following modified defect relation, which implies Theorem 10.13.

Theorem 10.14 ([39, Theorem 5.4.9]). *Let $x : M \to \mathbf{R}^4$ be a complete regular minimal surface in $\mathbf{R}^4$ with infinite total curvature and let $g = (g_1, g_2) : M \to P^1(\mathbf{C}) \times P^1(\mathbf{C})$ be the classical Gauss map of M. Take arbitrary mutually distinct $\alpha_1, \dots, \alpha_{q_1}$ and mutually distinct $\beta_1, \dots, \beta_{q_2}$ in $\bar{\mathbf{C}}$.*

(i) *If g_1 and g_2 are both nonconstant and $\sum_{i=1}^{q_1} D_{g_1}(\alpha_i) > 2$ and $\sum_{j=1}^{q_2} D_{g_2}(\beta_j) > 2$, then*

$$\frac{1}{\sum_{i=1}^{q_1} D_{g_1}(\alpha_i) - 2} + \frac{1}{\sum_{j=1}^{q_2} D_{g_2}(\beta_j) - 2} \geq 1.$$

(ii) *If g_2 is a constant, then $\sum_{j=1}^{q_1} D_{g_1}(\alpha_j) \leq 3$.*

To see this, take reduced representations (g_{k0}, g_{k1}) of g_k $(k = 1, 2)$ and set $h := (f_1 - \sqrt{-1} f_2)/(g_{10} g_{20})$, where $f_i := (\partial/\partial z) x_i$. Then we have

$$f_1 - \sqrt{-1} f_2 = h g_{10} g_{20}, \quad f_1 + \sqrt{-1} f_2 = -h g_{11} g_{21},$$
$$f_3 + \sqrt{-1} f_4 = h g_{11} g_{20}, \quad f_3 - \sqrt{-1} f_4 = -h g_{10} g_{21}.$$

Therefore, the induced metric on M is given by

$$\begin{aligned} ds^2 &= (|f_1 - \sqrt{-1} f_2|^2 + |f_1 + \sqrt{-1} f_2|^2 + |f_3 - \sqrt{-1} f_4|^2 \\ &\quad + |f_3 + \sqrt{-1} f_4|^2)|dz|^2 \\ &= |h|^2 (|g_{10}|^2 + |g_{11}|^2)(|g_{20}|^2 + |g_{21}|^2)|dz|^2 . \end{aligned}$$

Here, h is a nowhere zero holomorphic function. Consider the situation in (i). Then, the maps g_1 and g_2 of M into $P^1(\mathbf{C})$ are both nondegenerate. Moreover, in view of the above representation of ds^2, g satisfies the condition $(C)_{1\,1}$. We can apply Theorem 9.6 to assert (i). If g_2 is a constant, then $\rho_{g_1} \leq 1$. The assertion (ii) is an immediate consequence of Theorem 9.6 for the case $L = 1$.

References *

1. Ahlfors, L.V.: An extension of Schwarz's lemma. Trans. Am. Math. Soc. **43**, 359–364 (1938), Zbl. 18, 410
2. Ahlfors, L.V.: The theory of meromorphic curves. Acta Soc. Sci. Fenn. Nov. Ser. A I, No. 35 (1941), Zbl. 61, 152
3. Ahlfors, L.V., and Sario, L.: Riemann surfaces. Princeton Univ. Press, New Jersey 1960, Zbl. 196, 338
4. Beckenbach, E.F.: Defect relations for meromorphic minimal surfaces. Lecture Notes Math. **419**, Springer-Verlag, Berlin Heidelberg New York 1974, pp. 18–41, Zbl. 297.53002
5. Beckenbach, E.F., and Cootz, T.A.: The second fundamental theorem for meromorphic minimal surfaces. Bull. Am. Math. Soc. **76** (1970) 711–716, Zbl. 205, 256
6. Beckenbach, E.F., and Hutchison, G.A.: Meromorphic minimal surfaces. Pac. J. Math. **28**, 17–47 (1969), Zbl. 179, 112
7. Beckenbach, E.F., and Radó, T.: Subharmonic functions and minimal surfaces. Trans. Am. Math. Soc. **35**, 648–661 (1933), Zbl. 7, 129
8. Beckenbach, E.F., Eng, F.H., and Tafel, R.E.: Global properties of rational and logarithmico-rational minimal surfaces. Pac. J. Math. **50**, 355–381 (1974), Zbl. 282.53010
9. Bernstein, S.: Sur un théorème de géométrie et ses applications aux équations aux dérivées partielles du type elliptique. Comm. de la Soc. Math. de Kharkov (2éme sér.) **15**, 38–4 (1915–1917), Jbuch 48, 811 and 1401

*For the convenience of the reader, references to reviews in Zentralblatt für Mathematik (Zbl.), compiled by means of the MATH database, and Jahrbuch über die Fortschritte der Mathematik (Jbuch) have, as far as possible, been included in this bibliography.

10. Borel, E.: Sur les zéros des fonctions entières. Acta Math. **20**, 357–396 (1897), Jbuch 28, 360
11. Cartan, H.: Sur les zéros des combinaisons linéaires de p fonctions holomorphes données. Mathematica, **7**, 5–31 (1933), Zbl. 7, 415
12. Chen, W.: Cartan's conjecture: Defect relation for meromorphic maps from parabolic manifold to projective space. Ph.D. dissertation, Notre Dame University 1987
13. Chen, W.: Defect relations for degenerate meromorphic maps. Trans. Am. Math. Soc. **319**, 499–515 (1990), Zbl. 701.32013
14. Chern, S.S.: An elementary proof of the existence of isothermal parameters on a surface. Proc. Am. Math. Soc. **6**, 771–782 (1955), Zbl. 66, 154
15. Chern, S.S.: Complex manifolds without potential theory. Van Nostrand, 1967, Zbl. 158, 330
16. Chern, S.S., and Osserman, R.: Complete minimal surfaces in euclidean n-space. J. Anal. Math. **19**, 15–34 (1967), Zbl. 172, 228
17. Cowen, M., and Griffiths, P.: Holomorphic curves and metrics of negative curvature. J. Anal. Math. **29**, 93–153 (1976), Zbl. 352.32014
18. Earp, R.S., and Rosenberg, H.: On values of the Gauss map of complete minimal surfaces in $\mathbf{R}^3$. Comment. Math. Helv. **63**, 579–586 (1988), Zbl. 694.53007
19. Fang, Y.: On the Gauss map of complete minimal surfaces with finite total curvature. Indiana Univ. Math. J. **42**, No. 4, 1389–1411 (1993), Zbl. 794.53006
20. Finn, R.: On a class of conformal metrics, with application to differential geometry in the large. Comment. Math. Helvet. **40**, 1–30 (1965), Zbl. 192, 273
21. Forster, O.: Lectures on Riemann surfaces. Springer-Verlag, Berlin Heidelberg New York 1981, Zbl. 475.30002. (German original edition: Springer-Verlag, Berlin Heidelberg New York 1977, Zbl. 381.30021)
22. Fujimoto, H.: On meromorphic maps into the complex projective space. J. Math. Soc. Jap. **26**, 272–288 (1974), Zbl. 276.32013
23. Fujimoto, H.: The uniqueness problem of meromorphic maps into the complex projective space. Nagoya Math. J. **58**, 1–23 (1975), Zbl. 313.32005
24. Fujimoto, H.: A uniqueness theorem of algebraically non-degenerate meromorphic maps into $P^N(\mathbf{C})$. Nagoya Math. J. **64**, 117–147 (1976), Zbl. 329.32013
25. Fujimoto, H.: Remarks to the uniqueness problem of meromorphic maps into $P^N(\mathbf{C})$ I. Nagoya Math. J. **71**, 13–24 (1978), Zbl. 382.32016
26. Fujimoto, H.: Remarks to the uniqueness problem of meromorphic maps into $P^N(\mathbf{C})$ II. Nagoya Math. J. **71**, 25–41 (1978), Zbl. 382.32017
27. Fujimoto, H.: Remarks to the uniqueness problem of meromorphic maps into $P^N(\mathbf{C})$ III. Nagoya Math. J. **75**, 71–85 (1979), Zbl. 431.32021
28. Fujimoto, H.: Remarks to the uniqueness problem of meromorphic maps into $P^N(\mathbf{C})$ IV. Nagoya Math. J. **83**, 153–181 (1981), Zbl. 462.320.10
29. Fujimoto, H.: On the Gauss map of a complete minimal surface in $\mathbf{R}^m$. J. Math. Soc. Jap. **35**, 279–288 (1983), Zbl. 506.53005
30. Fujimoto, H.: Value distribution of the Gauss maps of complete minimal surfaces in $\mathbf{R}^m$. J. Math. Soc. Jap. **35**, 663–681 (1983), Zbl. 527.53004
31. H. Fujimoto, Non-integrated defect relationfor meromorphic maps of complete Kähler manifolds into $P^{N_1}(\mathbf{C}) \times \cdots \times P^{N_k}(\mathbf{C})$. Jap. J. Math. New Ser. **11**, 233–264 (1985), Zbl. 591.32029
32. Fujimoto, H.: On the number of exceptional values of the Gauss map of minimal surfaces. J. Math. Soc. Jap. **40**, 235–247 (1988), Zbl. 641.53009
33. Fujimoto, H.: Modified defect relations for the Gauss map of minimal surfaces. J. Diff. Geom. **29**, 245–262 (1989), Zbl. 676.53005
34. Fujimoto, H.: Modified defect relations for the Gauss map of minimal surfaces II. J. Diff. Geom. **31**, 365–385 (1990), Zbl. 719.53005

35. Fujimoto, H.: Modified defect relations for the Gauss map of minimal surfaces III. Nagoya Math. J. **124**, 13–40 (1991), Zbl. 739.49022
36. Fujimoto, H.: On the Gauss curvature of minimal surfaces. J. Math. Soc. Jap. **44**, 427–439 (1992), Zbl. 759.53004
37. Fujimoto, H.: Unicity theorems for the Gauss maps of complete minimal surfaces. J. Math. Soc. Jap. **45**, 481–487 (1993), Zbl. 796.53005
38. Fujimoto, H.: Unicity theorems for the Gauss maps of complete minimal surfaces II. Kodai Math. J. **16**, 335–354 (1993)
39. Fujimoto, H.: Value distribution theory of the Gauss map of minimal surfaces in $\mathbf{R}^m$. Aspect of Math. **E21**, Vieweg, Wiesbaden 1993
40. Fujimoto, H.: Nevanlinna theory for minimal surfaces of parabolic type. To appear in Kodai Math. J.
41. Hall, P.: A Picard theorem with an application to minimal surfaces. Trans. Am. Math. Soc. **314**, 597–603 (1989), Zbl. 682.53010
42. Hall, P.: A Picard theorem with an application to minimal surfaces II. Trans. Am. Math. Soc. **325**, 895–902 (1991), Zbl. 743.53002
43. Heinz, E.: Über die Lösungen der Minimalflächengleichung. Nachr. Akad. Wiss. Göttingen, 51–56 (1952), Zbl. 48, 154
44. Hoffman, D.A., and Osserman, R.: The geometry of the generalized Gauss map. Mem. Am. Math. Soc. **236** (1980) Zbl. 469.53004
45. Hopf, E.: On an inequality for minimal surfaces $z = z(x, y)$. J. Ration. Mech. Anal. **2**, 519–522 (1953), Zbl. 51, 126
46. Huber, A.: On subharmonic functions and differential geometry in the large. Comment. Math. Helvet. **32**, 13–72 (1957), Zbl. 80, 150
47. Jorge, L.P., and Meeks III, W.H.: The topology of complete minimal surfaces of finite total Gaussian curvature. Topology **22**, 203–221 (1983), Zbl. 517.53008
48. Kao, S.J.: On values of Gauss maps of complete minimal surfaces on annular ends. Math. Ann. **291**, 315–318 (1991), Zbl. 760.53005
49. Mo, X., and Osserman, R.: On the Gauss map and total curvature of complete minimal surfaces and an extension of Fujimoto's theorem. J. Diff. Geom. **31**, 343–355 (1990), 686.53010
50. Nevanlinna, R.: Einige Eindeutigkeitssätze in der Theorie der meromorphen Funktionen. Acta Math. **48**, 367–391 (1926), Jbuch 52, 323
51. Nevanlinna, R.: Le théorème de Picard-Borel et la théorie des fonctions méromorphes. Gauthier-Villars, Paris 1929, Jbuch 55, 773
52. Nitsche, J.C.C.: On an estimate for the curvature of minimal surfaces $z = z(x, y)$. J. Math. Mech. **7**, 767–769 (1958), Zbl. 85, 159
53. Nochka, E.I.: On the theory of meromorphic functions. Dokl. Akad. Nauk SSSR **269**, 547–552 (1983) Engl. transl.: Sov. Math. Dokl. **27** (2), 377–381 (1983), Zbl. 552.32024
54. Osserman, R.: An analogue of the Heinz-Hopf inequality. J. Math. Mech. **8**, 383–385 (1959), Zbl. 85, 159
55. Osserman, R.: On the Gauss curvature of minimal surfaces. Trans. Am. Math. Soc. **96**, 115–128 (1960), Zbl. 93, 343
56. Osserman, R.: Minimal surfaces in the large. Comment. Math. Helvet. **35**, 65–76 (1961), Zbl. 98, 349
57. Osserman, R.: Global properties of minimal surfaces in $\mathbf{E}^3$ and $\mathbf{E}^n$. Ann. Math. **80**, 340–364 (1964), Zbl. 134, 385
58. Osserman, R.: A survey of minimal surfaces. 2nd edition, Dover Publ. Inc., New York 1986
59. Ros, R.: The Gauss map of minimal surfaces. Preprint
60. Ru, M.: On the Gauss map of minimal surfaces immersed in $\mathbf{R}^n$. J. Diff. Geom. **34**, 411–423 (1991), Zbl. 733.53005

61. Ru, M.: On the Gauss map of minimal surfaces with finite total curvature. Bull. Aust. Math. Soc. **44**, 225–232 (1991), Zbl. 725.53010
62. Ru, M.: Gauss map of minimal surfaces with ramification. Trans. Am. Math. Soc. **39**, No. 2, 751–764 (1993), Zbl. 761.792.53003
63. Shabat, B.V.: Distribution of values of holomorphic mappings. Nauka, Moscow 1982. Engl. transl.: Transl. Math. Monographs **61**, Am. Math. Soc., 1985, Zbl. 537.32008, Zbl. 564.32016
64. Stoll, W.: Value distribution on parabolic spaces, Lecture Notes Math. **600**, Springer-Verlag, Berlin Heidelberg New York 1977, Zbl. 367.32001
65. Weitsman, A., and Xavier, F.: Some function theoretic properties of the Gauss map for hyperbolic complete minimal surfaces. Mich. Math. J. **34**, 275–283 (1987), Zbl. 631.53011
66. Weyl, H. and J.: Meromorphic functions and analytic curves. Princeton Univ. Press, Princeton, New Jersey 1943, Zbl. 61, 153
67. Wong, P.: Holomorphic curves in spaces of constant curvature. Lecture Notes Pure Appl. Math. **143**, 201–223 (1993), Zbl. 802.53020
68. Wu, H.: The equidistribution theory of holomorphic curves. Princeton Univ. Press, Princeton, New Jersey 1970
69. Xavier, F.: The Gauss map of a complete non-flat minimal surface cannot omit 7 points of the sphere. Ann. Math., II. Ser. **113**, 211–214 (1981), Zbl. 477.53007. Erratum, Ann. Math. **115** (2), 667 (1982), Zbl. 485.53006
70. Yau, S.T.: Some function-theoretic poperties of complete Riemannian manifolds and their applications to geometry. Indiana Univ. Math. J., **25**, 659–670 (1976), Zbl. 335.53041

III. Boundary Value Problems for Minimal Surfaces

Stefan Hildebrandt

The aim of this survey is to describe some basic results on boundary value problems for minimal surfaces $X: \Omega \to \mathbb{R}^3$ in three-dimensional Euclidean space. We are essentially concerned with questions of existence, regularity and geometric estimates for solutions of such problems. The material to be presented is divided as follows.

Contents

Chapter 1
Area and Minimal Surfaces

In this introductory chapter we discuss some geometric ideas connected with the notion of minimal surfaces. First we derive various forms of the first variation $\delta A(X,Y)$ of the area functional $A(X)$, and we note that the critical points X of $A(X)$ are exactly the surfaces of vanishing mean curvature. Secondly we prove that the free boundary condition connected with the area functional is the condition of orthogonality.

In Section 2 we show that each nonparametric minimal surface X can be transformed to a conformal immersion Z by means of a real analytic diffeomorphism, and this new parameter representation Z is the real part of an isotropic and holomorphic curve $\Phi : \Omega^* \to \mathbb{C}^3$. Thereby we in particular obtain that each nonparametric minimal surface is real analytic.

Combining these results with the uniformization theorem we see that every minimal surface in $\mathbb{R}^3$, i.e. every C^2-immersion of mean curvature zero, can be transformed by means of a suitable parameter transformation to another surface representation which is a conformal and harmonic mapping. This leads us to the notion of a generalized minimal surface which may have isolated branch points as singularities. For such surfaces we prove some representation formulae which play an important role in the explicit construction of specific examples of minimal surfaces.

1.1 The First Variation of Area. Minimal Surfaces

Let $X : \overline{\Omega} \to \mathbb{R}^3$ be a C^2–immersion of a smoothly bounded domain Ω of $\mathbb{R}^2$ into $\mathbb{R}^3$, the three-dimensional Euclidean space, with the scalar product $a \cdot b$ and the length $|a| = \sqrt{a \cdot a}$, $|b| = \sqrt{b \cdot b}$ of vectors $a, b \in \mathbb{R}^3$. We write

$$X(u,v) = (X^1(u,v), X^2(u,v), X^3(u,v))$$

where $w = (u,v) \in \overline{\Omega}$ denotes the parameter points in $\overline{\Omega}$ and $X^k(u,v)$ are the components of $X(u,v)$. The two partial derivatives of X with respect to u and v are X_u and X_v, and $\nabla X = (X_u, X_v)$ is the gradient of X with

$$|\nabla X|^2 = |X_u|^2 + |X_v|^2 .$$

Sometimes it is convenient to write u^1 and u^2 instead of u and v; then the partial derivative X_{u^α} of X with respect to u^α is abbreviated by $X_{,\alpha}$, and the second derivative $X_{u^\alpha u^\beta}$ is abbreviated by $X_{,\alpha\beta}$, etc.

Let $g_{\alpha\beta} = X_{,\alpha} \cdot X_{,\beta}$ be the *coefficients of the first fundamental form* of X, and $g = det(g_{\alpha\beta})$. We often use the Gauss notation

$$\mathcal{E} = |X_u|^2 , \quad \mathcal{F} = X_u \cdot X_v , \quad \mathcal{G} = |X_v|^2 ,$$

i.e.

$$\mathcal{E} = g_{11}\,, \quad \mathcal{F} = g_{12} = g_{21}\,, \quad \mathcal{G} = g_{22}\,,$$

and

$$\mathcal{W} = \sqrt{g} = \sqrt{\mathcal{E}\mathcal{G} - \mathcal{F}^2} = |X_u \wedge X_v|$$

where $X_u \wedge X_v$ denotes the exterior product of X_u and X_v. Then the *surface normal* or *Gauss map* of X is the map $N : \overline{\Omega} \to S^2 \subset \mathbb{R}^3$ into the 2-sphere $S^2 = \{a \in \mathbb{R}^3 : |a| = 1\}$ given by

$$N = \frac{1}{\mathcal{W}}\, X_u \wedge X_v$$

Let $b_{\alpha\beta} = N \cdot X_{,\alpha\beta} = -N_{,\alpha} \cdot X_{,\beta} = -N_{,\beta} \cdot X_{,\alpha}$ be the *coefficients of the second fundamental form* of X and $b = det(b_{\alpha\beta})$. We write also $\mathcal{L} = b_{11}, \mathcal{M} = b_{12} = b_{21}, \mathcal{N} = b_{22}$. Moreover $(g^{\alpha\beta})$ denotes the inverse of the matrix $(g_{\alpha\beta})$, and b_α^β stands for

$$b_\alpha^\beta = b_{\alpha\gamma} g^{\gamma\beta}$$

(summation convention: summation with respect to repeated Greek (or Latin) indices from 1 to 2 (or 3, respectively)).

Then the *mean curvature* H of X is given by

$$H = \frac{\mathcal{L}\mathcal{G} + \mathcal{N}\mathcal{E} - 2\mathcal{M}\mathcal{F}}{2(\mathcal{E}\mathcal{G} - \mathcal{F}^2)} = \frac{1}{2}\, b_{\alpha\beta}\, g^{\alpha\beta} = \frac{1}{2}\, (b_1^1 + b_2^2)\,,$$

and the *Gauss curvature* K of X is defined by

$$K = \frac{\mathcal{L}\mathcal{N} - \mathcal{M}^2}{\mathcal{E}\mathcal{G} - \mathcal{F}^2} = \frac{b}{g} = det(b_\alpha^\beta)\,.$$

We can represent $X_{,\alpha\beta}$ by means of the *Gauss equations* as

$$X_{,\alpha\beta} = \Gamma_{\alpha\beta}^\gamma X_{,\gamma} + b_{\alpha\beta} N$$

where

$$\Gamma_{\alpha\beta}^\gamma = g^{\gamma\sigma}\Gamma_{\alpha\sigma\beta}\,, \qquad \Gamma_{\alpha\sigma\beta} = \frac{1}{2}(g_{\alpha\sigma,\beta} + g_{\beta\sigma,\alpha} - g_{\alpha\beta,\sigma})$$

are the Christoffel symbols of second and first kind respectively. They are accompanied by the *Weingarten equations*

$$N_{,\alpha} = -b_\alpha^\beta X_{,\beta}\,, \qquad b_\alpha^\beta = g^{\beta\gamma} b_{\alpha\gamma}\,.$$

The area element dA of X is given by $dA = \mathcal{W} du dv$, and thus the area of the surface X is defined by

$$A(X) = \int_\Omega |X_u \wedge X_v| du dv = \int_X dA\,. \tag{1}$$

Now we consider a one-parameter family $Z : \overline{\Omega} \times (-\varepsilon_0, \varepsilon_0) \to \mathbb{R}^3, \varepsilon_0 > 0$, of surfaces

$$Z(w,\varepsilon) = X(w) + \varepsilon Y(w) + \varepsilon^2 R(w,\varepsilon), \qquad |R(w,\varepsilon)| \le \text{const},$$

such that Z and $\partial Z/\partial\varepsilon$ are of class C^1. Then we have

$$Z(w,0) = X(w), \qquad \frac{\partial Z}{\partial \varepsilon}(w,0) = Y(w).$$

We can write

$$Y(w) = \eta^\beta(w) X_{,\beta}(w) + \lambda(w) N(w)$$

with functions $\eta^1, \eta^2, \lambda \in C^1(\overline{\Omega})$. Then

$$Z_{,\alpha} = X_{,\alpha} + \varepsilon\,[\eta^\beta_{,\alpha} X_{,\beta} + \eta^\beta X_{,\alpha\beta} + \lambda_{,\alpha} N + \lambda N_{,\alpha}] + \varepsilon^2 R_{,\alpha}.$$

By virtue of Gauss's and Weingarten's equations we obtain

$$Z_{,\alpha} = X_{,\alpha} + \varepsilon\,[\xi^\gamma_\alpha X_{,\gamma} + \nu_\alpha N] + \varepsilon^2 R_{,\alpha}$$

with

$$\xi^\gamma_\alpha = \eta^\gamma{}_{,\alpha} + \Gamma^\gamma_{\alpha\beta}\eta^\beta - b_{\alpha\beta} g^{\beta\gamma}\lambda, \qquad \nu_\alpha = b_{\alpha\beta}\eta^\beta + \lambda_{,\alpha}.$$

Indicating by $\dots$ the terms of second and higher order in ε, we find

$$|Z_u|^2 |Z_v|^2 - (Z_u \cdot Z_v)^2 = \mathcal{W}^2 + 2\varepsilon \mathcal{W}^2(\xi^1_1 + \xi^2_2) + \dots$$

where

$$\xi^1_1 + \xi^2_2 = \eta^1_u + \eta^2_v - \lambda b_{\alpha\beta} g^{\alpha\beta} + \Gamma^\alpha_{\alpha\beta}\eta^\beta.$$

Since

$$b_{\alpha\beta} g^{\alpha\beta} = 2H, \qquad \Gamma^\alpha_{\alpha\beta} = \frac{1}{2g}\, g_{,\beta} = \frac{1}{\mathcal{W}}\,\mathcal{W}_{,\beta}$$

we infer that

$$\xi^1_1 + \xi^2_2 = \frac{1}{\mathcal{W}}[(\eta^1\mathcal{W})_u + (\eta^2\mathcal{W})_v] - 2H\lambda$$

whence

$$|Z_u \wedge Z_v| = \mathcal{W} + \varepsilon\,[(\eta^1\mathcal{W})_u + (\eta^2\mathcal{W})_v - 2H\mathcal{W}\lambda] + \dots. \tag{2}$$

Thus we obtain

Proposition 1. *The first variation*

$$\delta A(X,Y) = \frac{d}{d\varepsilon} A(Z(\cdot,\varepsilon))\Big|_{\varepsilon=0} \tag{3}$$

of the functional $A(X)$ at X in direction of $Y = \eta^\alpha X_{,\alpha} + \lambda N$ is given by

$$\delta A(X,Y) = \int_\Omega [(\eta^1\mathcal{W})_u + (\eta^2\mathcal{W})_v - 2H\mathcal{W}\lambda]\,du\,dv, \tag{4}$$

and an integration by parts yields

$$\delta A(X,Y) = -2\int_\Omega \lambda H \mathcal{W}\,du\,dv + \int_{\partial\Omega} \mathcal{W}(\eta^1 dv - \eta^2 du). \tag{4'}$$

The boundary integral in (4′) vanishes if $Y \in C^1_c(\Omega, \mathbb{R}^3)$, i.e. if Y has compact support in Ω, and thus the fundamental theorem of the calculus of variations yields

Proposition 2. *The first variation $\delta A(X,Y)$ of the area functional A at X vanishes for all $Y \in C^1_c(\Omega, \mathbb{R}^3)$ if and only if the mean curvature H of X is identically zero.*

This motivates the following

Definition. *A C^2-immersion $X : \Omega \to \mathbb{R}^3$ of a parameter domain $\Omega \subset \mathbb{R}^2$ is called a minimal surface if its mean curvature function H satisfies $H = 0$.*

Let $[a,b,c] := det(a,b,c) = a \cdot (b \wedge c)$ be the volume product of three vectors $a, b, c \in \mathbb{R}^3$. Then, for $Y = \eta^\alpha X_{,\alpha} + \lambda N$ we can write

$$\lambda \mathcal{W} = [Y, X_u, X_v]$$

and

$$\mathcal{W}(\eta^2 du - \eta^1 dv) = [Y, N, dX] \,.$$

Therefore formula (4′) can be written as

$$-\delta A(X,Y) = 2 \int_\Omega H[Y, X_u, X_v] du dv + \int_{\partial\Omega} [Y, N, dX] \,. \tag{5}$$

Let s be the arc-length parameter along the boundary $X\big|_{\partial\Omega} = \partial X$ of the surface X, $t = dX/ds$ the tangent and $n = N \wedge t$ the *side normal* of X along its boundary. Then (5) becomes

$$-\delta A(X,Y) = 2 \int_X (Y \cdot N) H dA + \int_{\partial X} (Y \cdot n) ds \tag{6}$$

If $H = 0$, i.e. if X is a minimal surface, we have in particular

$$\delta A(X,Y) = - \int_{\partial X} (Y \cdot n) ds \,.$$

Hence if Y is perpendicular to the side normal n along ∂X and $H = 0$, we have $\delta A(X,Y) = 0$. Conversely, if $\delta A(X,Y) = 0$ for all vector fields $Y \in C^1(\overline{\Omega}, \mathbb{R}^3)$ which are tangential to a given two-dimensional manifold S in $\mathbb{R}^3$ along ∂X, then $H = 0$ and n is perpendicular to S along ∂X. This leads us to another necessary condition for surfaces $X \in C^2(\overline{\Omega}, \mathbb{R}^3)$ which make the area functional A *stationary* with respect to all *variations* $Z(\cdot, \varepsilon) = X + \varepsilon Y + \varepsilon^2 R(\cdot, \varepsilon)$ *of* X such that $Z(w, \varepsilon)$ for $w \in \partial\Omega$ lies on a two-dimensional manifold $S \subset \mathbb{R}^3$ (which is often called a *supporting manifold* or simply a *support surface*).

Proposition 3. *Let $X : \overline{\Omega} \to \mathbb{R}^3$ be a C^2-immersion with boundary values $X|_{\partial\Omega}$ on a smooth support surface $S \subset \mathbb{R}^3$. Then we have*

$$\delta A(X,Y) := \frac{d}{d\varepsilon} A(Z(\cdot,\varepsilon))\Big|_{\varepsilon=0} = 0$$

for all smooth variations $Z(\cdot,\varepsilon) = X + \varepsilon Y + O(\varepsilon^2), |\varepsilon| < \varepsilon_o << 1$, *such that* $Z(w,\varepsilon) \in S$ *for all* $(w,\varepsilon) \in \partial\Omega \times (-\varepsilon_o, \varepsilon_o)$ *if and only if* X *is a minimal surface which meets* S *orthogonally at its boundary.*

In a situation as described in Proposition 3 we say that X *is a stationary surface with a free boundary on the support surface* S.

1.2 Conformal Representation of Minimal Surfaces

Suppose that $X : \overline{\Omega} \to \mathbb{R}^3$ is a surface in nonparametric form, that is, X is described as a graph of a function $z : \overline{\Omega} \to \mathbb{R}$ where Ω is a domain in $\mathbb{R}^2$. Such a surface is given in the form

$$X(x,y) = (x, y, z(x,y)), \qquad x, y \in \Omega .$$

A straightforward computation shows that

$$\mathcal{W} = \sqrt{1 + |\nabla z|^2} = \sqrt{1 + z_x^2 + z_y^2}$$

and

$$2\mathcal{W}^3 H = (1 + z_y^2) z_{xx} - 2 z_x z_y z_{xy} + (1 + z_x^2) z_{yy} =: T(z) .$$

Hence $H = 0$ is equivalent to $T(z) = 0$, i.e. to

$$(1 + z_y^2) z_{xx} - 2 z_x z_y z_{xy} + (1 + z_x^2) z_{yy} = 0 . \tag{1}$$

This equation is the so-called *minimal surface equation*, and graph z of a function $z \in C^2(\overline{\Omega})$ is a minimal surface if and only if z satisfies this equation. Because of

$$\operatorname{div}(\nabla z / \mathcal{W}) = T / \mathcal{W}^3 \tag{2}$$

we see that (1) is equivalent to

$$\operatorname{div} \frac{\nabla z}{\sqrt{1 + |\nabla z|^2}} = 0 \tag{3}$$

which is the Euler equation of the "nonparametric area functional"

$$A(X) = \int_\Omega \sqrt{1 + |\nabla z|^2}\, dx dy =: \mathcal{A}(z) \tag{4}$$

Thus, as to be expected, the nonparametric surface $X(x,y) = (x, y, z(x,y))$ is a minimal surface, i.e. a critical point of $A(X)$ if and only if z is a critical point of $\mathcal{A}(z)$, i.e. a solution of the minimal surface equation (1).

Consider an arbitrary nonparametric surface $X(x,y) = (x,y,z(x,y)), (x,y) \in \Omega$, with $z \in C^2(\Omega)$. The surface normal $N(x,y)$ of $X(x,y)$ is given by $N = (\xi, \eta, \zeta)$ with

$$\xi = -z_x/\mathcal{W}, \quad \eta = -z_y/\mathcal{W}, \quad \zeta = 1/\mathcal{W}$$

where $\mathcal{W} = \sqrt{1 + z_x^2 + z_y^2}$. A straightforward computation shows that the exterior differential of the vector-valued one-form

$$N \wedge dX = (\alpha, \beta, \gamma)$$

is given by

$$d\alpha = 2Hz_x\, dx \wedge dy\,, \quad d\beta = 2Hz_y\, dx \wedge dy\,, \quad d\gamma = -2H dx \wedge dy$$

where H is the mean curvature of X; therefore

$$d(N \wedge dX) = -2H\mathcal{W}N dx \wedge dy\,. \tag{5}$$

It follows that the 1-form $N \wedge dX$ is closed if and only if X is a minimal surface, and we obtain

Proposition 1. *A nonparametric surface $X(x,y) = (x,y,z(x,y))$ defined by means of a function $z \in C^2(\Omega)$ on a simply connected domain Ω of $\mathbb{R}^2$ is a minimal surface if and only if the associated 1-form $N \wedge dX$ is exact, i.e. if and only if there is a mapping $X^* \in C^2(\Omega, \mathbb{R}^3)$ such that*

$$-dX^* = N \wedge dX\,. \tag{6}$$

If we write $X^* = (a,b,c)$ and $N \wedge dX = (\alpha, \beta, \gamma)$, equation (6) is equivalent to

$$-da = \alpha\,, \quad -db = \beta\,, \quad -dc = \gamma\,. \tag{6'}$$

Let us apply this result to a nonparametric minimal surface $X(x,y) = (x,y,z(x,y))$ where $z \in C^2(\Omega)$ is defined on an open convex set Ω of $\mathbb{R}^2$. Then we obtain that there is a function $a \in C^2(\Omega)$ such that $da = -\alpha$, i.e.

$$da = \frac{1}{\mathcal{W}}\left[z_x z_y dx + (1 + z_y^2)dy\right], \quad \mathcal{W} = \sqrt{1 + z_x^2 + z_y^2}\,. \tag{7}$$

Consider the mapping $\varphi : \Omega \to \mathbb{R}^2$ defined by $\varphi(x,y) = (x, a(x,y))$, i.e. by the pair of equations

$$u = x\,, \quad v = a(x,y)$$

Since $a_y = (1 + z_y^2)/\mathcal{W} > 0$ and Ω is convex, the mapping φ is one-to-one, and its Jacobian $J = a_y$ is positive on Ω. Hence φ is a C^2-diffeomorphism of Ω onto some domain Ω^* of $\mathbb{R}^2$. Its inverse $\psi : \Omega^* \to \Omega$ of class C^2 is given by

$$x = u\,, \quad y = f(u,v)$$

with some function $f \in C^2(\Omega^*)$, and we have

$$\begin{bmatrix} 1 & , & 0 \\ f_u & , & f_v \end{bmatrix} = \begin{bmatrix} 1 & , & 0 \\ -a_x/a_y & , & 1/a_y \end{bmatrix}$$

because of $D\psi(u,v) = [D\varphi(x,y)]^{-1}$. On account of (7) it follows that

$$f_u = -z_x z_y/(1+z_y^2)\,, \qquad f_v = \mathcal{W}/(1+z_y^2)\,. \tag{8}$$

Transforming $X(x,y)$ to the new parameter representation $Z(u,v)$ by $Z := X \circ \psi$, that is,

$$Z(u,v) = (u, f(u,v), g(u,v)) \text{ with } g(u,v) := z(u, f(u,v))$$

we obtain by a short computation

$$|Z_u|^2 = |Z_v|^2\,, \quad Z_u \cdot Z_v = 0\,, \tag{9}$$

and $|Z_u|^2 = \mathcal{W}^2/(1+z_y^2) > 0$. Thus u, v are conformal parameters for the representation Z, and so $Z = X \circ \psi$ is a *conformally represented minimal surface which is equivalent to the nonparametric minimal surface* X.

Now we use the equations $-db = \beta$, $-dc = \gamma$, which state that

$$db = -\frac{1}{\mathcal{W}}\left[(1+z_x^2)dx + z_x z_y dy\right], \quad dc = \frac{1}{\mathcal{W}}\left[-z_y dx + z_x dy\right]. \tag{10}$$

Introducing the surface $Z^* : \Omega^* \to \mathbb{R}^3$ by

$$\begin{aligned} &Z^*(u,v) := (v, f^*(u,v), g^*(u,v))\,, \\ &f^*(u,v) := b(u, f(u,v))\,, \quad g^*(u,v) := c(u, f(u,v)) \end{aligned}$$

we infer from (8) and (10) that

$$f_u = f_v^*\,, \qquad f_v = -f_u^*\,, \qquad g_u = g_v^*\,, \qquad g_v = -g_u^*\,, \tag{11}$$

which is equivalent to the vectorial Cauchy-Riemann equations

$$Z_u = Z_v^*\,, \qquad Z_v = -Z_u^*\,. \tag{12}$$

Thus $f + i f^*$ and $g + i g^*$ are holomorphic functions of the complex variable $w = u + i v$ in Ω^*, if we now interpret Ω^* as a domain in the complex plane $\mathbb{C}$. Hence the diffeomorphisms ψ and φ are real analytic, and therefore $X = Z \circ \varphi$ is real analytic. Thus we have proved:

Theorem 1. *Every solution $z \in C^2(\Omega)$ of the minimal surface equation (1) is real analytic.*

Furthermore the above discussion yields the following result.

Theorem 2. *Let $X(x,y) = (x, y, z(x,y))$ be a nonparametric minimal surface of class $C^2(\Omega, \mathbb{R}^3)$ defined on a convex domain Ω of $\mathbb{R}^2$. Then there is a real analytic diffeomorphism $\varphi : \Omega \to \Omega^*$ of Ω onto some simply connected domain Ω^* with a real analytic inverse $\psi : \Omega^* \to \Omega$ such that*

$Z(u,v) := X(\psi(u,v))$ satisfies the conformality relations (9) and is the real part of a holomorphic curve

$$\Phi(w) = Z(u,v) + iZ^*(u,v)\,, \qquad w = u + iv \in \Omega^* \tag{13}$$

in $\mathbb{C}^3$ which is isotropic, i.e.

$$\Phi'(w) \cdot \Phi'(w) = 0\,, \qquad ' = \frac{d}{dw}\,. \tag{14}$$

If the diffeomorphism $\varphi : (x,y) \mapsto (u,v)$ is chosen as $u = x, v = a(x,y)$ where $a(x,y)$ is a solution of (7), then $X^* := Z^* \circ \varphi$ is related to X by the formula $dX^* = -N \wedge dX$ where N is the surface normal of X.

Remark. Note that (14) follows from (9) and (12) because of

$$\Phi'(w) \cdot \Phi'(w) = |Z_u|^2 - |Z_v|^2 - 2i\, Z_u \cdot Z_v\,.$$

1.3 Definition of Generalized Minimal Surfaces. Representation Formulae

By a celebrated theorem of Lichtenstein every $C^{2,\alpha}$-immersion $X : \Omega \to \mathbb{R}^3, 0 < \alpha < 1$, of a domain $\Omega \subset \mathbb{R}^2$ is equivalent to a conformal immersion $Z : \Omega^* \to \mathbb{R}^3$, i.e. there is a $C^{2,\alpha}$-diffeomorphism $\tau : \Omega^* \to \Omega$ such that $Z = X \circ \tau$ satisfies

$$|Z_u|^2 = |Z_v|^2\,, \qquad Z_u \cdot Z_v = 0\,.$$

For minimal surfaces it is much easier to prove that they are equivalent to conformal surface representations since we can apply Theorem 2 of the preceding section. In fact, any minimal immersion is locally equivalent to a nonparametric surface, i.e. to a minimal graph above one of the three coordinate planes of $\mathbb{R}^3$, and therefore it is locally equivalent to a conformal immersion. Then the uniformization theorem implies that X is globally equivalent to a conformal immersion, i.e. we have

Theorem 1. *Any C^2-immersion $X : \Omega \to \mathbb{R}^3$ of vanishing mean curvature, i.e. any minimal surface, is equivalent to an immersed surface represented by conformal parameters.*

For this reason we consider in the sequel only C^2-surfaces $X : \Omega \to \mathbb{R}^3$ satisfying the conformality relations

$$|X_u|^2 = |X_v|^2\,, \qquad X_u \cdot X_v = 0\,. \tag{1}$$

We first note the following result.

Theorem 2. *If $X \in C^2(\Omega, \mathbb{R}^3)$ is a conformal immersion with the mean curvature function $H(u,v)$, then*

$$\Delta X = 2H X_u \wedge X_v \tag{2}$$

where Δ denotes the ordinary Laplacian, i.e. $\Delta X = X_{uu} + X_{vv}$.

Proof. Equations (1) mean $\mathcal{E} = \mathcal{G}, \mathcal{F} = 0$, and therefore

$$H = \frac{\mathcal{L} + \mathcal{N}}{2\mathcal{E}}.$$

Since

$$\mathcal{L} = \langle X_{uu}, N \rangle, \quad \mathcal{N} = \langle X_{vv}, N \rangle$$

we then obtain

$$\Delta X \cdot N = 2\mathcal{E} H. \tag{3}$$

On the other hand, by differentiating (1) with respect to u and v we obtain four equations from which we can infer that ΔX is perpendicular to both X_u and X_v, and since $\mathcal{E} = |X_u \wedge X_v|$ on account of (1), we also have $X_u \wedge X_v = \mathcal{E} N$. Therefore (2) is a consequence of (3).

Corollary. *A conformal immersion $X \in C^2(\Omega, \mathbb{R}^3)$ is a minimal surface if and only if $\Delta X = 0$, i.e. if and only if the mapping $X : \Omega \to \mathbb{R}^3$ is harmonic.*

This leads us to the following

Definition. *A locally nonconstant mapping $X : \Omega \to \mathbb{R}^3$ of class C^2 of an open set $\Omega \subset \mathbb{R}^2$ is called a "generalized minimal surface", or simply a "minimal surface", if it satisfies the conformality relations (1) as well as $\Delta X = 0$. (Note that we allow for points $w = (u, v)$ where $\mathcal{W}$ vanishes).*

Here we usually assume that Ω is a connected open set in $\mathbb{R}^2$, i.e. a domain in $\mathbb{R}^2$. For the purposes of differential geometry it is necessary to consider even minimal surfaces $\mathcal{X} : M \to \mathbb{R}^3$ which are defined on Riemann surfaces M as parameter domains; but we want to confine ourselves to planar domains M as parameter domains, since many of the results presented in this article can easily be carried over to the general situation.

Consider now a minimal surface $X : \Omega \to \mathbb{R}^3$ defined on a *simply connected* domain Ω of $\mathbb{R}^2$($\mathbb{R}^2$ will be identified with $\mathbb{C}$). Then we define the *adjoint surface* $X^* : \Omega \to \mathbb{R}^3$ to X as a solution to the Cauchy-Riemann equations

$$X_u = X_v^*, \qquad X_v = -X_u^*. \tag{4}$$

From

$$\Delta X = 0, \quad |X_u|^2 = |X_v|^2, \quad X_u \cdot X_v = 0$$

and (4) we infer that also

$$\Delta X^* = 0, \quad |X_u^*|^2 = |X_v^*|^2, \quad X_u^* \cdot X_v^* = 0,$$

and clearly X^* is nonconstant. Thus the adjoint X^* to X is a minimal surface. Then

$$f(w) := X(u, v) + iX^*(u, v), \quad w = u + iv \in \Omega, \tag{5}$$

is a holomorphic mapping $f : \Omega \to \mathbb{C}^3$ which can be viewed as a holomorphic curve in $\mathbb{C}^3$. Its complex derivative $f' = df/dw$ is given by

$$f' = X_u + iX_u^* = X_u - iX_v \tag{6}$$

whence

$$f' \cdot f' = |X_u|^2 - |X_v|^2 - 2i\, X_u \cdot X_v\,.$$

Thus the conformality relations (1) yield the *isotropy condition*

$$f' \cdot f' = 0\,. \tag{7}$$

Let ∂_u and ∂_v be the operators $\partial/\partial u$ and $\partial/\partial v$, and denote by $\partial_w = \partial/\partial w, \partial_{\overline{w}} = \partial/\partial \overline{w}$ the Wirtinger operators

$$\partial_w = \frac{1}{2}(\partial_u - i\partial_v)\,, \quad \partial_{\overline{w}} = \frac{1}{2}(\partial_u + i\partial_v)\,.$$

Then $\Delta X = 0$ and (1) are equivalent to

$$X_{w\overline{w}} = 0 \tag{8}$$

and

$$X_w \cdot X_w = 0 \tag{9}$$

respectively. We also note that

$$\mathcal{W} = |X_u|^2 = \frac{1}{2}|\nabla X|^2 = \frac{1}{2}|f'|^2 = 2|X_w|^2 \tag{10}$$

Thus the zeros of $\mathcal{W}$ in Ω are the zeros of f', and of X_u. These zeros are called *branch points of* X; they are clearly isolated in Ω since X is nonconstant. Branch points are the only geometric singularities of "generalized" minimal surfaces $X : \Omega \to \mathbb{R}^3$, and X is a minimal immersion if and only if it has no branch points.

Let us investigate the *asymptotic behaviour* of a minimal surface $X : \Omega \to \mathbb{R}^3$ in the neighbourhood of a branch point $w_0 \in \Omega$. Locally we have $X(w) = \operatorname{Re} f(w)$ where f is a holomorphic and isotropic curve $f : B_R(w_0) \to \mathbb{C}^3$ where $B_R(w_0)$ denotes the disk $B_R(w_0) = \{w \in \mathbb{C} : \mid w - w_0 \mid < R\}$. Since w_0 is a branch point, we have $f'(w_0) = 0$. As $X(w)$ is nonconstant, there is an integer $m \geq 1$ such that $f^{(k)}(w_0) = 0$ for $1 \leq k \leq m$, but $f^{(m+1)}(w_0) \neq 0$. Taylor's expansion yields

$$f'(w) = \frac{1}{m!} f^{(m+1)}(w_0)(w - w_0)^m + \dots\ .$$

Set $X_0 := X(w_0)$ and

$$A = \frac{1}{2}(\alpha - i\beta) := \frac{1}{2m!} f^{(m+1)}(w_0), \;\; B := \frac{2}{m+1} A\,.$$

Then we conclude from

$$f'(w) = X_u(w) - iX_v(w) = 2X_w(w)$$

that

$$X_w(w) = A(w - w_0)^m + O(|\, w - w_0\,|^{m+1}) \text{ as } w \to w_0 \tag{11}$$

and

$$X(w) = X_0 + \operatorname{Re}\{B(w - w_0)^{m+1} + O(|\, w - w_0\,|^{m+2})\} \tag{12}$$

The conformality relation $X_w \cdot X_w = 0$ implies that the vector $A \in \mathbb{C}^3 - \{0\}$ satisfies $A \cdot A = 0$, and therefore the two vectors $\alpha, \beta \in \mathbb{R}^3$ fulfil $|\alpha|^2 = |\beta|^2$ and $\alpha \cdot \beta = 0$, and $A \neq 0$ yields $|\alpha| = |\beta| > 0$. Then we infer from (11) that

$$X_u(w) \wedge X_v(w) = (\alpha \wedge \beta) \,|\, w - w_0\,|^{2m} + O(|\, w - w_0\,|^{2m+1}) \text{ as } w \to w_0\,, \tag{13}$$

and we conclude that the surface normal $N(w)$ tends to a limit N_0 as $w \to w_0$; in fact,

$$\lim_{w \to w_0} N(w) = \frac{\alpha \wedge \beta}{|\,\alpha \wedge \beta\,|} =: N_0\,. \tag{14}$$

If we rotate the axes of the coordinate system in $\mathbb{R}^3$ such that $e_1 := \alpha/|\alpha|$, $e_2 := \beta/|\beta|, e_3 := N_0$ point in the directions of the new coordinate axes, we obtain for $X(w) = (X^1(w), X^2(w), X^3(w))$ the *normal form*

$$\begin{aligned} X^1(w) + iX^2(w) &= X_0^1 + iX_0^2 + a(w - w_0)^{m+1} + O(|\, w - w_0\,|^{m+2})\,, \\ X^3(w) &= X_0^3 + O(|\, w - w_0\,|^{m+2}) \end{aligned} \tag{15}$$

where $X_0 = (X_0^1, X_0^2, X_0^3)$ and $a = |\alpha|/(m+1)$. This shows that $X(w)$ behaves in the neighbourhood of the branch point w_0 like a branch point of m-th order of a Riemann surface. Therefore we denote the integer m associated with w_0 as the *order of the branch point* w_0.

By stereographic projection of S^2 on $\hat{\mathbb{C}} = \mathbb{C} \cup \{\infty\}$ one can prove that $N(w)$ is actually real analytic, and then a straightforward computation yields that

$$-\Delta N = N|\nabla N|^2 \tag{16}$$

as well as

$$|N_u|^2 = |N_v|^2\,, \quad N_u \cdot N_v = 0 \tag{17}$$

on Ω. However the above reasoning is rather useful as it can be used to discuss the asymptotic behaviour of minimal surfaces at boundary branch points where one cannot apply Taylor's expansion but only a weaker kind of asymptotic expansion due to Hartman and Wintner.

Let us briefly touch on one of the celebrated Weierstrass–Enneper representation formulas suggested by (5) and (7). We indicate a global approach that was pointed out by Osserman.

First we infer from (5) and (7) that every holomorphic map $\Phi : \Omega \to \mathbb{C}^3$, $\Omega \subset \mathbb{C}$, with $\Phi \cdot \Phi = 0$ and $\Phi(w) \not\equiv 0$ defines a minimal surface $X : \Omega \to \mathbb{R}^3$ by means of the formula

$$X(w) = X_0 + \text{Re} \int_{w_0}^{w} \Phi(\zeta) d\zeta \ , \qquad w \in \Omega \, , \tag{18}$$

if $w_0 \in \Omega, X_0 \in \mathbb{R}^3$, and Ω is simply connected. Suppose now that $\Phi = (\Phi_1, \Phi_2, \Phi_3)$. Then we write the equation $\Phi \cdot \Phi = 0$ as $\Phi_1^2 + \Phi_2^2 + \Phi_3^2 = 0$, which is equivalent to

$$(\Phi_1 - i\Phi_2)(\Phi_1 + i\Phi_2) + \Phi_3^2 = 0 \, . \tag{19}$$

Suppose now that $\Phi_1 - i\Phi_2 \not\equiv 0$. Then we consider the functions

$$\mu = \Phi_1 - i\Phi_2 \, , \quad \nu = \Phi_3/\mu \, .$$

By (19) we have $\Phi_1 + i\Phi_2 = -\mu\nu^2$, and therefore

$$(\Phi_1, \Phi_2, \Phi_3) = \left(\frac{1}{2}\mu(1 - \nu^2), \ \frac{i}{2}\mu(1 + \nu^2), \ \mu\nu \right) . \tag{20}$$

A close inspection of these formulas yields:

Lemma. *If μ is holomorphic and ν meromorphic in $\Omega, \mu(w) \not\equiv 0$, and μ has a zero of order $\geq 2n$ where ν has a pole of order n, then the triple $\Phi = (\Phi_1, \Phi_2, \Phi_3)$ defined by (20) is holomorphic and satisfies $\Phi \cdot \Phi = 0$ and $\Phi(w) \not\equiv 0$. Conversely each holomorphic triple $\Phi : \Omega \to \mathbb{C}^3$, satisfying $\Phi{\cdot}\Phi = 0$ and $\Phi(w) \not\equiv 0$ can be written in the form (20) if and only if $\Phi_1 - i\Phi_2 \not\equiv 0$.*

This leads to

Theorem 3 (Enneper–Weierstrass representation formula). *For every non-planar minimal surface $X(w)$ on a simply connected domain Ω in $\mathbb{C}$ there is a holomorphic function μ and a meromorphic function ν in Ω with $\mu \not\equiv 0, \nu \not\equiv 0$ such that $\mu\nu^2$ is holomorphic and*

$$X(w) = X_0 + \text{Re} \int_{w_0}^{w} \Phi(\zeta) d\zeta \, , \qquad w, w_0 \in \Omega \, , \tag{21}$$

where Φ is given by (20) and $X_0 = X(w_0)$.

Conversely, two functions μ, ν as above define by (21) a minimal surface $X : \Omega \to \mathbb{R}^3$ provided that Ω is simply connected.

If we in particular choose

$$\Phi(w) = ((1 - w^2)F(w), i(1 + w^2)F(w), 2wF(w)) \tag{22}$$

where $F(w) \not\equiv 0$ is holomorphic in Ω, then (21) defines a minimal surface $X : \Omega \to \mathbb{R}^3$ with the Gauss map

$$N(w) = \frac{1}{1 + u^2 + v^2} \ (2u, 2v, u^2 + v^2 - 1) \, , \quad w = u + iv \, , \tag{23}$$

and the line element

$$ds^2 = \mathcal{E}(w)(du^2 + dv^2) \, , \quad \mathcal{E}(w) = |F(w)|^2 (1 + u^2 + v^2)^2 \tag{24}$$

whose Gauss curvature in $\Omega' = \{w \in \Omega : F(w) \neq 0\}$ is given by

$$K(w) = -4 \mid F(w) \mid^{-2} (1+ \mid w \mid^2)^{-4} \tag{25}$$

The coefficients $\mathcal{L}, \mathcal{M}, \mathcal{N}$ of the second fundamental form satisfy $\mathcal{L} + \mathcal{N} = 0$ and can be obtained from the holomorphic function $l(w) = \mathcal{L}(w) - i\mathcal{M}(w) = -2F(w)$. The branch points of X are the zeros of $F(w)$. Let $dw = du + idv$. Then the directions (du, dv) of the asymptotic lines are characterized by $\operatorname{Re} F(w)(dw)^2 = 0$, while $\operatorname{Im} F(w)(dw)^2 = 0$ describes the lines of curvature.

Chapter 2
Boundary Value Problems for Minimal Surfaces

In the first two sections we treat the simplest problem of this kind, which is known as Plateau's problem. This is the question to span a closed simple curve by a disk-type minimal surface of least area. The third section shows how the approach to Plateau's problem can be modified in order to treat also partially free problems. In addition, we briefly discuss some other boundary value problems, and in the last section we add some remarks concerning uniqueness and nonuniqueness of solutions to such problems.

In his treatise *Statique expérimentale et théorétique des liquides soumis aux seules forces moléculaires* from 1873, Plateau described a multitude of experiments connected with the phenomenon of capillarity. Among other things, Plateau noted that every contour consisting of a single closed wire, whatever its geometric form, bounds at least one soap film. Now the mathematical model of a thin wire is a closed Jordan curve of finite length. Moreover, the mathematical objects modelling soap films are two-dimensional surfaces in $\mathbb{R}^3$. To every such surface, the phenomenological theory of capillarity attaches a potential energy that is proportional to its surface area. Hence, by Johann Bernoulli's principle of virtual work, *soap films in stable equilibrium correspond to surfaces of minimal area.*

Turning this argument around, it stands to reason that every rectifiable closed Jordan curve bounds at least one surface of least area and that all possible solutions to Plateau's problem can be realized by soap film experiments.

However, as R. Courant [12] has remarked, *empirical evidence can never establish mathematical existence – nor can the mathematician's demand for existence be dismissed by the physicist as useless rigor. Only a mathematical existence proof can ensure that the mathematical description of a physical phenomenon is meaningful.*

The mathematical question that we have above denoted as Plateau's problem was a great challenge to mathematicians, and its solution turned out to be a formidable task. During the nineteenth century, Plateau's problem was solved for many special contours Γ, but a sufficiently general solution was

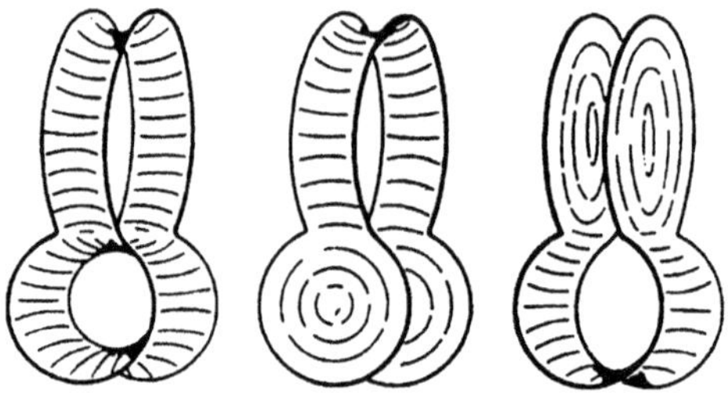

Fig. 1

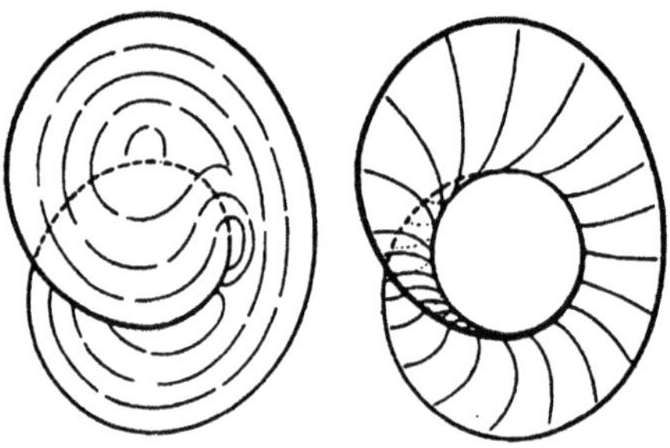

Fig. 2

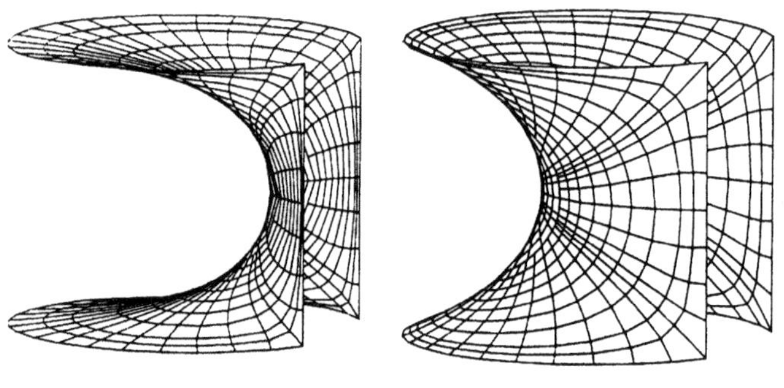

Fig. 3

only obtained in 1930 by J. Douglas [10,11] and simultaneously by T. Radó [5,6]. A considerable simplification of their methods was found by R. Courant [1,2] and, independently, by L. Tonelli [1]. In the Sections 1 and 2 we want to describe the Courant-Tonelli approach to Plateau's problem.

Recall that regular surfaces of least area are minimal surfaces, that is, their mean curvature vanishes throughout. Thus we can formulate a somewhat more general version of Plateau's problem: *Given a closed rectifiable Jordan curve*

Γ, *find a minimal surface spanned by* Γ. Then the least area problem for Γ is more stringent than the Plateau problem: the first question deals with the (absolute or relative) minimizers of area, whereas the second is concerned with the stationary points of the area functional.

Note that for a fixed boundary contour Γ the solutions to Plateau's problem are by no means uniquely determined. Moreover, there may exist solutions of different genus within the same boundary curve, and there may exist both orientable and non-orientable minimal surfaces within the same boundary frame. This is illustrated by the minimal surfaces depicted in Fig. 2.

Even if we fix the topological type of the solutions to Plateau's problem, the unique solvability is, in general, not ensured. For instance, Figs. 1 and 3 depict some boundary configurations which can span several minimal surfaces of the topological type of the disk.

2.1 Plateau's Problem

If one tries to formulate and to solve Plateau's problem, cumbersome difficulties turn up. Among other problems one has to face the fact that there exist mathematical solutions to Plateau's problem which cannot be realized in experiment by soap films. This is, of course, to be expected for stationary solutions which are not minimizing, because they correspond to unstable soap films, and these will be destroyed by tiniest perturbations of the soap lamellae caused by, say, a slight shaking of the boundary frame or by a breath of air.

However it can also happen that (mathematical) solutions of Plateau's problem have branch points, and that they have self-intersections. Both phenomena are physically unrealistic as can be demonstrated by experiments.

Solutions of Plateau's problem, which are absolute minimizers of area, cannot have interior branch points according to a result of Osserman-Gulliver-Alt. The proof of this result is rather difficult and lengthy; thus it cannot be proved here. Instead we refer to the original papers by Osserman [7], Gulliver [1], Alt [1,2], and to a recent diploma-thesis by D. Wienholtz [1]. Tromba [10] has found a comparatively simple argument showing that *generically* minimizers are free of interior branch points.

Unfortunately no similarly satisfactory result is known about the absence of self-intersections. In fact it may happen (cf. Almgren-Thurston [1]) that (mathematical) solutions to the "strict" Plateau problem, i.e. minimizers of area among all disk-type surfaces spanning a prescribed closed Jordan curve Γ, can have self-intersections, even if Γ is unknotted. In other words, minimizers are immersions but they might not be embeddings. One of the few known positive results is the following:

If Γ is a closed Jordan curve that lies on a convex surface, then Γ bounds a minimal surface without self-intersections.

We shall use the classical approach which consists in minimizing area among surfaces given as mappings from a two-dimensional parameter domain into $\mathbb{R}^3$, this way fixing the topological type of the admissible surfaces. However, as we have already seen, it is by no means clear what the topological type of the surface of least area in a given configuration Γ will be. In fact, there may be rectifiable boundaries for which the area minimizing solution of Plateau's problem is of infinite genus. An example for this phenomenon is depicted in Fig. 4.

Fig. 4

Let us now restrict ourselves to surfaces $X \in C^0(\overline{B}, \mathbb{R}^3)$ which are parametrized on the closure of the unit disk $B = \{w \in \mathbb{C} : |w| < 1\}$, and which map the circle ∂B topologically onto a prescribed closed Jordan curve Γ in $\mathbb{R}^3$. Such a surface is said to be a solution of Plateau's problem for Γ if its restriction to B is a minimal surface. Since minimal surfaces are the critical points of the area functional

$$A_B(X) = \int_B |X_u \wedge X_v| dudv\,, \tag{1}$$

one is tempted to look for solutions of Plateau's problem by minimizing $A_B(X)$ in the class of all surfaces $X \in C^0(\overline{B}, \mathbb{R}^3)$ mapping ∂B homeomorphically onto Γ.

But this method does not work very well in the "classical context" of surfaces since the area functional $A_B(X)$ is rather degenerate. For instance it remains invariant with respect to all self-mappings of B which are diffeomorphisms. Moreover it does not change if one attaches "hairs" to a surface X. In order to avoid these difficulties, one uses an artifice similar to that applied in studying geodesics where one replaces the length functional by the 1-dimensional Dirichlet integral. This idea is based on the fact that the critical points of the 1-dimensional Dirichlet integral are also critical points of the length functional which are parametrized proportionally to the arc length, and vice versa. An analogous relation holds between the stationary surfaces of the two-dimensional *Dirichlet integral*

$$D(X) = D_B(X) = \frac{1}{2} \int_B (|X_u|^2 + |X_v|^2) dudv \tag{2}$$

and the area functional $A_B(X)$. This can be seen as follows: For arbitrary vectors $p, q \in \mathbb{R}^3$, we have

$$|p \wedge q| \le |p||q|,$$

and therefore

$$|p \wedge q| \le \frac{1}{2}|p|^2 + \frac{1}{2}|q|^2. \tag{3}$$

The equality sign in (3) holds if and only if $p \perp q$ and $|p| = |q|$. Suppose now that $X \in C^1(B, \mathbb{R}^3)$ has a finite Dirichlet integral $D_B(X)$. Then we obtain the inequality

$$A_B(X) \le D_B(X), \tag{4}$$

and the equality sign is satisfied if and only if the conformality relations

$$|X_u|^2 = |X_v|^2, \quad X_u \cdot X_v = 0$$

are fulfilled on B. In other words, *area functional and Dirichlet integral coincide on the conformally parametrized surfaces X, and, in general, the Dirichlet integral furnishes a majorant for the area functional.*

Moreover, every smooth regular surface $X : B \to \mathbb{R}^3$ can, by Lichtenstein's theorem, be reparametrized by a regular change of parameters $\tau : B \to B$ such that $Y := X \circ \tau$ satisfies the conformality relations

$$|Y_u|^2 = |Y_v|^2, \qquad Y_u \cdot Y_v = 0,$$

and we obtain

$$D_B(Y) = A_B(Y) = A_B(X).$$

This observation makes it plausible that, within a class $\mathcal{C}$ of surfaces which is invariant with respect to parameter changes, minimizers of $D_B(X)$ will also be minimizers of $A_B(X)$, and more generally, that stationary points of $D_B(X)$ will be stationary points of $A_B(X)$.

Certainly the class $\mathcal{C}$ defined by Plateau's boundary condition $X : \partial B \to \Gamma$ has this invariance property. Thus we are led to the idea to minimize Dirichlet's integral instead of the area functional since this way we would equally well obtain a minimizer for $A_B(X)$.

Let us now work out this idea. We set $B = \{w \in \mathbb{C} : |w| < 1\}$ and $C = \partial B = \{w \in \mathbb{C} : |w| = 1\}$. Recall that a closed Jordan curve in $\mathbb{R}^3$ is a subset of $\mathbb{R}^3$ which is homeomorphic to ∂B, i.e. which is the topological image of S^1.

Definition 1. *Given a closed Jordan curve Γ in $\mathbb{R}^3$, we say that $X : \overline{B} \to \mathbb{R}^3$ is a solution of Plateau's problem for the boundary contour Γ (or: a minimal surface spanning Γ) if it fulfils the following three conditions:*

(i) $X \in C^0(\overline{B}, \mathbb{R}^3) \cap C^2(B, \mathbb{R}^3)$;

(ii) *The surface X satisfies*

$$\Delta X = 0\,, \quad |X_u|^2 = |X_v|^2\,, \quad X_u \cdot X_v = 0 \;\; in \;\; B\,; \tag{5}$$

(iii) *The restriction* $X|_C$ *of* X *to the boundary* C *of the parameter domain* B *is a homeomorphism of* C *onto* Γ.

If it is necessary to be more precise, we shall denote a minimal surface X described by this definition as a *disk-type solution of Plateau's problem for the contour* Γ, or as a *minimal disk spanning* Γ.

Condition (iii) is often expressed by saying that $X|_C$ is a continuous, strictly monotonic mapping of C onto Γ.

Clearly this condition is not closed with respect to uniform convergence on C since uniform limits of strictly monotonic functions can be merely weakly monotonic, that is, they may have arcs of constancy on C. To be precise, we give the following

Definition 2. *Suppose that* Γ *is a closed Jordan curve in* $\mathbb{R}^3$, *and let* $\varphi : C \to \Gamma$ *be a homeomorphism from* C *onto* Γ. *Then a continuous mapping* $\psi : C \to \Gamma$ *of* C *onto* Γ *is said to be weakly monotonic if there is a nondecreasing continuous function* $\tau : [0, 2\pi] \to \mathbb{R}$ *with* $\tau(0) = 0, \tau(2\pi) = 2\pi$ *such that*

$$\psi(e^{i\theta}) = \varphi(e^{i\tau(\theta)}) \quad for \;\; 0 \le \theta \le 2\pi\,.$$

In other words, ψ is weakly monotonic if the image points $\psi(w)$ traverse Γ in a constant direction when w moves along C in a constant direction. The image points may "stand still" but never move backwards if w moves monotonically on C, and $\psi(w)$ moves once around Γ if w travels once around C.

One sees at once that the following holds true:

Lemma 1. *Let* $\{\psi_n\}$ *be a sequence of weakly monotonic, continuous mappings of* C *onto a closed Jordan curve* Γ, *and suppose that the mappings* ψ_n *converge uniformly on* C *to some mapping* $\psi : C \to \mathbb{R}^3$. *Then* ψ *is a weakly monotonic continuous mapping of* C *onto* Γ.

Now we want to set up the variational problem that will lead us to a solution of Plateau's problem. First we define the *class* $\mathcal{C}(\Gamma)$ *of admissible functions*. We will choose $\mathcal{C}(\Gamma)$ as a subset of the Sobolev space $H_2^1(B, \mathbb{R}^3)$, consisting of measurable functions $X : B \to \mathbb{R}^3$ such that X, X_u, X_v are square integrable on B. (Here X_u and X_v denote the weak derivatives of X with respect to the variables u and v.) Recall that every function $X \in H_2^1(B, \mathbb{R}^3)$ has a "trace" $X|_C$ on the boundary $C = \partial B$ which is of class $L_2(C, \mathbb{R}^3)$.

Definition 3. *Given a closed Jordan curve* Γ *in* $\mathbb{R}^3$, *a mapping* $X : B \to \mathbb{R}^3$ *is said to be of class* $\mathcal{C}(\Gamma)$ *if* $X \in H_2^1(B, \mathbb{R}^3)$, *and if its trace* $X|_C$ *can be represented by a weakly monotonic, continuous mapping* $\psi : C \to \Gamma$ *of* C *onto* Γ *(i.e., every* $L_2(C)$*-representative of* $X|_C$ *coincides with* ψ *except for a subset of zero* 1*-dimensional Hausdorff measure).*

Now we define the variational problem $\mathcal{P}(\Gamma)$ associated with Plateau's problem for the curve Γ as the following task:

Minimize Dirichlet's integral $D(X)$, defined by (2), in the class $\mathcal{C}(\Gamma)$.

In other words, setting

$$e(\Gamma) := \inf\{D(X) : X \in \mathcal{C}(\Gamma)\}, \tag{6}$$

we are to find a surface $X \in \mathcal{C}(\Gamma)$ such that $D(X) = e(\Gamma)$ is satisfied.

In order to solve the minimum problem $\mathcal{P}(\Gamma)$, we have to find a minimizing sequence $\{X_n\}$ whose boundary values $X_n|_C$ contain a subsequence which is uniformly convergent on C. The selection of such a minimizing sequence will be achieved by the following artifice:

Fix three distinct points w_1, w_2, w_3 on C and three distinct points Q_1, Q_2, Q_3 on Γ, and consider those mappings $X \in \mathcal{C}(\Gamma)$ which satisfy the *three-point condition*

$$X(w_k) = Q_k, \quad k = 1, 2, 3. \tag{7}$$

The set of such mappings X will be denoted by $\mathcal{C}^*(\Gamma)$. Set

$$e^*(\Gamma) := \inf\{D(X) : X \in \mathcal{C}^*(\Gamma)\}. \tag{8}$$

We clearly have

$$e(\Gamma) \le e^*(\Gamma).$$

Moreover, if $X \in \mathcal{C}(\Gamma)$, then there exist three distinct points $\zeta_1, \zeta_2, \zeta_3$ on C such that

$$X(\zeta_k) = Q_k, \quad k = 1, 2, 3.$$

Let σ be a conformal selfmapping of B that maps w_1, w_2, w_3 onto $\zeta_1, \zeta_2, \zeta_3$. Then the mapping $Y := X \circ \sigma$ is of class $\mathcal{C}^*(\Gamma)$ and satisfies $D(Y) = D(X)$, because of the conformal invariance of the Dirichlet integral. Hence we even obtain

$$e(\Gamma) = e^*(\Gamma). \tag{9}$$

Consequently, any solution X of the *restricted minimum problem*

$$\mathcal{P}^*(\Gamma) : \text{"Minimize } D(X) \text{ in the class } \mathcal{C}^*(\Gamma)\text{"} \tag{10}$$

is also a solution of the original minimum problem $\mathcal{P}(\Gamma)$. Hence we shall try to solve $\mathcal{P}^*(\Gamma)$ instead of $\mathcal{P}(\Gamma)$, in this way obtaining a convenient compactness property of the boundary values of any minimizing sequence, as we shall see.

Before we can start with our minimizing process, one final difficulty remains to be solved. Since $\mathcal{P}^*(\Gamma)$ would not have a solution if $\mathcal{C}^*(\Gamma)$ were empty, let us now study under which circumstances $\mathcal{C}^*(\Gamma)$ is certainly nonempty.

Let $\varphi : C \to \Gamma$ be a homeomorphism representing Γ, and let

$$\varphi(e^{i\theta}) = \frac{A_0}{2} + \sum_{n=1}^{\infty}\{A_n \cos n\theta + B_n \sin n\theta\}$$

be its Fourier expansion, $A_n, B_n \in \mathbb{R}^3$, which is convergent in $L_2([0,2\pi],\mathbb{R}^3)$. We can assume that φ satisfies the prescribed three-point condition, i.e.

$$\varphi(w_k) = Q_k\,, \quad k = 1,2,3\,.$$

Let ρ, θ be polar coordinates about the origin of the w-plane, that is, $w = \rho e^{i\theta}$, and set

$$X(w) := \frac{A_0}{2} + \sum_{n=1}^{\infty} \rho^n (A_n \cos n\theta + B_n \sin n\theta)\,.$$

This mapping is harmonic and assumes the boundary values φ, and a straightforward computation yields that

$$D(X) = \frac{\pi}{2} \sum_{n=1}^{\infty} n(|A_n|^2 + |B_n|^2)\,.$$

Consequently the map $X : \overline{B} \to \mathbb{R}^3$ belongs to the class $H_2^1(B,\mathbb{R}^3)$ if and only if

$$\sum_{n=1}^{\infty} n(|A_n|^2 + |B_n|^2) < \infty\,. \tag{11}$$

If this is true, then $\mathcal{C}^*(\Gamma)$ is nonempty.

Condition (11) is satisfied if and only if $\phi(\theta) := \varphi(e^{i\theta})$ has "half a derivative" which is square integrable. This is for instance the case if φ is Lipschitz continuous, and thus we conclude that $\mathcal{C}(\Gamma)$ and $\mathcal{C}^*(\Gamma)$ are nonempty if Γ is rectifiable.

2.2 Existence of Solutions to Plateau's Problem

We want to prove the following basic existence theorems due to J. Douglas and T. Radó; our reasoning follows the proofs of R. Courant and L. Tonelli.

Theorem 1. *If $\mathcal{C}(\Gamma)$ is nonempty, then the minimum problem $\mathcal{P}(\Gamma)$ has at least one solution which is continuous on $\overline{B}$ and harmonic in B. In particular, $\mathcal{P}(\Gamma)$ has such a solution for every rectifiable curve Γ.*

The crucial argument in the proof of Theorem 1 is based on the

Courant-Lebesgue Lemma. *Suppose that X is of class $C^0(\overline{B},\mathbb{R}^3) \cap C^1(B,\mathbb{R}^3)$ and satisfies $D(X) \le M$ for some M with $0 \le M < \infty$. Then, for every $z_0 \in C$ and for each $\delta \in (0,1)$, there exists a number $\rho \in (\delta, \sqrt{\delta})$ such that the distance of the images $X(z), X(z')$ of the two intersection points z and z' of C with the circle $\partial B_\rho(z_0)$ can be estimated by*

$$|\, X(z) - X(z')\,| \le \left\{ \frac{4M\pi}{\log 1/\delta} \right\}^{1/2}. \tag{1}$$

In conjunction with the Arzelà-Ascoli theorem, the Courant-Lebesgue lemma furnishes a compactness result for the boundary values $X\big|_{\partial B}$ of every family of functions $X \in \mathcal{C}(\Gamma)$ with bounded Dirichlet integrals $D(X)$. To prove this lemma, we set $B := \{w : |w| < 1\}, C := \partial B, S_r(z_0) := B \cap B_r(z_0), C_r(z_0) := \overline{B} \cap \partial B_r(z_0), B_r(z_0) := \{w : |w - z_0| < r\}$.

Lemma 1. *Let $X \in H_2^1(B, \mathbb{R}^3), Z(r,\theta) = X(z_0 + re^{i\theta})$ and $p(r) := \int_{\theta_1(r)}^{\theta_2(r)} |Z_\theta(r,\theta)|^2 \, d\theta$, where $\{z_0 + re^{i\theta} : \theta_1(r) \le \theta \le \theta_2(r)\}$ describes the arc $C_r(z_0)$ for $z_0 \in \overline{B}$, i.e. $0 \le \theta_1(r) \le \theta_2(r) \le 2\pi$. Let $\mathcal{I}$ be a measurable subset of $(0,1)$ such that $0 < \int_{\mathcal{I}} dr/r < \infty$ and $\int_{\mathcal{I}} p(r)/r \le M < \infty$. Then the set $\mathcal{I}_M := \{\rho \in \mathcal{I} : p(\rho) \int_{\mathcal{I}} dr/r \le M\}$ has a positive 1-dimensional Lebesgue measure $\mathcal{L}^1(\mathcal{I}_M) > 0$, and for almost all $\rho \in \mathcal{I}_M$ and all θ, θ' with $\theta_1(\rho) \le \theta \le \theta' \le \theta_2(\rho)$ we obtain*

$$\int_\theta^{\theta'} |Z_\theta(\rho,\theta)| \, d\theta \le \left(M \Big/ \int_{\mathcal{I}} \frac{dr}{r}\right)^{1/2} |\theta - \theta'|^{1/2} . \tag{2}$$

Proof. For $\rho \in \mathcal{I}_M$ and $\theta_1(\rho) \le \theta \le \theta' \le \theta_2(\rho)$ we obtain

$$\begin{aligned}\int_\theta^{\theta'} |Z_\theta(\rho,\theta)| \, d\theta &\le \left(\int_\theta^{\theta'} |Z_\theta(\rho,\theta)|^2 \, d\theta\right)^{1/2} |\theta - \theta'|^{1/2} \\ &\le \left(M \Big/ \int_{\mathcal{I}} \frac{dr}{r}\right)^{1/2} |\theta - \theta'|^{1/2},\end{aligned}$$

and $\mathcal{L}^1(\mathcal{I}_M) > 0$ is an obvious consequence of our assumptions. □

Lemma 2. *Every $X \in H_2^1(B, \mathbb{R}^3)$ possesses a representative $Z(r,\theta)$ of $X(z_0 + re^{i\theta}), z_0 \in C$, which is absolutely continuous with respect to θ for a.a. $r \in (0,2)$ and which has the following property:*

For every $\delta \in (0, R^2), 0 < R < 1$, there is a measurable subset $\mathcal{J}$ of the interval $(\delta, \sqrt{\delta})$ with $\mathcal{L}^1(\mathcal{J}) > 0$ such that

$$|Z(\rho,\theta) - Z(\rho,\theta')| \le \int_\theta^{\theta'} |Z_\theta(\rho,\theta)| \, d\theta \le \eta(\delta,R) \, |\theta - \theta'|^{1/2} \tag{3}$$

holds for a.a. $\rho \in \mathcal{J}$ and $\theta_1(\rho) \le \theta \le \theta' \le \theta_2(\rho)$, where

$$\eta(\delta, R) := \left\{\frac{4}{\log 1/\delta} D(X, S_R(z_0))\right\}^{1/2}, \tag{4}$$

and $D(X, \Omega) := D_\Omega(X)$.

Proof. Let $p(r)$ be defined as in Lemma 1. Then we obtain

$$\begin{aligned}\int_0^r \frac{p(\rho)}{\rho} d\rho &\le \int_0^r \int_{\theta_1(\rho)}^{\theta_2(\rho)} \left\{|Z_\rho(\rho,\theta)|^2 + \frac{1}{\rho^2} |Z_\theta(\rho,\theta)|^2\right\} \rho d\theta d\rho \\ &= 2D(X, S_r(z_0)) .\end{aligned}$$

For $M := 2D(X, S_R(z_0))$ and $\mathcal{J} = (\delta, \sqrt{\delta})$, we infer from Lemma 1 that there is some ρ with $\delta < \rho < \sqrt{\delta} \le R$ such that

$$\begin{aligned}\int_\theta^{\theta'} | Z_\theta(\rho,\theta) \,|\, d\theta &\le \left\{ M \Big/ \int_\delta^{\sqrt{\delta}} \frac{dr}{r} \right\}^{1/2} |\theta - \theta'|^{1/2} \\ &= \left\{ 4D(X, S_R(z_0)) \, \frac{1}{\log 1/\delta} \right\}^{1/2} |\theta - \theta'|^{1/2} \\ &= \eta(\delta, R)|\theta - \theta'|^{1/2},\end{aligned}$$

and from

$$Z(\rho,\theta') - Z(\rho,\theta) = \int_\theta^{\theta'} Z_\theta(\rho,\theta) d\theta$$

we infer that

$$| Z(\rho,\theta') - Z(\rho,\theta) | \le \int_\theta^{\theta'} | Z_\theta(\rho,\theta) \,|\, d\theta \le \eta(\theta, R)|\theta - \theta'|^{1/2}. \qquad \square$$

The Courant-Lebesgue lemma is now an immediate consequence of Lemma 2.

Let Q_1, Q_2, Q_3 be three distinct points on the Jordan curve Γ and w_1, w_2, w_3 three distinct points on $C = \partial B$. Denote by $\mathcal{C}^*(\Gamma)$ the set of surfaces $X \in \mathcal{C}(\Gamma)$ satisfying the 3-point condition

$$X(w_k) = Q_k\,, \qquad k = 1, 2, 3. \tag{5}$$

Since Γ is the topological image of C, there exists a number $\lambda(\varepsilon) > 0$ for every $\varepsilon > 0$ such that the following holds true: *Any pair of points* $P, Q \in \Gamma$ *with*

$$0 <| P - Q |< \lambda(\varepsilon) \tag{6}$$

decomposes Γ *into two arcs* $\Gamma_1(P,Q)$ *and* $\Gamma_2(P,Q)$ *such that*

$$\operatorname{diam} \Gamma_1(P,Q) < \varepsilon \tag{7}$$

holds. Hence, if $0 < \varepsilon < \varepsilon_0 := \min_{j \ne k} | Q_j - Q_k |$, then $\Gamma_1(P,Q)$ can contain at most one of the points Q_j appearing in the three-point condition (5).

Let now X be an arbitrary element of $\mathcal{C}^*(\Gamma)$ satisfying the assumptions of the Courant-Lebesgue lemma, and let δ_0 be a fixed number with $2\sqrt{\delta_0} < \min_{j \ne k} |w_j - w_k|$. For $\varepsilon \in (0, \varepsilon_0)$ we choose a $\delta = \delta(\varepsilon) > 0$ such that $\delta < \delta_0$ and

$$\left(4\pi M / \log \frac{1}{\delta} \right)^{1/2} < \lambda(\varepsilon).$$

Consider an arbitrary point z_0 on C, and let $\rho \in (\delta, \sqrt{\delta})$ be some number such that the images $P := X(z), Q := X(z')$ of the two intersection points z, z' of C and $\partial B_\rho(z_0)$ satisfy

$$| P - Q | \leq \left\{ \frac{4M\pi}{\log 1/\delta} \right\}^{1/2} .$$

Then $| P - Q | < \lambda(\varepsilon)$, whence

$$\operatorname{diam} \Gamma_1(P, Q) < \varepsilon$$

holds on account of (7). Because of $\varepsilon < \varepsilon_0$ the arc $\Gamma_1(P, Q)$ contains at most one of the points Q_j. On the other hand, since $X \in \mathcal{C}^*(\Gamma)$, we infer from (5), $\delta < \delta_0$ and $2\sqrt{\delta_0} < \min_{j \neq k} | w_j - w_k |$ that $X(C \cap \overline{B_\rho(z_0)})$ contains at most one of the points Q_j; thus it coincides with $\Gamma_1(P, Q)$. Consequently we have

$$|X(w) - X(w')| < \varepsilon \quad \text{for all } w, w' \in C \cap B_\rho(z_0) ,$$

which implies

$$|X(w) - X(w')| < \varepsilon \quad \text{for all } w, w' \in C \text{ with } |w - w'| < \delta/2 . \tag{8}$$

Now we choose a sequence $\{X_n\}$ of mappings $X_n \in \mathcal{C}^*(\Gamma)$ such that $D(X_n) \to e^*(\Gamma)$ as $n \to \infty$. We can assume that $X_n \in C^0(\overline{B}, \mathbb{R}^3) \cap C^2(B, \mathbb{R}^3)$ and $\triangle X_n = 0$ in B for $n = 1, 2, \ldots$, because otherwise we could replace X_n by the solution Z_n of the boundary value problem

$$\triangle Z_n = 0 \quad \text{in } B , \quad Z_n = X_n \quad \text{on } C$$

which is continuous on $\overline{B}$ and of class $C^2 \cap H_2^1(B, \mathbb{R}^3)$. It is well known that this problem has exactly one solution. This solution minimizes $D(X)$ among all $X \in H_2^1(B, \mathbb{R}^3)$ with $X - X_n \in \overset{\circ}{H}{}_2^1(B, \mathbb{R}^3)$. Consequently, $D(Z_n) \leq D(X_n)$, and by construction we have $Z_n \in \mathcal{C}^*(\Gamma)$ whence $e^*(\Gamma) \leq D(Z_n)$. Thus we obtain

$$e^*(\Gamma) \leq D(Z_n) \leq D(X_n) \to e^*(\Gamma) ,$$

and therefore

$$\lim_{n \to \infty} D(Z_n) = e^*(\Gamma) .$$

Hence we have found a minimizing sequence $\{Z_n\}$ for $\mathcal{P}^*(\Gamma)$ consisting of harmonic mappings Z_n which are continuous on $\overline{B}$.

We now claim that the boundary values $X_n|_C$ of the terms of any minimizing sequence $\{X_n\}$ for $\mathcal{P}^*(\Gamma)$ are equicontinuous on C. Since $D(X_n)$ tends to $e^*(\Gamma)$ there is a number $M > 0$ such that $D(X_n) < M$ for all $n \in \mathbb{N}$. Therefore we can apply (8) to $X = X_n$, $n \in \mathbb{N}$, and in conjunction with $X_n(C) = \Gamma$ we infer by means of the Arzelà-Ascoli theorem that $\{X_n|_C\}$ contains a subsequence converging uniformly on C to some $\varphi \in C^0(C, \mathbb{R}^3)$, and by renumbering we can achieve that

$$\lim_{n \to \infty} \sup_{w \in C} | X_n(w) - \varphi(w) | = 0 .$$

It follows that φ is a weakly monotonic mapping of C onto Γ. Since the functions X_n are continuous on $\overline{B}$ and harmonic in B, it follows that X_n tends to some function X, uniformly on $\overline{B}$, which is continuous on $\overline{B}$, harmonic in B, satisfies (5), and has the boundary values φ. Consequently, X is of class $\mathcal{C}^*(\Gamma)$, and therefore $e^*(\Gamma) \le D(X)$. Moreover, a classical result for harmonic functions implies that grad X_n tends to grad X as $n \to \infty$, uniformly on every $B' \subset\subset B$, whence $\lim_{n\to\infty} D_{B'}(X_n) = D_{B'}(X)$ and therefore $\liminf_{n\to\infty} D_B(X_n) \ge D_{B'}(X)$ if $B' \subset\subset B$.

Thus we finally obtain

$$e^*(\Gamma) = \lim_{n\to\infty} D(X_n) \ge D(X) \ge e^*(\Gamma),$$

i.e. $D(X) = e^*(\Gamma)$. Therefore $X \in \mathcal{C}^*(\Gamma)$ is a minimizer of Dirichlet's integral in the class $\mathcal{C}^*(\Gamma)$, and since $e(\Gamma) = e^*(\Gamma)$, the mapping X is also a minimizer of Dirichlet's integral within $\mathcal{C}(\Gamma)$. □

Theorem 1 states that $D(X)$ has at least one minimizer, and this minimizer is harmonic. By a standard argument (Weyl's lemma) one easily infers that, in fact, every solution of $\mathcal{P}(\Gamma)$ is a harmonic mapping, i.e. we have

Theorem 2. *Every minimizer X of the Dirichlet integral within the class $\mathcal{C}(\Gamma)$ is continuous in $\overline{B}$ and harmonic in B.*

The same reasoning as in the proof of Theorem 1 yields

Theorem 3. *Let $\{\Gamma_n\}$ be a sequence of closed Jordan curves in $\mathbb{R}^3$ which converge in the sense of Fréchet to some closed Jordan curve Γ (notation: $\Gamma_n \to \Gamma$ as $n \to \infty$), and let $\{X_n\}$ be a sequence of mappings $X_n \in \mathcal{C}(\Gamma_n)$ with uniformly bounded Dirichlet integral, i.e.,*

$$D(X_n) \le M, \qquad n \in \mathbb{N}.$$

Then their boundary values $\varphi_n := X_n|_C$ are equicontinuous if they satisfy a uniform three-point condition

$$\varphi_n(w_j) = Q_j^{(n)}, \qquad j = 1, 2, 3,$$

with some points $w_j \in C$ and $Q_j^{(n)} \in \Gamma_n, j = 1, 2, 3$, such that $\lim_{n\to\infty} Q_j^{(n)} = Q_j$ holds, where Q_1, Q_2, Q_3 denote three different points on the limit curve Γ. If, moreover, the mappings X_n are continuous on $\overline{B}$ and harmonic in B, then we can extract a subsequence $\{X_{n_p}\}$ that converges uniformly on $\overline{B}$ to some mapping $X \in \mathcal{C}(\Gamma)$ which is continuous on $\overline{B}$ and harmonic in B.

If $\{X_n\}$ is a sequence of minimal surfaces $X_n \in \mathcal{C}(\Gamma_n)$ where Γ_n are rectifiable Jordan curves with $L(\Gamma_n)$ (= length of Γ_n) $\le l$ for all $n \in \mathbb{N}$, then we have $D(X_n) \le l^2/4\pi$ because of the isoperimetric inequality for minimal surfaces (see Section 4.2). Thus we obtain

Theorem 4. *Let $\Gamma, \Gamma_1, \Gamma_2, \dots$ be closed Jordan curves in $\mathbb{R}^3$ with $\Gamma_n \to \Gamma$ as $n \to \infty, L(\Gamma_n) \le l$, and $\lim_{n\to\infty} e(\Gamma_n) = e(\Gamma)$. Furthermore, let $X_n \in \mathcal{C}(\Gamma_n)$*

be a sequence of solutions for $\mathcal{P}(\Gamma_n)$ whose boundary values $\varphi_n = X_n|_C$ satisfy a uniform three-point condition such as in Theorem 3. Then we can extract a subsequence $\{X_{n_p}\}$ which converges uniformly on $\overline{B}$ to some solution X of $\mathcal{P}(\Gamma)$ as $p \to \infty$, and $D(X_{n_p}) \to D(X)$.

Now we want to prove that any solution X of the minimum problem $\mathcal{P}(\Gamma)$ satisfies the conformality relations

$$|X_u|^2 = |X_v|^2, \qquad X_u \cdot X_v = 0.$$

To this end, we exploit the minimum property of X by changing the independent variables u, v in the direction of an arbitrarily prescribed vector field $\lambda(u,v) = (\mu(u,v), \nu(u,v))$ on $\overline{B}$. Such variations of X will be called *inner variations.*

In order to make this variational technique precise, we start with an arbitrary vector field $\lambda = (\mu, \nu)$ on $\overline{B}$ which is of class $C^1(\overline{B}, \mathbb{R}^2)$. Without restriction we can assume that λ is defined on all of $\mathbb{R}^2$ and is of class $C^1(\mathbb{R}^2, \mathbb{R}^2)$. With λ we associate some 1-parameter family of mappings $\tau_\varepsilon : \mathbb{R}^2 \to \mathbb{R}^2$ which satisfies

$$\tau_\varepsilon(w) = \tau(w, \varepsilon) = w - \varepsilon\lambda(w) + o(\varepsilon) \quad \text{as} \quad \varepsilon \to 0,$$

$w = (u,v)$. For instance, we could take $\tau_\varepsilon(w) = w - \varepsilon\lambda(w)$. The function $\tau(w,\varepsilon)$ is of class C^1 on $\mathbb{R}^2 \times \mathbb{R}$. Choose some open set B_0 with $B \subset\subset B_0$. Then it is easy to see that $\tau_\varepsilon : B_0 \to \tau_\varepsilon(B_0)$ furnishes a C^1-diffeomorphism of B_0 onto its image $\tau_\varepsilon(B_0)$ provided that $|\varepsilon| < \varepsilon_0$ for some sufficiently small $\varepsilon_0 > 0$, because $\tau_\varepsilon(w)$ is just a perturbation of the identity map $\tau_0(w) = w$.

Clearly the inverse mappings $\sigma_\varepsilon = \tau_\varepsilon^{-1}$ exist on a common domain of definition Ω satisfying $B_\varepsilon^* \subset\subset \Omega \subset\subset B_0$, where we have set $B_\varepsilon^* = \tau_\varepsilon(B)$. We write $\omega = \tau_\varepsilon(w) = \tau(w,\varepsilon)$ and $w = \sigma_\varepsilon(\omega) = \sigma(\omega, \varepsilon)$. The function $\sigma(\omega,\varepsilon)$ is of class C^1 on $\Omega \times (-\varepsilon_0, \varepsilon_0)$ and satisfies both

$$\sigma(\omega,\varepsilon) = \omega + \varepsilon\lambda(\omega) + o(\varepsilon) \quad \text{and} \quad \tau(\sigma(\omega,\varepsilon)) = \omega \quad \text{for all } (\omega,\varepsilon) \in \Omega \times (-\varepsilon_0, \varepsilon_0).$$

Restricting the region of definition of $\tau_\varepsilon = \tau(\cdot,\varepsilon)$ and $\sigma_\varepsilon = \sigma(\cdot,\varepsilon)$ to $\overline{B}$ and $\overline{B_\varepsilon^*}$, respectively, the mapping τ_ε is a diffeomorphism of $\overline{B}$ onto $\overline{B_\varepsilon^*}$, with the inverse σ_ε, and we have in particular

$$B_0^* = B, \quad \sigma(w,0) = w, \quad \frac{\partial}{\partial\varepsilon}\sigma(w,\varepsilon)|_{\varepsilon=0} = \lambda(w) \quad \text{for } w \in \overline{B}$$

and

$$\frac{\partial}{\partial\varepsilon} \det D\tau_\varepsilon \Big|_{\varepsilon=0} = -(\mu_u + \nu_v) = -\operatorname{div}\lambda.$$

Consider now an arbitrary $X \in C^1(\overline{B}, \mathbb{R}^3)$ and form the family of mappings $Z_\varepsilon := X \circ \sigma_\varepsilon$, $\sigma_\varepsilon : \overline{B_\varepsilon^*} \to \overline{B}$. For an arbitrary integral $\mathcal{F}(X)$ of the kind

$$\mathcal{F}_B(X) = \mathcal{F}(X,B) := \int_B F(X, X_u, X_v)\,du dv \tag{9}$$

with a C^1-Lagrangian $F(x,p,q)$ we define *the first inner* variation of $\mathcal{F}_B$ at X in direction of $\lambda = (\mu,\nu)$ as

$$\partial\mathcal{F}_B(X,\lambda) := \frac{d}{d\varepsilon}\mathcal{F}(X\circ\sigma_\varepsilon, B_\varepsilon^*)\Big|_{\varepsilon=0}. \tag{10}$$

It turns out that

$$\begin{aligned}\partial\mathcal{F}_B(X,\lambda) = \int_B [F_p(\ldots)\cdot(X_u\mu_u + X_v\nu_v)\\ + F_q(\ldots)\cdot(X_u\mu_v + X_v\nu_v) - F(\ldots)(\mu_u+\nu_v)]dudv\end{aligned} \tag{11}$$

where the arguments $(\ldots)$ of F, F_p, F_q are to be taken as (X, X_u, X_v).

For the Dirichlet integral $D(X) = \frac{1}{2}\int_B(|X_u|^2 + |X_v|^2)dudv$ and any $\lambda = (\mu,\nu) \in C^1(\overline{B},\mathbb{R}^2)$, the first inner variation $\partial D(X,\lambda)$ is given by

$$2\partial D(X,\lambda) = \int_B [a(\mu_u - \nu_v) + b(\mu_v+\nu_u)]dudv\,, \tag{12}$$

where

$$a := |X_u|^2 - |X_v|^2\,, \quad b := 2X_u\cdot X_v\,. \tag{13}$$

In case of the generalized Dirichlet integral

$$E(X) = \frac{1}{2}\int_B g_{ik}(X)\{X_u^iX_u^k + X_v^iX_v^k\}dudv \tag{14}$$

and for any $\lambda = (\mu,\nu)\in C^1(\overline{B},\mathbb{R}^2)$, we obtain

$$2\partial E(X,\lambda) = \int_B [(a(\mu_u-\nu_v) + b(\mu_v+\nu_u)]dudv \tag{15}$$

with

$$a := g_{ik}(X)X_u^iX_u^k - g_{ik}(X)X_v^iX_v^k\,, \quad b := 2g_{ik}(X)X_u^iX_v^k\,. \tag{16}$$

Now we can prove

Theorem 5. *If $X \in H_2^1(B,\mathbb{R}^3)$ satisfies*

$$\partial D(X,\lambda) = 0 \quad \text{for all } \lambda\in C^1(\overline{B},\mathbb{R}^2)\,, \tag{17}$$

then X fulfils the conformality relations

$$|X_u|^2 = |X_v|^2\,, \quad X_u\cdot X_v = 0\,. \tag{18}$$

Conversely if $X \in H_2^1(B,\mathbb{R}^3)$ satisfies (18), then (17) holds true.

Proof. Choose arbitrary functions $\rho,\sigma \in C_c^\infty(B)$ and determine functions $h,k\in C^\infty(\overline{B})$ with

$$\Delta h = \rho\,, \quad \Delta k = \sigma \text{ on } B\,, \quad h = 0\,, \quad k = 0 \text{ on } \partial B\,.$$

Then the functions $\mu := h_u + k_v$, $\nu := -h_v + k_u$ are of class $C^\infty(\overline{B})$ and satisfy $\mu_u - \nu_v = \rho$, $\mu_v + \nu_u = \sigma$. We infer from (12), (13), (17) that

$$\int_B (a\rho + b\sigma) dudv = 0 \quad \text{for all } \rho, \sigma \in C_c^\infty(B)$$

whence $a = 0$ and $b = 0$ a.e. on B. The converse can be seen in a similar way. □

Consider the inner variations $\sigma_\varepsilon : \overline{B_\varepsilon^*} \to \overline{B}$ as above. Since B and B_ε^* are diffeomorphic, there is a conformal map $\kappa_\varepsilon : B \to B_\varepsilon^*$ of B onto B_ε^*, on account of Riemann's mapping theorem, and by a classical result of Carathéodory one can extend κ_ε to a homeomorphism of $\overline{B}$ onto $\overline{B_\varepsilon}$. Let $Z_\varepsilon := X \circ \tau_\varepsilon$ and $Y_\varepsilon := Z_\varepsilon \circ \kappa_\varepsilon$. Since Dirichlet's integral is invariant with respect to conformal maps, we have

$$D(Y_\varepsilon, B) = D(Z_\varepsilon, B_\varepsilon^*)$$

If X is a solution of $\mathcal{P}(\Gamma)$, it follows $X \in \mathcal{C}(\Gamma)$ and $D(X) \le D(Y_\varepsilon)$ whence $D(X,B) \le D(Z_\varepsilon, B_\varepsilon^*)$ for $|\ \varepsilon\ | << 1$, and therefore $\partial D(X,\lambda) = 0$ for all $\lambda \in C^1(\overline{B}, \mathbb{R}^2)$. By virtue of Theorem 5 we arrive at

Theorem 6. *Every solution X of $\mathcal{P}(\Gamma)$ is of class $C^0(\overline{B}, \mathbb{R}^3) \cap C^2(B, \mathbb{R}^3)$ and satisfies $\triangle X = 0$ and (18), i.e. every solution X of $\mathcal{P}(\Gamma)$ is a minimal surface.*

Theorem 7. *Every minimal surface X of class $\mathcal{C}(\Gamma)$ yields a topological mapping of C onto Γ.*

Proof. It suffices to prove that $X_{|C}$ is one-to-one. Suppose that this were not true. Since $X_{|C}$ is weakly monotonic, we could then find an arc $C_0 = \{e^{i\theta} : \theta_1 < \theta < \theta_2\}$ which is mapped onto a single point $P \in \mathbb{R}^3$:

$$X(e^{i\theta}) = P \quad \text{for all } \theta \in (\theta_1, \theta_2)\,.$$

By Schwarz's reflection principle we could extend $X(w)$ as a harmonic mapping across C. Differentiating (25) in the tangential direction, we would then obtain

$$\frac{\partial}{\partial\theta} X(e^{i\theta}) = 0$$

and, applying the conformality relations, it would follow that grad X vanishes identically on C. This would imply grad $X \equiv 0$ on B, or $X(w) \equiv P$, a contradiction to $X \in \mathcal{C}(\Gamma)$. □

From Theorems 1 , 6 and 7 we infer the following

Main Theorem. *Let Γ be a closed Jordan curve in $\mathbb{R}^3$ and suppose that $\mathcal{C}(\Gamma)$ is nonempty. Then the minimum problem $\mathcal{P}(\Gamma)$: "$D(X) \to$ min in the class $\mathcal{C}(\Gamma)$" has at least one solution. Every solution X of $\mathcal{P}(\Gamma)$ is continuous in $\overline{B}$, harmonic in B, satisfies the conformality relations* (18) *in B, and maps*

C topologically onto Γ. In particular, every closed rectifiable curve Γ in $\mathbb{R}^3$ bounds at least one minimal surface of the type of the disk.

So far we have proved that every closed rectifiable curve Γ in $\mathbb{R}^3$ bounds at least one minimal surface X of class $\mathcal{C}(\Gamma)$, and this solution of the Plateau problem has been obtained by minimization of the Dirichlet integral among all (disk-type) surfaces of class $\mathcal{C}(\Gamma)$. Since any minimizer X is automatically continuous on $\overline{B}$, the solution of Plateau's problem can as well be achieved by minimizing $D(X)$ within the class

$$\overline{\mathcal{C}}(\Gamma) := \mathcal{C}(\Gamma) \cap C^0(\overline{B}, \mathbb{R}^3) .$$

Although every minimizer X satisfies

$$D(X) = A(X) ,$$

it is by no means clear that a minimizer of the Dirichlet integral in $\overline{\mathcal{C}}(\Gamma)$ also minimizes the area functional among all surfaces in $\overline{\mathcal{C}}(\Gamma)$. For this we need to know that

$$\overline{a}(\Gamma) = \overline{e}(\Gamma) , \tag{19}$$

where $\overline{a}(\Gamma)$ and $\overline{e}(\Gamma)$ denote the infimum of $A(X)$ and $D(X)$ respectively, among all $X \in \overline{\mathcal{C}}(\Gamma)$. However, the inequality

$$A(X) \leq D(X)$$

only implies that

$$\overline{a}(\Gamma) \leq \overline{e}(\Gamma) .$$

In fact, the proof of the equality sign requires a nontrivial result on global conformal mappings which we will not supply in these notes. We shall only quote a basic result due to Morrey, from which we can derive the equation (19). Roughly speaking, this result says that one can introduce "nearly conformal" parameters on every reasonable surface X.

Lemma on ε-Conformal Mappings. *Let X be of class $C^0(\overline{B}, \mathbb{R}^3) \cap H_2^1(B, \mathbb{R}^3)$. Then, for every $\varepsilon > 0$, there exists a homeomorphism τ_ε of $\overline{B}$ onto itself which is of class H_2^1 on $\overline{B}$ and satisfies both*

$$Z_\varepsilon := X \circ \tau_\varepsilon \in C^0(\overline{B}, \mathbb{R}^3) \cap H_2^1(B, \mathbb{R}^3)$$

and

$$D(Z_\varepsilon) \leq A(X) + \varepsilon .$$

(For a proof, we refer to Morrey [1], pp. 141–143, and [3], pp. 814–815).

In order to prove (19), let X be an arbitrary surface of $\overline{\mathcal{C}}(\Gamma)$. Then, by Morrey's lemma, we can find homeomorphisms τ_n of $\overline{B}$ onto itself such that $Z_n := X \circ \tau_n \in \overline{\mathcal{C}}(\Gamma)$ and $D(Z_n) \leq A(X_n) + 1/n$. Since $\overline{e}(\Gamma) \leq D(Z_n)$, it follows that $\overline{e}(\Gamma) \leq A(X)$, whence $\overline{e}(\Gamma) \leq \overline{a}(\Gamma)$ and (19) is proved; therefore $A(X) = D(X)$. We thus obtain

Theorem 8. *Every solution $X \in \mathcal{C}(\Gamma)$ of the minimum problem $\mathcal{P}(\Gamma)$ is a surface of least area in $\mathcal{C}(\Gamma) \cap C^0(\overline{B}, \mathbb{R}^3)$.*

Historical remarks, references to the literature, and further results.

Solutions of Plateau's problem for specific boundaries, mainly of polygonal type, were already given by H. A. Schwarz, Riemann and Weierstrass. An extensive presentation of this work can be found in Darboux's *Leçons* [1]. The first general existence proof for the *nonparametric Plateau problem* was given by Haar [1] in 1927. An attempt to solve Plateau's problem for general unknotted and piecewise smooth Jordan curves is due to Garnier (1928, cf. [2]). General and convincing proofs of Plateau's problem were published independently and nearly simultaneously by J. Douglas [10], [11] and T. Radó [5],[6] about 1930.

Douglas based his approach on the so-called *Douglas* functional

$$A(\xi) = \frac{1}{4\pi} \int_0^{2\pi} \int_0^{2\pi} \frac{|\, \xi(\theta) - \xi(\varphi) \,|^2}{4 \sin^2 \frac{1}{2}(\theta - \varphi)} d\theta d\varphi$$

with $\xi(\theta) := X(e^{i\theta}) = X(\cos\theta, \sin\theta)$, $X : B \to \mathbb{R}^3, B = \{w \in \mathbb{C} : |\, w \,| < 1\}$, which for harmonic mappings $X : B \to \mathbb{R}^3$ coincides with the Dirichlet integral. This way the Plateau problem is reduced to a one-dimensional question, and thus it seems to be easier to handle. On the other hand Dirichlet's integral appears to be a more natural tool which lends itself easily to interesting generalizations such as free boundary problems or Plateau's problem for surfaces of prescribed mean curvature.

Radó's method is much closer to the one used in Sections 2.1 and 2.2, which was invented independently by Courant [1] and Tonelli [1] in 1936; important ideas of their proofs are due to Radó.

Detailed presentations of results on Plateau's problem and related questions can be found in the survey of Radó [9], the monographs of Courant [12], Nitsche [18], Dierkes/Hildebrandt/Küster/Wohlrab [1], Jost [12] and in the notes of Struwe [4].

Concerning the existence of unstable minimal surfaces in a prescribed contour we first mention the work Shiffman, Morse-Tompkins and in particular Courant. A detailed presentation is given in Courant [12]. The work of Shiffman and Courant was further developed and extended by E. Heinz [4–9] and Sauvigny [1–4]. By now a kind of Morse theory for minimal surfaces has been developed; we shall only mention some results which apply to the setting of Sections 2.1 and 2.2, while we do not refer to numerous important results obtained in the context of geometric measure theory. Important initial steps in this direction were the results [1–4] of Böhme and the work of Böhme–Tomi [1] and Tromba [1,2]. Of particular importance is the so-called *index theorem* of Böhme-Tromba [2] for $\pi|_{\mathcal{M}} : \mathcal{M} \to \mathcal{A}, \pi$ being the projection map $\pi : \mathcal{N} \to \mathcal{A}, \mathcal{M}$ the set of minimal surfaces viewed as a stratified subset in the bundle $\mathcal{N} = \mathcal{D} \times \mathcal{A}$ where $\mathcal{D}$ consists of Sobolev representations of S^1

which are of degree one, and $\mathcal{A}$ consists of Sobolev representations of smooth simple curves. This result, for instance, shows that the number of solutions bounded by a closed simple curve is generically finite, although there might be irregular Jordan curves (with at least one "very singular" point) bounding infinitely many minimal surfaces of the type of the disk. Generalizations of the index theorem to minimal surfaces of higher topological type were developed by Schüffler [3], Schüffler-Tomi, and Thiel; recently a very general index theorem was established by Tomi and Tromba [5]. Various Morse-theoretic results were found by Tromba [5–8, 10], Struwe [2,3], and Jost-Struwe [1].

How many minimal surfaces can be spanned in a given closed Jordan curve? The answer to this question is in general not known, not even if we fix the topological type of the solutions of Plateau's problem. As we have considered only disk-type minimal surfaces, we want to consider the more modest question of:

How many minimal surfaces of the type of the disk are bounded by a given closed Jordan curve Γ?

The situation would be simple if we could prove that Γ bounds only one disk-type minimal surface $X \in \mathcal{C}(\Gamma)$ (up to reparametrizations $X \circ \tau$ of X by conformal selfmappings $\tau : B \to B$ of the parameter domain B; such reparametrizations would not be counted as different from X and could be excluded by fixing a three-point condition for the surfaces $X \in \mathcal{C}(\Gamma)$ which are prospective solutions).

However, examples (cf. Figures 1 and 3) warn us not to expect uniqueness for disk-type solutions of Plateau's problem. Thus we may ask whether additional geometric conditions for Γ are known which ensure this uniqueness. Essentially, two results are known:

Theorem of Radó [4,8]. *If Γ has a one-to-one parallel projection onto a planar convex curve γ, then Γ bounds at most one disk-type minimal surface.*

The same result holds true if Γ admits a one-to-one *central projection* onto a planar convex curve γ. Nitsche [18], cf. §§ 401 and 586, has proved that Radó's theorem remains true if one allows "vertical" segments for Γ which are mapped to simple points on γ. However, the proof of this fact is not merely a simple generalization of Radó's reasoning, but needs a new argument furnished by Nitsche.

Theorem of Nitsche [16]. *If Γ is real analytic and has a total curvature less than 4π, then Γ bounds only one disk-type minimal surface.*

For polygonal Γ of total curvature less than 4π, this result was earlier conjectured by R. Schneider [2] whose sketch of a proof contained some of the ideas used in Nitsche's proof.

In general, however, nothing is known about the number of solutions of Plateau's problem which are of class $\mathcal{C}(\Gamma)$. Actually, the situation seems to be rather unpromising on account of the following result due to Böhme [5]:

For each positive integer N and for each $\varepsilon > 0$, there exists a real analytic Jordan curve Γ in $\mathbb{R}^3$ with total curvature less than $4\pi + \varepsilon$ which bounds at least N minimal surfaces of class $\mathcal{C}(\Gamma)$, i.e., of the type of the disk.

In 1965, Nitsche [4] raised the question whether every "sufficiently regular" say, real analytic) Jordan curve bounds at most finitely many minimal surfaces of the type of the disk. So far, the answer to this outstanding problem is unknown, but there is a striking partial result due to Tomi [3,5]: *There are only finitely many disk-type solutions of Plateau's problem which are absolute minimizers of the area functional in $\mathcal{C}(\Gamma)$, provided that Γ is a regular real-analytic Jordan curve.* Nitsche [21,22] pointed out that *every sufficiently regular Jordan curve of total curvature less than or equal to 6π bounds at most finitely many minimal surfaces of the type of the disk.*Other finiteness results are due to Beeson [3–5].

In certain cases one can view the changes of minimal surfaces in terms of their boundaries as a bifurcation problem. This was carried out in papers by Nitsche [6], Ruchert [1], Beeson and Tromba [1], and finally Büch [1].

It was a long-standing question whether the area-minimizing solution of Plateau's problem obtained by Douglas and Radó is a *regular* surface, that is, an immersion. This was eventually confirmed in a series of papers by Osserman [7], Gulliver [1], Alt [1,2], and Gulliver-Osserman-Royden [1]. The breakthrough was achieved by Osserman [7] who, by an ingenious idea, was able to rule out the existence of *true branch points* for minimizers. A "true branch point" of a minimal surface $X : B \to \mathbb{R}^3$ is characterized by the fact that there are several geometrically different sheets of the surface lying over the tangent plane at w_0. These sheets intersect transversally along smooth curves in $\mathbb{R}^3$ emanating from $X(w_0)$. A "false (interior) branch point" is a singular point $w_0 \in B$ which has a neighbourhood U in B such that $X(U)$ turns out to be (the trace of) an embedded surface. In other words, false branch points cannot be detected by looking at the image of a minimal surface; they are just the result of a "false parametrization".

Osserman's reasoning did not rule out the existence of false branch points for a Douglas-Radó solution. This second part of the regularity proof was, more or less simultaneously, achieved by Gulliver and Alt in the papers cited above. Another treatment can be found in the paper of Gulliver-Osserman-Royden. It is still an open problem whether there can be branch points at the boundary ∂B; however, Gulliver and Lesley [1] were able to prove that the Douglas-Radó solution is free of boundary branch points if Γ is a regular, real-analytic Jordan curve. (See D. Wienholtz [2] for new results.) Thus we now have the following sharpened version of the

Fundamental Existence Theorem. *Every closed rectifiable Jordan curve Γ in $\mathbb{R}^3$ bounds an area minimizing surface $X : B \to \mathbb{R}^3$ of the type of the disk, and all solutions of this type are regular surfaces, i.e. they are free of branch points $w_0 \in B$. If Γ is regular and real analytic, then they have no branch points on ∂B, either.*

The quite intricate arguments leading to the nonexistence of true interior branch points for area-minimizing solutions of Plateau's problem were recently inspected and completed by D. Wienholtz [1] in a carefully written diploma-thesis.

A minimal surface without branch points is an immersion, but it might still not be an embedding. However, self-intersecting minimal surfaces can never appear as soap films, according to the experiments of Plateau. In fact stable soap films either appear as surfaces of higher topological type (thereby avoiding selfintersections which necessarily have to appear for disk-type surfaces bounded by a knotted curve), or they arrange themselves as "systems" of minimal surfaces with the characteristic 120^0 angles at their common liquid edges. If we want to find embedded minimizers, we have therefore to restrict ourselves to unknotted Jordan curves as boundaries, or to allow a higher topological type for the parameter domain. In the first direction we have the following result proved in a sequence of papers by Gulliver-Spruck [1], Tomi-Tromba [1], Almgren-Simon [1], and Meeks-Yau [3]: *Let K be a strictly convex body in $\mathbb{R}^3$ whose boundary ∂K is of class C^2, and suppose that Γ is a closed rectifiable Jordan curve contained in ∂K. Then there exists an embedded minimal surface of the type of the disk which is bounded by Γ.*

The optimal result is due to Meeks-Yau who proved that any disk-type minimizer is embedded. Moreover they showed in [4] that K can be replaced by a compact region in $\mathbb{R}^3$ whose boundary is C^2-smooth and has a nonnegative mean curvature with respect to the inward normal. Then any minimizing disk-type surface in Γ with its boundary on ∂K has to be embedded. In addition they connected the problem of embeddedness with the uniqueness question.

In the other direction we have the following fundamental result by Hardt and Simon [1]: *Each closed Jordan curve $\Gamma \subset \mathbb{R}^3$ of class $C^{1,\alpha}$ bounds at least one embedded orientable minimal surface.*

Note that here nothing is said about the topological type of the minimal surface in question.

2.3 The Partially Free Boundary Problem, and Other Boundary Value Problems

Let us see how the method of Section 2.2 is to be modified in order to treat another boundary value problem for minimal surfaces, the so-called *semi-free* or *partially free problem.*

Consider a boundary configuration $< \Gamma, S >$ consisting of a closed set S in $\mathbb{R}^3$ (e.g., a smooth surface S with or without boundary, see Fig. 5), and a Jordan arc Γ, the endpoints P_1 and P_2 of which lie on S, $P_1 \neq P_2$, while all other points of Γ are disjoint from S.

Let us denote the arcs of ∂B lying in the half-planes $\{\operatorname{Im} w \geq 0\}$ and $\{\operatorname{Im} w \leq 0\}$ by C and I respectively. The class $\mathcal{C}(\Gamma, S)$ of admissible surfaces

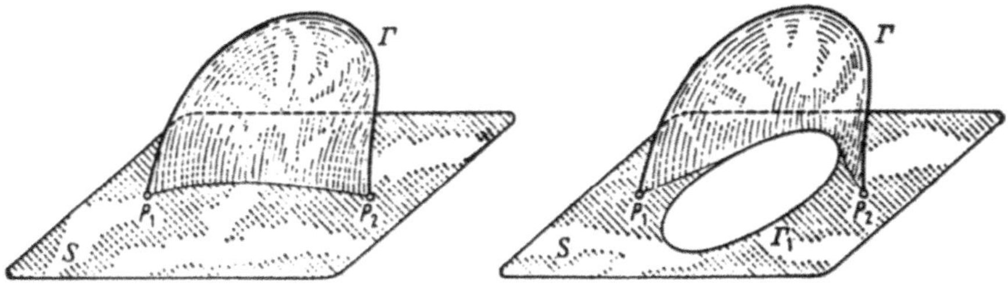

Fig. 5

for the semi-free problem is the set of all maps $X \in H_2^1(B, \mathbb{R}^3)$ whose L_2-traces on C and I satisfy

(i) $X(w) \in S$ for $\mathcal{H}^1$ – almost all $w \in I$;
(ii) $X|_C$ maps C continuously and in a weakly monotonic way onto Γ such that $X(1) = P_1$ and $X(-1) = P_2$.

The corresponding variational problem $\mathcal{P}(\Gamma, S)$ reads:

$$D(X) \to \min \quad \text{in the class } \mathcal{C}(\Gamma, S).$$

Again, as in the study of the Plateau problem, it is desirable to introduce a three-point-condition. Since we have already fixed the images of two boundary points, the image of only one more point needs to be prescribed: Let P_3 be some point of Γ different from P_1 and P_2, and let $\mathcal{C}^*(\Gamma, S)$ denote the class of all those surfaces $X \in \mathcal{C}(\Gamma, S)$ mapping $i = \sqrt{-1} \in C$ to P_3. The corresponding variational problem $\mathcal{P}^*(\Gamma, S)$ then requires:

$$D(X) \to \min \quad \text{in } \mathcal{C}^*(\Gamma, S).$$

Then, corresponding to the Main Theorem of the previous section, we can formulate the following result.

Theorem 1. *If $\mathcal{C}(\Gamma, S)$ is nonempty, then there is a solution of $\mathcal{P}(\Gamma, S)$. Each solution X of $\mathcal{P}(\Gamma, S)$ is a minimal surface of class $C^0(B \cup C', \mathbb{R}^3)$ for every closed arc C' contained in the interior of C.*

(Note that $\mathcal{C}(\Gamma, S)$ is nonempty if Γ is rectifiable and P_1, P_2 can be connected by a rectifiable arc on S).

Proof. Let $\{X_n\}$ be a minimizing sequence for $\mathcal{P}(\Gamma, S)$; we can assume that $X_n \in \mathcal{C}^*(\Gamma, S), n \in \mathbb{N}$. Then

$$\lim_{n \to \infty} D(X_n) = e(\Gamma, S) = e^*(\Gamma, S)$$

where $e(\Gamma, S)$ and $e^*(\Gamma, S)$ are the infima of $D(X)$ on $\mathcal{C}(\Gamma, S)$ and $\mathcal{C}^*(\Gamma, S)$ respectively. Then $D(X_n) \leq \text{const}$, and a suitable variant of Poincaré's inequality together with $\sup_{w \in C} |X_n(w)| \leq \text{const}$ yields

$$|X_n|_{H_2^1(B)} \leq \text{const} \quad \text{for } n \in \mathbb{N}.$$

Hence (by renumbering), we can assume that $\{X_n\}$ converges weakly to some $X \in H_2^1(B, \mathbb{R}^3)$. Since Dirichlet's integral is lower semicontinuous with respect to weak convergence in $H_2^1(B, \mathbb{R}^3)$, we have $\liminf_{n\to\infty} D(X_n) \geq D(X)$, and therefore

$$D(X) \leq e(\Gamma, S). \tag{1}$$

Exploiting the Courant-Lebesgue lemma of Section 2.2 one can prove that $\mathcal{C}^*(\Gamma, S)$ *is closed with respect to weak convergence of sequences in* $H_2^1(B, \mathbb{R}^3)$, and thus we obtain $X \in \mathcal{C}^*(\Gamma, S)$ whence $D(X) \geq e(\Gamma, S)$. In conjunction with (1) we arrive at $D(X) = e(\Gamma, S)$, and so X is a solution of $\mathcal{P}(\Gamma, S)$. Weyl's lemma shows that X is harmonic, and the minimum property of X leads to

$$\partial D(X, \lambda) = 0 \quad \text{for all } \lambda \in C^1(\overline{B}, \mathbb{R}^2),$$

for any solution X of $\mathcal{P}(\Gamma, S)$ whence $|X_u|^2 = |X_v|^2$ and $X_u \cdot X_u = 0$ on account of Theorem 5 of Section 2.2. The continuity of X in $B \cup \operatorname{int} C$ follows from a well known result for harmonic functions. □

Remark 1. For the semi-free problems we often choose the parameter domain B as a semidisk $\{w : |w| < 1, \operatorname{Im} w > 0\}$, $I := \{u \in \mathbb{R} : |u| < 1\}$, and $C = \partial B - I$.

Free boundary value problems for minimal surfaces were first solved by H.A. Schwarz. He considered boundary configurations consisting of finitely many planar surfaces and straight arcs, so-called Schwarzian chains. By means of *reflection principles* solutions of such problems can be used to generate periodic minimal surfaces. General free boundary problems were first studied by Courant [6]; his results are presented in some detail in Courant's monograph [12] and in Dierkes/Hildebrandt/Küster/Wohlrab [1], chapter 5, as well as in Nitsche [18]. We have seen that for partially free problems, containing also prescribed arcs as part of the boundary configuration, solutions can be found by a minimizing procedure based on the Courant-Lebesgue lemma. If, however, the boundary configuration is "completely free", minimizing sequences might shrink to a point, and one is left with nothing. It is Courant's merit to have seen how one can establish the existence of nontrivial minimizers if the support surfaces S of the free boundary contain closed loops which cannot be contracted to a point within S. Still more difficult is the case when S is a topological sphere, say, the boundary of a compact convex body. Then there exist no nontrivial minimal surfaces bounded by S which minimize area; instead, one can only hope for *stationary solutions* within S which are to be constructed by minimax-methods. However, this idea cannot easily be realized since the Palais-Smale condition is not satisfied, and certain approximation procedures are needed to make the method valid. The first result in this direction was found by Struwe [1] (cf. also Imbusch [1]); special cases were earlier treated by Schwarz, and by B. Smyth [1]. Embedded stationary solutions within a closed convex surface S of class C^4 were first proved to exist by Grüter and Jost [1]. The existence of several stationary embedded solutions within S was established in the penetrating papers [4,5,8,10] by Jost.

Here we must refrain from presenting any more details on minimal surfaces with free boundaries. We only want to define what we mean by a *stationary surface within a given boundary configuration*. For simplicity we explain the matter for a configuration $< \Gamma, S >$ such as considered in the beginning of this section. We assume that S is a smooth surface in $\mathbb{R}^3$ with a smooth boundary ∂S, which could be empty. Let the class $\mathcal{C}(\Gamma, S)$ of admissible surfaces within $< \Gamma, S >$ be defined as before. Finally let Γ be a smooth arc with its endpoints P_1, P_2 on S.

Definition 1. *A stationary minimal surface in* $< \Gamma, S >$ *is an element* $X \in \mathcal{C}(\Gamma, S)$ *which is a minimal surface of class* $C^1(B \cup I, \mathbb{R}^3) \cap C^2(B, \mathbb{R}^3)$, $I = \partial B \cap \{\operatorname{Im} w < 0\}$ *=free boundary, and along* $I_1 := \{w \in I : X(w) \in \mathit{int}\ S\}$, *the normal derivative* $\partial X / \partial \nu$ *with respect to the exterior normal* ν *to* ∂B *is perpendicular to* S. *For any* $w \in I_2 := I - I_1 = \{w \in I : X(w) \in \partial S\}$ *and every interior tangent vector* V *to* S *at* $X(w)$ *we have* $V \cdot \frac{\partial X}{\partial \nu}(w) \geq 0$.

Note that the free boundary condition means that X is perpendicular to S along $X|_I$ if the support surface S has no boundary.

It turns out that this definition is equivalent to the following one, provided that X is not too badly behaved at its boundary.

Definition 2. *An element* $X \in \mathcal{C}(\Gamma, S)$ *is called a critical (or stationary) point of Dirichlet's integral in the class* $\mathcal{C}(\Gamma, S)$ *if*

$$\lim_{\varepsilon \to +0} \frac{1}{\varepsilon}\{D(X_\varepsilon) - D(X)\} \geq 0$$

holds for all "admissible variations" $X_\varepsilon, |\varepsilon| < \varepsilon_0$, *of* X. *A family* $\{X_\varepsilon\}_{|\varepsilon|<\varepsilon_0}$ *of surfaces* $X_\varepsilon \in \mathcal{C}(\Gamma, S)$ *is said to be an admissible variation of* X, *if it is one of the following two types.*

Type I (inner variations): $X_\varepsilon = X \circ \sigma_\varepsilon$ *where* $\{\sigma_\varepsilon\}_{|\varepsilon|<\varepsilon_0}, \varepsilon_0 > 0$, *is a differentiable family of diffeomorphisms* $\sigma_\varepsilon : \overline{B}^*_\varepsilon \to \overline{B}$ *which are defined as inverse mappings of the diffeomorphisms* $\tau_\varepsilon : \overline{B} \to \overline{B}^*_\varepsilon$ *defined by*

$$\tau_\varepsilon(w) = w - \varepsilon\lambda(w)\,,\ \lambda \in C^1(\mathbb{R}^2, \mathbb{R}^2)\,.$$

Type II (outer variations): $X_\varepsilon = X + \varepsilon\phi(\cdot, \varepsilon)$ *for all* $\varepsilon \in (0, \varepsilon_0)$ *and some* $\varepsilon_0 > 0$, *where the following holds:*

(α) *the Dirichlet integrals of the mappings* $\phi(\cdot, \varepsilon)$ *are uniformly bounded, i.e.,* $D(\phi(\cdot, \varepsilon)) \leq \text{const}$ *for all* $\varepsilon \in (0, \varepsilon_0)$;

(β) *the functions* $\phi(\cdot, \varepsilon)$ *converge pointwise a.e. in* B *to some function* $\phi_0 \in H_2^1 \cap L_\infty(B, \mathbb{R}^3)$.

Concerning the equivalence of these two definitions we have the following result:

Assume that S and ∂S are of class C^1 and that Γ is rectifiable. Then every stationary minimal surface in $\langle \Gamma, S \rangle$ is a stationary point of Dirichlet's integral in $\mathcal{C}(\Gamma, S)$. If S and ∂S are of class C^4, then also the converse holds true, that is, every stationary point of Dirichlet's integral in $\mathcal{C}(\Gamma, S)$ furnishes a stationary minimal surface in $\langle \Gamma, S \rangle$.

This result is certainly not optimal, and it would be worthwhile to investigate what are the weakest assumptions on Γ and S such that Definitions 1 and 2 are equivalent.

In a similar way one can define *stationary surfaces* within other free boundary configurations and show the equivalence of the definitions corresponding to Definitions 1 and 2.

Of particular interest is the *generalized Plateau problem* (or *Douglas problem*), and the so-called *thread problem*. In the first case the boundary consists of several disjoint closed Jordan curves. Here the conformal type of the parameter domain of any solution is not a priori determined but must be found together with the solution. In the second problem some parts of the boundary are freely movable arcs of fixed length, and thus the problem can be considered as a generalization of the isoperimetric problem.

Concerning the generalized Plateau problem we refer to Courant [12], Nitsche [18], Jost [12], and Dierkes/Hildebrandt/Küster/Wohlrab [1], and to the references to the literature quoted in these texts. In DHKW, vol. II, chapter 10 one also finds a presentation of the thread problem which was first solved by H. W. Alt [3]. The generalized Plateau problem was first attacked by Douglas, Courant, and Shiffman. Modern existence proofs were given by Tromba, Tomi–Tromba, and Jost. The recent survey by Struwe [7] provides a brief, but very clear report on the generalized Plateau problem, on the structure of the set of solutions and on Morse and index theory. A very instructive report on the construction of embedded minimal surfaces bounded by given boundary configuration is given in the survey of Jost [9]. Methods of geometric measure theory are of fundamental importance in this context.

Chapter 3
Boundary Regularity and Geometric Estimates for Minimal Surfaces

In the setting of our article, regularity questions for minimal surfaces are essentially questions of boundary regularity which consist of two problems: (i) Does a minimal surface $X : B \to \mathbb{R}^3$ assume smooth boundary values in a smooth way, provided that its boundary lies on a smooth boundary configuration? (ii) What are the "geometric" singularities of a minimal surface? Concerning question (i) we prove that, under certain precautions, the answer is "yes". In Section 3.2 we present the pertinent results for solutions of Plateau's problem, and in Section 3.3 the boundary behaviour of solutions to free boundary problems is sketched.

The only geometric singularities of minimal surfaces in the interior are branch points. It will be shown in Section 3.4 that the same holds true at the boundary. This follows from an asymptotic expansion, first noted by E. Heinz, which is obtained by a technique due to Hartman and Wintner. By means of the Gauss-Bonnet theorem, the number of branch points in the interior and on the boundary of the parameter domain B can be estimated from above in geometric terms of the boundary configuration which the minimal surface is spanning (see Section 3.5). The results on boundary regularity obtained in this way enable us to establish various geometric estimates for minimal surfaces such as isoperimetric inequalities and length estimates of the free traces. Together with "enclosure theorems", derived e.g. from the maximum principle, such kinds of estimates furnish a certain rough picture of a minimal surface spanning a pre-assigned boundary and lead to various compactness results for minimal surfaces. We prove a selection of such geometric estimates in Section 3.5.

3.1 Solutions of Differential Inequalities

In this section we sketch some auxiliary results, due to Heinz, which are useful for studying the boundary behaviour of minimal surfaces. Detailed proofs can be found in DHKW [1], Sections 7.1 and 7.2, or in Heinz's original papers.

We consider solutions $X \in C^2(\Omega, \mathbb{R}^N)$ of

$$|\Delta X| \le a|\nabla X|^2, \qquad a = \text{const}, \tag{1}$$

which can equivalently be written as

$$|X_{w\bar{w}}| \le a|X_w|^2. \tag{1'}$$

Let $|X|_{0,\Omega} = \sup_\Omega |X(w)|$. Then we have

Lemma 1. *Let $X \in C^2(\Omega, \mathbb{R}^N)$ be a solution of (1) in the open set Ω of $\mathbb{R}^2$ which satisfies $| X |_{0,\Omega} \leq M$. Then we obtain*

$$\Delta |X|^2 \geq 2(1 - aM)|\nabla X|^2 \tag{2}$$

in Ω. In particular, if $aM < 1$, then $| X |^2$ is subharmonic in Ω.

Proof. Follows immediately from the identity

$$\Delta |X|^2 = 2|\nabla X|^2 + 2X \cdot \Delta X\,. \tag{3}$$

□

Lemma 2. *Suppose that $X \in C^0(\overline{B}_R(w_0), \mathbb{R}^N) \cap C^2(B_R(w_0), \mathbb{R}^N)$ satisfies (1) in $B_R(w_0)$. Assume also that $|X(w)| \leq M$ for $w \in \overline{B}_R(w_0)$ and $aM < 1$ are satisfied. Then for any $\rho \in (0, R)$ we have*

$$\int_{B_\rho(w_0)} | \nabla X |^2 \, dudv \leq \frac{1}{\log \frac{R}{\rho}} \, \frac{2\pi M}{1 - aM} \max_{w \in \partial B_R(w_0)} | X(w) - X(w_0) | \tag{4}$$

and

$$\int_{B_\rho(w_0)} | \nabla X |^2 \, dudv \leq \frac{1}{\log \frac{R}{\rho}} \, \frac{4\pi M^2}{1 - aM}\,. \tag{5}$$

Proof. One represents $x(w) := |X(w)|^2$ by means of Green's function, uses $|x(w) - x(w_0)| \leq 2M|X(w) - X(w_0)|$, and applies (2). □

Proposition 1. *There is a continuous function $\kappa(t), 0 \leq t < 1$, with the following property: For any solution $X \in C^2(B_R(w_0), \mathbb{R}^N)$ of the differential inequality (1) in $B_R(w_0)$ satisfying*

$$|X(w)| \leq M \qquad \textit{for } w \in B_R(w_0) \tag{6}$$

for some constant M with $aM < 1$, the estimates

$$|\nabla X(w_0)| \leq \kappa(aM) \frac{M}{R} \tag{7}$$

and

$$|\nabla X(w_0)| \leq \frac{\kappa(aM)}{R} \sup_{w \in B_R(w_0)} |X(w) - X(w_0)| \tag{8}$$

hold true.

Proof. Fix any $R' \in (0, R)$, and consider the nonnegative function

$$f(w) := (R' - |w - w_0|)|\nabla X(w)|$$

on $\overline{B}_{R'}(w_0)$ which vanishes on $\partial B_{R'}(w_0)$. Then there is some point $w_1 \in B_{R'}(w_0)$ where $f(w)$ assumes its maximum K, i.e.,

$$f(w_1) = K := \max\{f(w) : w \in \overline{B}_{R'}(w_0)\}\,.$$

Set $r = |w - w_1|$ for $w \in \overline{B}_{R'}(w_0)$ and $\rho := R' - |w_1 - w_0|$. Clearly, we have $0 < \rho \leq R'$. For any $\theta \in (0,1)$ we can represent X_u by

$$X_u(w_1) = \frac{1}{\pi\rho^2\theta^2}\int_{B_{\rho\theta}(w_1)} X_u\,dudv - \frac{1}{2\pi}\int_{B_{\rho\theta}(w_1)} (u-u_1)\left(\frac{1}{r^2} - \frac{1}{\rho^2\theta^2}\right)\Delta X\,dudv\,,$$

and an analogous formula holds for $X_v(w_1)$. By means of Schwarz's inequality we infer that

$$| \nabla X(w_1) | \leq \frac{1}{\sqrt{\pi}\rho\theta}\left\{\int_{B_{\rho\theta}(w_1)} | \nabla X |^2\,dudv\right\}^{1/2} + \frac{1}{2\pi}\int_{B_{\rho\theta}(w_1)} \frac{| \Delta X |}{r}\,dudv\,.$$

Applying Lemma 2, (5) to $\rho\theta$ and ρ instead of ρ and R, we also obtain

$$\int_{B_{\rho\theta}(w_1)} | \nabla X |^2\,dudv \leq \frac{c(a,M)}{\log\frac{1}{\theta}}\,.$$

Taking $| \Delta X | \leq a\,| \nabla X |^2$ into account, we arrive at

$$| \nabla X(w_1) | \leq \frac{\sqrt{c}}{\sqrt{\pi}\rho\theta}\left(\log\frac{1}{\theta}\right)^{-1/2} + \frac{a}{2\pi}\int_{B_{\rho\theta}(w_1)} \frac{1}{r}\,| \nabla X |^2\,dudv$$

and

$$\frac{a}{2\pi}\int_{B_{\rho\theta}(w_1)} \frac{1}{r}\,| \nabla X |^2\,dudv \leq a\rho\theta \sup_{B_{\rho\theta}(w_1)} | \nabla X |^2\,.$$

Note that $|\nabla X(w_1)| = K/\rho$, and that for $r = |w - w_1| < \rho\theta$ we infer that $|\nabla X(w)| \leq K/(1-\theta)\rho$ for all $w \in B_{\rho\theta}(w_1)$. Therefore

$$\frac{a}{2\pi}\int_{B_{\rho\theta}(w_1)} \frac{1}{r}\,| \nabla X |^2\,dudv \leq \frac{a\theta K^2}{(1-\theta)^2\rho}$$

whence

$$K \leq \frac{\sqrt{c/\pi}}{\theta\left(\log\frac{1}{\theta}\right)^{1/2}} + \frac{a\theta K^2}{(1-\theta)^2}\,.$$

Set

$$\alpha(\theta) := \frac{a\theta}{(1-\theta)^2}\,, \quad \beta(\theta) := \frac{\sqrt{c/\pi}}{\theta\left(\log\frac{1}{\theta}\right)^{1/2}}\,.$$

Then we have

$$\alpha K^2 - K + \beta \geq 0\,,$$

and we infer that there is a number $\theta_0 = \theta_0(a,M) \in (0,1)$ such that $4\alpha(\theta)\beta(\theta) \leq 3/4$ if $0 < \theta \leq \theta_0$, i.e. $\sqrt{1 - 4\alpha(\theta)\beta(\theta)} \geq 1/2$. We conclude that either (i) $K \leq m^-(\theta)$, or (ii) $K \geq m^+(\theta)$ for every $\theta \in (0,\theta_0)$, where $m^{\pm}(\theta) = [1 \pm \sqrt{1 - 4\alpha(\theta)\beta(\theta)}]/2\alpha(\theta)$. Furthermore, m^- and m^+ are continuous on $(0,\theta_0]$ and satisfy $m^- < m^+$ as well as $m^+(\theta) \to \infty$ for $\theta \to +0$.

Therefore (ii) is excluded for θ close to zero, and a continuity argument yields $K \le m^-(\theta_0)$. We also note that $m^-(\theta_0) \le (4/3)\beta(\theta_0)$. Consequently

$$K \le \frac{4}{3}\beta(\theta_0) = \frac{4\sqrt{c/\pi}}{3\theta_0\left(\log\frac{1}{\theta_0}\right)^{1/2}} := c^*(a,M).$$

Because of

$$R' \mid \nabla X(w_0) \mid = f(w_0) \le f(w_1) = K \le c^*(a,M)$$

we arrive at

$$|\nabla X(w_0)| \le c^*(a,M)/R' \ \text{ for any } R' \in (0,R)$$

whence

$$|\nabla X(w_0)| \le c^*(a,M)/R. \tag{7*}$$

This estimate is close to (7). We now introduce the function $\kappa(t) := c^*(t,1)$. A close inspection of the previous computations shows that $\kappa(t)$ can be assumed to be an increasing continuous function on the interval $(0,1)$.

In order to prove (7) we assume that $M > 0$ because this inequality trivially holds true if $M = 0$. Then $Z(w) := M^{-1}X(w)$ satisfies both $|Z(w)| \le 1$ and

$$\mid \Delta Z \mid \le aM \mid \nabla Z \mid^2 .$$

Applying the estimate (7*) to Z, we arrive at

$$\mid \nabla Z(w_0) \mid \le \kappa(aM)/R.$$

Multiplying this inequality by M, we obtain (7).

Estimate (8) is now an easy consequence of (7). To see this we introduce the quantity

$$m := \sup\{|X(w) - X(w_0)| \ : w \in B_R(w_0)\}.$$

If $m = 0$ or $m = \infty$, the estimate (8) is true for trivial reasons. If $M \le m < \infty$, (8) follows directly from (7). If $0 < m < M$, we introduce $Z := m^{-1}[X - X(w_0)]$ and obtain as before

$$|\nabla Z(w_0)| \le \kappa(am)/R \le \kappa(aM)/R,$$

and this implies (8). □

Corollary 1. *Suppose that $X \in C^2(B_R(w_0), \mathbb{R}^N)$ is a solution of (1) in $B_R(w_0)$ satisfying*

$$|X(w)| \le M \ \textit{for} \ w \in B_R(w_0)$$

and for some constant M with $aM < 1$. Then we have

$$\mid \nabla X(w) \mid \le \kappa(aM)\frac{M}{\rho} \ \textit{for all} \ w \in B_{R-\rho}(w_0), \quad 0 < \rho < R. \tag{9}$$

Proposition 2. *Let $X \in C^0(\overline{B}_R(w_0), \mathbb{R}^N) \cap C^2(B_R(w_0), \mathbb{R}^N)$ be a solution of the differential inequality (1) in $B_R(w_0)$, and suppose that $|X(w)| \leq M$ holds for all $w \in \overline{B}_R(w_0)$ and for some number M with $2aM < 1$. Moreover, set $x(w) := | X(w) |^2$, and let $H \in C^0(\overline{B}_R(w_0), \mathbb{R}^N) \cap C^2(B_R(w_0), \mathbb{R}^N)$ and $h \in C^0(\overline{B}_R(w_0)) \cap C^2(B_R(w_0))$ be the solutions of the boundary value problems*

$$\Delta H = 0 \quad \text{in } B_R(w_0), \quad H = X \quad \text{on } \partial B_R(w_0), \tag{10}$$

$$\Delta h = 0 \quad \text{in } B_R(w_0), \quad h = x \quad \text{on } \partial B_R(w_0). \tag{11}$$

Then for any $w \in B_R(w_0)$ and $w^* \in \partial B_R(w_0)$ we have the inequality

$$| X(w) - X(w^*) | \leq \frac{a}{2(1-2aM)} | h(w) - h(w^*) | + \frac{1-aM}{1-2aM} | H(w) - H(w^*) | . \tag{12}$$

Proof. Relations (1) and (2) imply

$$| \Delta X | \leq a \Delta x / 2(1 - aM) .$$

We infer that for any constant $E \in \mathbb{R}^N, | E | = 1$, the function

$$z(w) := \frac{a}{2(1-aM)} [x(w) - h(w)] + [H(w) - X(w)] \cdot E$$

vanishes on $\partial B_R(w_0)$ and satisfies $\Delta z \geq 0$. Then the maximum principle yields

$$[H - X] \cdot E \leq \frac{a}{2(1-aM)} [h - x] \text{ for all } E \text{ with } | E | = 1$$

whence

$$| H - X | \leq \frac{a}{2(1-aM)} [h - x] . \tag{13}$$

Since $|H| \leq M$, we infer

$$h - x \leq 2M[|H| - |X|] + h - |H|^2$$

and then (13) leads to

$$| H - X | \leq \frac{a}{2(1-2aM)} (h - |H|^2) \text{ on } \overline{B}_R(w_0) . \tag{14}$$

For $w^* \in \partial B_R(w_0)$ we have

$$|X(w^*)|^2 = x(w^*) = |H(w^*)|^2 = h(w^*) ,$$

and therefore

$$\begin{aligned} |X(w) - X(w^*)| &\leq |X(w) - H(w)| + |H(w) - H(w^*)| \\ &\leq \frac{a}{2(1-2aM)} (h(w) - |H(w)|^2) + |H(w) - H(w^*)| . \end{aligned}$$

Because of

$$\begin{aligned} h(w) - |H(w)|^2 &= h(w) - h(w^*) + |H(w^*)|^2 - |H(w)|^2 \\ &\le |h(w) - h(w^*)| + 2M|H(w) - H(w^*)|\,, \end{aligned}$$

we may now conclude that (12) holds for any $w \in \overline{B}_R(w_0)$ and for any $w^* \in \partial B_R(w_0)$. □

Remark 1. Note that the differential inequality (1) remains invariant with respect to conformal transformations of the parameter domain. Thus we can carry over Proposition 2 from $B_R(w_0)$ to any bounded domain Ω in $\mathbb{C}$ which is of the conformal type of the disk and has a closed Jordan curve as its boundary.

Proposition 3. *For any $a \ge 0, R > 0, M \ge 0$, and $k \ge 0$ with $2aM < 1$, there is a number $c = c(a, R, M, k) \ge 0$ having the following property: Let $X \in C^0(\overline{B}_R(w_0), \mathbb{R}^N) \cap C^2(B_R(w_0), \mathbb{R}^N)$ be a solution of (1) in $B_R(w_0)$ satisfying $|X(w)| \le M$ for all $w \in \overline{B}_R(w_0)$. Suppose also that the boundary values $\mathcal{X}(\varphi) := X(w_0 + \mathrm{Re}^{i\varphi})$ are of class $C^2(\mathbb{R})$ and satisfy $|\mathcal{X}''(\varphi)| \le k$ for all $\varphi \in \mathbb{R}$. Then we have*

$$|\nabla X(w)| \le c(a, R, M, k) \text{ for all } w \in B_R(w_0)\,. \tag{15}$$

Proof. It suffices to treat the case $w_0 = 0$ and $R = 1$; that is, we consider the parameter domain $B = B_1(0)$. Let $w = re^{i\theta}, 0 \le r < 1$, be an arbitrary point of B. By formula (8) of Proposition 1 we have

$$|\nabla X(w)| \le \frac{c(a, M)}{1 - r} \sup\{|X(w') - X(w)| : w' \in B_{1-r}(w)\}\,. \tag{16}$$

Moreover, for $w, w' \in B$ and $w^* \in \partial B$ it follows from Proposition 2 that

$$\begin{aligned} |X(w) - X(w')| &\le |X(w) - X(w^*)| + |X(w') - X(w^*)| \\ &\le \frac{a}{2(1 - 2aM)}\{|h(w) - h(w^*)| + |h(w') - h(w^*)|\} \\ &\quad + \frac{(1 - aM)}{1 - 2aM}[|H(w) - H(w^*)| + |H(w') - H(w^*)|] \end{aligned} \tag{17}$$

holds true where H and h are harmonic in B and have the boundary values X and $x := |X|^2$ respectively on ∂B. Moreover, we have $|\nabla H| \le ck$ whence

$$|H(w) - H(w^*)| + |H(w') - H(w^*)| \le 3ck(1 - r) \tag{18}$$

for $w = re^{i\theta}, w^* = e^{i\theta}, w' \in B_{1-r}(w)$.

The boundary values $\eta(\varphi)$ of $x(w_0 + re^{i\varphi})$ satisfy

$$|\eta''| \le 2|\mathcal{X}'|^2 + 2Mk$$

It is not difficult to see that $|\mathcal{X}'| \le 2\pi k$, whence

$$|\eta''| \le 8\pi^2 k^2 + 2Mk\,.$$

From this one can deduce that

$$|\nabla h(w)| \le c^*(1+k^2) \text{ for all } w \in B$$

whence

$$|h(w_1) - h(w_2)| \le c^*(1+k^2)|w_1 - w_2| \text{ for all } w_1, w_2 \in \overline{B}$$

and consequently

$$|h(w) - h(w^*)| + |h(w') - h(w^*)| \le 3c^*(1+k^2)(1-r) \tag{19}$$

for $w = re^{i\theta}, w^* = e^{i\theta}, w' \in B_{1-r}(w)$.

Combining (17), (18) and (19), we arrive at

$$|X(w) - X(w')| \le c(a, M, k)(1-r) \qquad \text{for } w = re^{i\theta}, 0 < r < 1 \text{ and } w' \in B_{1-r}(w)\,,$$

and this implies

$$|\nabla X(w)| \le c(a, M, k) \text{ for all } w \in B\,,$$

taking (16) into account. □

Theorem 1. *Suppose that the assumptions of Proposition 3 are satisfied. Then X is of class $C^{1,\mu}(\overline{B}_R(w_0), \mathbb{R}^N)$ for all $\mu \in (0,1)$ and we have the following estimate for the Hölder seminorm of ∇X:*

$$[\nabla X]_{\mu, \overline{B}_R(w_0)} \le c(a, R, M, k, \mu)\,. \tag{20}$$

The proof follows from Proposition 3 in conjunction with a well-known potential-theoretic reasoning. □

The most important result for the following section is

Theorem 2. *For $w_0 \in \partial B$, we introduce the set $S_\rho(w_0) := B \cap B_\rho(w_0)$. Assume that, for some $\rho \in (0,1), X \in C^0(\overline{S}_\rho(w_0), \mathbb{R}^N) \cap C^2(S_\rho(w_0), \mathbb{R}^N)$ is a solution of the differential inequality (1) in $S_\rho(w_0)$ that vanishes on $\partial S_\rho(w_0) \cap \partial B$. Then we obtain $X \in C^{1,\mu}(\overline{S}_{\rho'}(w_0), \mathbb{R}^N)$ for every $\mu \in (0,1)$ and every $\rho' \in (0, \rho)$.*

Proof. It suffices to show that for any $w^* = e^{i\theta} \in \partial S_\rho(w_0) \cap \partial B$ there is a $\delta > 0$ such that $X \in C^{1,\mu}(\overline{S}_\delta(w^*), \mathbb{R}^N)$, where $S_\delta(w^*)$ denotes the circular two-gon $B \cap B_\delta(w^*)$. We may also assume that $a > 0$.

Thus, having fixed an arbitrary $w^* = e^{i\theta} \in \partial B \cap \partial S_\rho(w_0)$, we first choose an $\varepsilon > 0$ such that $S_{3\varepsilon}(w^*) \subset S_\rho(w_0)$ holds and that

$$\sup\{|X(w)| \;:\; w \in S_{3\varepsilon}(w^*)\} \le \frac{1}{4a}\,.$$

Then the mapping

$$Z(w) := 4aX(w), \quad w \in S_{3\varepsilon}(w^*),$$

satisfies the inequalities

$$|Z| \leq 1 \quad \text{and} \quad |\Delta Z| \leq \frac{1}{4}|\nabla Z|^2$$

in $S_{3\varepsilon}(w^*)$. We now consider the functions $H(w)$ and $h(w)$ which are harmonic in $S_{3\varepsilon}(w^*)$ and have the boundary values X and $|X|^2$ respectively on $\partial S_{3\varepsilon}$. As h and H vanish on the circular arc

$$C := \partial B \cap \partial S_{3\varepsilon}(w^*),$$

we can extend h and H to harmonic functions in $B_{3\varepsilon}(w^*)$ by a reflection in C, applying Schwarz's reflection principle. Hence there is a number $c(\varepsilon)$ such that

$$|\nabla H(w)| + |\nabla h(w)| \leq c(\varepsilon) \quad \text{for all } w \in B_{2\varepsilon}(w^*)$$

whence

$$|H(w_1) - H(w_2)| + |h(w_1) - h(w_2)| \leq c(\varepsilon)|w_1 - w_2| \text{ for all } w_1, w_2 \in B_{2\varepsilon}(w^*). \tag{21}$$

Fix some $w = re^{i\varphi} \in S_\varepsilon(w^*)$. Then we have $|w| = r > 1-\varepsilon$ and, for $|w-w'| < 1-r$, we have $|w' - w^*| \leq |w' - w| + |w - w^*| < \varepsilon + \varepsilon = 2\varepsilon$, and consequently

$$B_{1-r}(w) \subset S_{2\varepsilon}(w^*) \subset S_{3\varepsilon}(w^*) \subset S_\rho(w_0).$$

By Proposition 2 and the subsequent Remark 1 we obtain for $\Omega := S_{3\varepsilon}(w^*)$, $w' \in \Omega$ and $e^{i\varphi} \in \partial B \cap \partial S_{3\varepsilon}(w^*)$ that

$$|Z(w')| = |Z(w') - Z(e^{i\varphi})| \leq \frac{1}{4}|h(w') - h(e^{i\varphi})| + \frac{3}{2}|H(w') - H(e^{i\varphi})|.$$

In connection with (21) we infer for any $w' \in B_{1-r}(w)$ that

$$|Z(w')| \leq \frac{7}{4}c(\varepsilon)|w' - e^{i\varphi}| \leq \frac{7}{2}c(\varepsilon)(1-r) < 4c(\varepsilon)(1-r).$$

In other words, we have

$$\sup\{|Z(w')| : w' \in B_{1-r}(w)\} \leq 4c(\varepsilon)(1-r)$$

for any $w \in S_\varepsilon(w^*)$ with $|\, w \,| = r$.

Moreover, we infer from Proposition 1, (8) that

$$|\nabla Z(w)| \leq \frac{2\kappa(1/4)}{1-r} \sup\{|\, Z(w') \,| : w' \in B_{1-r}(w)\}$$

for any $w \in S_\varepsilon(w^*)$ with $|\, w \,| = r$. This implies

$$|\nabla Z(w)| \leq c^*(\varepsilon) \quad \text{for all } w \in S_\varepsilon(w^*).$$

Since $X = \frac{1}{4a}Z$, we conclude that

$$|\nabla X|_{0,S_\varepsilon(w^*)} \le \text{const}\,, \quad \text{and} \quad |\Delta X|_{0,S_\varepsilon(w^*)} \le \text{const}\,.$$

Now we choose a cut-off function $\eta \in C_c^\infty(\mathbb{R}^2)$ with $\eta(w) = 1$ for $w \in B_\delta(w^*), \delta := \frac{\varepsilon}{2}$, and with $\eta(w) = 0$ for $|w - w^*| \ge \frac{3}{4}\varepsilon$. Then the mapping $Y := \eta X$ on $S_\varepsilon(w^*)$ satisfies

$$\Delta Y = \eta \Delta X + 2\nabla\eta \cdot \nabla X + \Delta\eta\, X$$

and therefore

$$|\Delta Y(w)| \le \text{const} \quad \text{for all } w \in S_\varepsilon(w^*) \tag{22}$$

$$Y(w) = 0 \quad \text{on } \partial S_\varepsilon(w^*) \tag{23}$$

$$Y(w) = 0 \quad \text{for all } w \in S_\varepsilon(w^*) \text{ with } \frac{3}{4}\varepsilon < |w - w^*| < \varepsilon\,. \tag{24}$$

Consider a conformal mapping τ of the unit disk B onto the two-gon $S_\varepsilon(w^*)$. We can extend τ to a homeomorphism of $\overline{B}$ onto $\overline{S}_\varepsilon(w^*)$, and it can be assumed that $\zeta = \pm 1$ are mapped onto the two vertices of the two-gon. By the reflection principle the mapping $\tau(\zeta)$ is holomorphic on $\overline{B} - \{-1, 1\}$. Then it follows from (22) and (24) that the mapping $Y^*(\zeta) := Y(\tau(\zeta))$ is of class $C^0(\overline{B}, \mathbb{R}^N) \cap C^2(B, \mathbb{R}^N)$ and satisfies

$$|\Delta Y^*(\zeta)| \le \text{const} \quad \text{for all } \zeta \in B\,.$$

Moreover, we infer from (23) that

$$Y^*(\zeta) = 0 \quad \text{for all } \zeta \in \partial B\,.$$

Then, by a well-known potential-theoretic reasoning, we obtain that $Y^* \in C^{1,\mu}(\overline{B}, \mathbb{R}^N)$ for any $\mu \in (0,1)$. It follows that $Y \in C^{1,\mu}(\overline{S}_\varepsilon(w^*), \mathbb{R}^N)$, and therefore $X \in C^{1,\mu}(\overline{S}_\delta(w^*), \mathbb{R}^N)$ since $Y(w) = X$ for all $w \in \overline{S}_\delta(w^*), \delta = \frac{\varepsilon}{2}$.

□

3.2 Boundary Regularity of Solutions to Plateau's Problem

Here we investigate the boundary behaviour of minimal surfaces bounded by smooth arcs. As all results are of a local nature, it suffices to formulate them on simply connected parameter domains B, say, for surfaces $X : B \to \mathbb{R}^3$ where $B = \{w \in \mathbb{C} : |w| < 1\}$. Our main result is

Theorem 1. *Consider a minimal surface $X : B \to \mathbb{R}^3$ of class $C^0(B \cup \gamma, \mathbb{R}^3) \cap C^2(B, \mathbb{R}^3)$ which maps an open subarc γ of ∂B onto an open Jordan arc Γ of $\mathbb{R}^3$ which is a regular curve of class $C^{m,\mu}$ for some integer $m \ge 1$ and some $\mu \in (0,1)$. Then X is of class $C^{m,\mu}(B \cup \gamma, \mathbb{R}^3)$. Moreover, if Γ is a regular real analytic Jordan arc, then X can be extended as a minimal surface across γ.*

We only prove here the result under the assumption that $\Gamma \in C^{m,\mu}$ with $m \geq 2, \mu \in (0,1)$. The case $\Gamma \in C^{1,\mu}$ has been handled by Nitsche [7,11] and [18, Kap. V, 2.1], Kinderlehrer [1], and Warschawski [6].

Our method to prove Theorem 1 is due to E. Heinz [3]. As it also covers the boundary behaviour of surfaces of prescribed mean curvature, we shall prove the following more general result.

Theorem 2. *Let $X \in C^0(B \cup \gamma, \mathbb{R}^3) \cap C^2(B, \mathbb{R}^3)$ be a solution of the equations*

$$\Delta X = 2H(X) X_u \wedge X_v \tag{1}$$

$$|X_u|^2 = |X_v|^2, \qquad X_u \cdot X_v = 0 \tag{2}$$

in B which maps an open subarc $\gamma = \{e^{i\theta} : \theta_1 < \theta < \theta_2\}$ of ∂B into some closed regular Jordan arc Γ of $\mathbb{R}^3$, i.e. $X(w) \in \Gamma$ for all $w \in \gamma$. Then the following holds:

(i) *If $\mathcal{H}(w) := H(X(w))$ is of class $L_\infty(B)$ and if $\Gamma \in C^2$, then we obtain that $X \in C^{1,\mu}(B \cup \gamma, \mathbb{R}^3)$ for any $\mu \in (0,1)$.*
(ii) *If H is of class $C^{0,\mu}$ on $\mathbb{R}^3$, and if $\Gamma \in C^{2,\mu}, 0 < \mu < 1$, then $X(w)$ is of class $C^{2,\mu}(B \cup \gamma, \mathbb{R}^3)$.*

Proof. (i) It suffices to show that for any $w_0 \in \gamma$ there is some $\delta > 0$ such that $X \in C^{1,\mu}(\overline{S}_\delta(w_0), \mathbb{R}^3), 0 < \mu < 1$, provided that $\mathcal{H}(w) := H(X(w))$ is of class $L_\infty(B)$ and that $\Gamma \in C^2$. Here $S_\delta(w_0)$ denotes as usual the two-gon $B \cap B_\delta(w_0)$.

Thus we fix some $w_0 \in \gamma$. By a suitable motion in $\mathbb{R}^3$ we may assume that $X(w_0) = 0$ and that Γ has a vertical tangent at 0. Then Γ can be represented in a neighbourhood of 0 by

$$x^1 = g^1(t), \quad x^2 = g^2(t), \quad x^3 = t, \quad |t| < 2t_0,$$

where the functions $g^1(t)$ and $g^2(t)$ are of class C^2 and satisfy

$$g^k(0) = 0, \quad \dot{g}^k(0) = 0, \quad k = 1,2.$$

Moreover, we can assume that $w_0 = 1$ and $\gamma = \{w \in \partial B : |w - 1| < R_0\}$ for some $R_0 \in (0,1)$. We set $X(w) = (X^1(w), X^2(w), X^3(w))$. For sufficiently small $t_0 > 0$ and $R \in (0, R_0]$ we obtain

$$|\dot{g}^1(t)|^2 + |\dot{g}^2(t)|^2 \leq 1/8 \quad \text{if } |t| < t_0, \quad |X^3(w)| < t_0 \quad \text{for } w \in \overline{S}_R(1)$$

We introduce the auxiliary map $Y(w) = (Y^1(w), Y^2(w))$ by

$$Y^k(w) := X^k(w) - g^k(X^3(w)), \quad w \in \overline{S}_R(1), \quad k = 1,2.$$

Clearly Y is continuous on $\overline{S}_R(1)$ and of class C^2 in $S_R(1)$, and $Y(w) = 0$ on $\partial B \cap \partial S_R(1)$. From (1) and

$$\Delta Y^k = \Delta X^k - \dot{g}^k(X^3)\Delta X^3 - \ddot{g}^k(X^3) \mid \nabla X^3 \mid^2, \quad k = 1,2,$$

we infer that

$$| \Delta Y | \leq \alpha | \nabla X |^2$$

holds for some constant $\alpha > 0$. Since

$$X_w^k = Y_w^k + g^k(X^3)X_w^3 ,$$

we have

$$| X_w^1 |^2 + | X_w^2 |^2 \leq 2 | Y_w^1 |^2 + 2 | Y_w^2 |^2 + \frac{1}{4} | X_w^3 |^2 ,$$

and $X_w \cdot X_w = 0$ yields

$$| X_w^3 |^2 \leq | X_w^1 |^2 + | X_w^2 |^2 ;$$

then we arrive at $| \nabla X |^2 \leq 8 | \nabla Y |^2$ whence

$$| \Delta Y | \leq 8\alpha | \nabla Y |^2 \quad \text{on } S_R(1) , \tag{3}$$

and we already know that

$$Y = 0 \text{ on } \partial B \cap \partial S_R(1) . \tag{4}$$

By Theorem 2 of Section 3.1 we obtain $Y \in C^{1,\mu}(\overline{S}_\varepsilon(1), \mathbb{R}^2)$ for any $\varepsilon \in (0, R)$ and $\mu \in (0,1)$.

Setting $p^k := \dot{g}^k / q$, $q := 1 + | \dot{g}^1 |^2 + | \dot{g}^2 |^2$, we obtain

$$\left[X_w^3 + \sum_{k=1}^{2} p^k(X^3)Y_w^k \right]^2 = \left\{ \sum_{k=1}^{2} p^k(X^3)Y_w^k \right\}^2 - \frac{1}{q(X^3)} \sum_{k=1}^{2} (Y_w^k)^2 , \tag{5}$$

making use once more of the relation $X_w \cdot X_w = 0$. Since the right-hand side of (5) is continuous on $\overline{S}_\varepsilon(1)$, we see that [...] and therefore also X_w is continuous, i.e. $X \in C^1(\overline{S}_\varepsilon(1), \mathbb{R}^3)$ for any $\varepsilon \in (0, R)$. If $w = re^{i\varphi}$, we have

$$wX_w = \frac{1}{2}(rX_r - iX_\varphi), \quad wY_w = \frac{1}{2}(rY_r - iY_\varphi) .$$

Let us multiply (5) by w^2 with $w \in \partial S_\varepsilon(1) \cap \partial B$. Because of (4) and the bounds $|\dot{g}^k|^2 < 1/8$, the right-hand side of the resulting equation is real and nonpositive, and we conclude that

$$X_r^3 + \sum_{k=1}^{2} p^k(X^3)Y_r^k = 0 \text{ on } \gamma' := \partial S_\varepsilon(1) \cap \partial B .$$

Thus

$$| \Delta X^3 | + | \nabla X^3 | \leq \text{const on } S_\varepsilon(1) , \quad X_r^3 \in C^{0,\mu}(\gamma') .$$

Let $y(w) = \eta(w)X^3(w), w \in \overline{S}_\varepsilon(1)$, with a suitable cut-off function η around $w = 1$. Then $y_r \in C^{0,\mu}(\gamma')$. A reflection argument in conjunction with the Korn-Privalov theorem leads to $y \in C^{1,\mu}(\overline{S}_\delta(1))$ whence $X^3 \in C^{1,\mu}(\overline{S}_\delta(1))$

for $0 < \delta < \varepsilon/2$, and therefore $X \in C^{1,\mu}(\overline{S}_\delta(1), \mathbb{R}^3)$ for any $\mu \in (0,1)$. Thus part (i) of Theorem 2 is proved.

(ii) The further proof is routine. One has alternatingly to consider the two problems (with a suitable Q):

$$\Delta Y = Q \text{ in } S_R(1), \quad Y = 0 \text{ on } \partial B \cap S_R(1)$$

and

$$\Delta X^3 = 2H(X)(X_u^1 X_v^2 - X_v^1 X_u^2) \text{ in } S_R(1)$$

$$X_r^3 = -\sum_{k=1}^{2} p^k(X^3) Y_r^k \text{ on } \partial B \cap S_R(1).$$

□

Finally, we state Lewy's regularity theorem.

Theorem 3. *Let $X \in C^0(B \cup I, \mathbb{R}^3) \cap C^2(B, \mathbb{R}^3)$ be a minimal surface which maps I into a real-analytic and regular Jordan arc Γ in $\mathbb{R}^3$. Then X can be extended analytically across I as a minimal surface.*

Proof. Let $X^*(w)$ be an adjoint minimal surface to $X(w)$ which is assumed to satisfy $\Delta X = 0$ and (2) in B, and let

$$f(w) = X(w) + iX^*(w) = (f^1(w), f^2(w), f^3(w))$$

be the holomorphic curve in $\mathbb{C}^3$ with $X = \operatorname{Re} f$ and $X^* = \operatorname{Im} f$ satisfying

$$f'(w) \cdot f'(w) = 0.$$

We can assume that B is the semi-disk $\{w \in \mathbb{C} : |w| < 1, \operatorname{Im} w > 0\}$, and that I is the interval $\{u \in \mathbb{R} : -1 < u < 1\}$ on the real axis.

By Theorem 2 we know already that $X \in C^2(B \cup I, \mathbb{R}^3)$ holds true. We have to show that for any $u_0 \in I$ there is some $\delta > 0$ such that $f(w)$ can be extended as a holomorphic curve from $S_\delta(u_0) := B \cap B_\delta(u_0)$ to $B_\delta(u_0)$. Without loss of generality we can assume that $u_0 = 0$. Set $B_\delta := B_\delta(0)$ and $S_\delta = S_\delta(0) = B \cap B_\delta$. As in the proof of Theorem 2 we can arrange for the following:

$$X^1(0) = X^2(0) = X^3(0) = 0, \text{ i.e., } X(0) = 0.$$

In the neighbourhood of the origin 0 we can Γ represent in the form

$$x^1 = g^1(t), \quad x^2 = g^2(t), \quad x^3 = t, \qquad t \in I_{2R_0}$$

where $I_\delta := \{t \in \mathbb{R} : | t | < \delta\}$, and $g^1(t)$ and $g^2(t)$ are real analytic functions on $I_{2R_0}, R_0 > 0$. Hence, choosing R_0 sufficiently small, we can assume that $g^1(\zeta)$ and $g^2(\zeta)$ are holomorphic functions of $\zeta \in B_{2R_0}$; hence

$$g(\zeta) := (g^1(\zeta), g^2(\zeta), \zeta), \qquad |\zeta| < 2R_0$$

is a holomorphic curve on B_{2R_0}. In addition, we may assume that

$$g^1(0) = g^2(0) = \dot{g}^1(0) = \dot{g}^2(0) = 0$$

and

$$\left|\frac{dg^1}{d\zeta}(\zeta)\right|^2 + \left|\frac{dg^2}{d\zeta}(\zeta)\right|^2 \leq \frac{1}{2} \text{ for } |\zeta| < 2R_0$$

are satisfied, and that

$$|f^3(w)| < R_0 \text{ holds for } w \in \overline{S}_R$$

where R is a sufficiently small positive number.

Consider now the holomorphic function

$$F(w,\zeta) := \frac{g'(\zeta) \cdot f'(w)}{g'(\zeta) \cdot g'(\zeta)}$$

of $(w,\zeta) \in S_R \times B_{R_0}$, and note that F is of class C^1 on $\overline{S}_R \times \overline{B}_{R_0}$ (of course, differentiability in the second statement is real differentiability).

We claim that the differential equation

$$X_u^3(u) = F(u, X^3(u)) \quad \text{for } u \in I_R \tag{6}$$

holds true. In fact, the boundary condition $X(I) \subset \Gamma$ together with the above normalization implies that

$$X(u) = g(X^3(u)) \quad \text{for } u \in I_R\,, \tag{7}$$

hence

$$X_u(u) = g'(X^3(u))X_u^3(u) \quad \text{for } u \in I_R\,. \tag{8}$$

Therefore $X_u^3(u) = 0$ for some $u \in I_R$ yields $X_u(u) = 0$, and (2) gives $X_v(u) = 0$ or $X_u^*(u) = 0$ and consequently $f'(u) = 0$; therefore we also have $F(u, X^3(u)) = 0$, and (6) is trivially satisfied. Thus we may now assume that $X_u^3(u) \neq 0$. Because of (8) and (2), we obtain on I_R:

$$\begin{aligned} X_u^3 g'(X^3) \cdot f' &= g'(X^3)X_u^3 \cdot f' = X_u \cdot f' \\ &= X_u \cdot (X_u + iX_u^*) = |X_u|^2 - iX_u \cdot X_v = |X_u|^2 \\ &= g'(X^3) \cdot g'(X^3)(X_u^3)^2 \end{aligned}$$

whence

$$X_u^3(u) = \frac{g'(X^3) \cdot f'}{g'(X^3) \cdot g'(X^3)}(u) = F(u, X^3(u))\,,$$

and (6) is verified. Thus $\zeta(u) = X^3(u)$ is a solution of the integral equation

$$\zeta(u) = \int_0^u F(\underline{u}, \zeta(\underline{u}))d\underline{u}\,. \tag{9}$$

It can easily be shown that there is some constant $M > 0$ such that

$$|F(w,\zeta) - F(w,\zeta')| \leq M|\zeta - \zeta'|$$

holds for all $w \in \overline{S}_R$ and $\zeta, \zeta' \in \overline{B}_{R_0}$. Then it follows from a standard fixed point argument that there is a number $\delta \in (0, R)$ such that the integral equation

$$z(w) = \int_0^w F(\underline{w}, z(\underline{w}))d\underline{w}\,, \qquad w \in \overline{S}_\delta\,,$$

has exactly one solution $z(w), w \in \overline{S}_\delta$, in the Banach space $\mathcal{A}(\overline{S}_\delta)$ of functions $z : \overline{S}_\delta \to \mathbb{C}$ which are holomorphic in S_δ and continuous on $\overline{S}_\delta$. (As usual, the proof of this fact can easily be carried out by Picard's iteration method.) Similarly one sees that the real integral equation (9) has (for $u \in \overline{I}_\delta$) exactly one solution $\zeta(u), u \in \overline{I}_\delta$, whence we obtain $\zeta(u) = X^3(u)$ and $\zeta(u) = z(u)$ for $| u | \le \delta$, that is,

$$z(u) = X^3(u) \quad \text{for } u \in \overline{I}_\delta\,.$$

Consequently $z(w)$ is real-valued on I_δ, and by Schwarz's reflection principle we can extend $z(w)$ to a holomorphic function in B_δ.

Now we consider the mapping $\phi : \overline{S}_\delta \to \mathbb{C}^3$, defined by

$$\phi(w) := f(w) - g(z(w))\,,$$

which is continuous on $\overline{S}_\delta$, holomorphic in S_δ, and purely imaginary on I_δ, since we have

$$\phi(u) = f(u) - g(z(u)) = X(u) + iX^*(u) - X(u) = iX^*(u)$$

on account of (7). Applying the reflection principle once again, we can extend $\phi(w)$ to a holomorphic function on B_δ, and therefore also

$$f(w) = \phi(w) + g(z(w))$$

is extended to a holomorphic mapping on B_δ. □

Historical Remarks and References to the Literature

The first results concerning the boundary behaviour of minimal surfaces are the *reflection principles* of Schwarz. They guarantee that a minimal surface can be extended analytically (as a minimal surface) across any straight part Γ_0 of its boundary by reflection in Γ_0, or across any part of its boundary where the surface meets some plane E_0 perpendicularly by reflection in E_0.

Another early result, based on a celebrated theorem by F. and M. Riesz from 1916, is Tsuji's theorem, stating that any minimal surface X of class $H_2^1(B, \mathbb{R}^3)$ has boundary values of class $H_1^1(\partial B, \mathbb{R}^3)$ if $\int_{\partial B} |dX| < \infty$ (see Tsuji [1]).

Theorem 3 is Lewy's celebrated regularity theorem from 1951. Lewy did not use Tsuji's result nor any other intermediate result like Theorem 1, as such results were unknown at that time ; cf. Lewy [2].

Preliminary results in direction of Theorem 1 were derived by Hildebrandt [1] in 1966, thereby proving that every $\Gamma \in C^{m,\mu}, m \ge 4, \mu \in (0,1)$ bounds at least one minimal surface X of class $C^{m,\mu}(\overline{B}, \mathbb{R}^3)$. In 1968, Hildebrandt [3] proved Theorem 1 for $m \geq 4$. Briefly thereafter, Heinz and Tomi [1] succeeded in establishing the result under the hypothesis $m \geq 3$, and both Nitsche [7] and Kinderlehrer [1] provided the final result for $m \geq 1$. Warschawski [6] verified that X has Dini-continuous first derivatives on $\overline{B}$, if the first derivatives of Γ with respect to arc length are Dini continuous; cf. also Lesley [1].

The method used in proving Theorem 2 and thereby Theorem 1 (for $m \geq 2$) is due to Heinz [2].

3.3 Boundary Regularity of Minimal Surfaces at Free Boundaries

The study of the boundary behaviour of minimal surfaces at free boundaries is somewhat more involved, and so we shall only state some of the major results, without proving them. There are several new phenomena which complicate the study of minimal surfaces with a free boundary. Firstly, Courant pointed out that there might be smooth (noncompact) support surfaces S bounding stationary minimal surfaces whose trace is discontinuous (see DHKW [1], vol.2, pp. 43–45). This phenomenon can, however, be excluded (at least for minimizers of the area) if we assume that S suffices a *chord-arc condition*, i.e. a kind of Lipschitz condition. We, in fact, have the following result:

(i) *Suppose that X minimizes Dirichlet's integral in the class $\mathcal{C}(\Gamma, S)$. Assume also that the support surface S satisfies a chord-arc condition. Then X is of class $C^{0,\mu}(B \cup I, \mathbb{R}^3)$ for some $\mu \in (0,1)$.*

Presently it is not known whether an analogous result holds for *stationary* minimal surfaces within a configuration whose support surfaces merely satisfy a chord-arc condition. However for smooth support surfaces one obtains satisfactory results, as we have

(ii) *Let S be an admissible support surface of class C^2, and suppose that X is a stationary point of Dirichlet's integral in the class $\mathcal{C}(\Gamma, S)$. Then there is some $\alpha \in (0,1)$ such that $X \in C^{0,\alpha}(B \cup I, \mathbb{R}^3)$.*

It turns out that for non-minimizing stationary surfaces one can in general not derive a priori estimates depending only on geometric data of their boundary configurations. Thus it is not surprising that one has to choose a completely different method to treat non-minimizers. In fact, the approach is indirect and will be based on *monotonicity formulas*.

One of the main results is

(iii) *Let S be an admissible support surface with $\partial S = \emptyset$ which is of class $C^{m,\beta}, m \geq 3, \beta \in (0,1)$. Then any stationary point X of Dirichlet's integral* *in* $\mathcal{C}(\Gamma, S)$ *is of class* $C^{m,\beta}(B \cup I, \mathbb{R}^3)$. *If* S *is real analytic then* X *is real analytic on* $B \cup I$ *and can be continued analytically across* I.

The analyticity result in (iii) is most easily proved by a method due to H. Lewy [1]. It is closely related to the proof of Theorem 3 in Section 3.2.

For support surfaces S with nonempty boundary, one can prove:

(iv) *If* S *is an admissible support surface of class* C^4 *with* $\partial S \in C^4$, *and if* X *is a stationary point of Dirichlet's integral in* $\mathcal{C}(\Gamma, S)$, *then* X *is of class* $C^{1,1/2}(B \cup I, \mathbb{R}^3)$.

One can show that the Hölder exponent $\mu = 1/2$ for ∇X is, in general, best possible, as one can infer from specific examples.

For a detailed discussion of the results (i)–(iv) we refer the reader to Chapter 7 in DHKW [1], vol. 2. There one also finds a brief historical account (pp. 130–132).

It might be useful to give a precise definition of the chord-arc condition used in the result (i).

Definition. *A set* S *in* $\mathbb{R}^3$ *is said to fulfil a chord-arc condition with constants* M *and* $\delta, M \geq 1$ *and* $\delta > 0$, *if it is closed and if any two points* P *and* Q *of* S *with a distance* $|P - Q|$ *less than or equal to* δ *can be connected in* S *by a rectifiable arc* Γ^* *whose length* $L(\Gamma^*)$ *satisfies*

$$L(\Gamma^*) \leq M|P - Q|\,.$$

It is also of great interest to determine the topological and analytical nature of the set of coincidence ("touching set") of a minimal surface subject to an obstacle constraint. In this regard the most penetrating result is due to Kinderlehrer [6] (cf. also [8]).

3.4 Asymptotic Expansions of Minimal Surfaces at Boundary Branch Points

We have seen that a minimal surface $X : B \to \mathbb{R}^3$ can be extended analytically and as a minimal surface across those parts of ∂B which are mapped by X into an analytic arc or which correspond to a free trace on an analytic support surface. At a branch point of such a part of ∂B, the minimal surface X, therefore, possesses an asymptotic expansion as described in Section 1.3. In this section we want to state an analogous expansion of X at boundary branch points, assuming merely that Γ or S are of some appropriate class C^m. Using an asymptotic expansion due to Hartman-Wintner (as modified by E. Heinz), we obtain

Theorem 1. *Let* S *be an admissible support surface of class* C^3 *and* X *be a stationary point of Dirichlet's integral in the class* $\mathcal{C}(\Gamma, S)$. *Assume also that*

$w_0 \in I$ is a boundary branch point of X. Then there exists an integer $\nu \geq 1$ and a vector $A \in \mathbb{C}^3$ with $A \neq 0$ and

$$A \cdot A = 0 \tag{1}$$

such that

$$X_w(w) = A(w - w_0)^\nu + o(| w - w_0 |^\nu) \text{ as } w \to w_0 . \tag{2}$$

We call ν the order of the branch point w_0.

From this expansion we can draw the same geometric conclusions as in Section 1.3. To this end we write

$$A = \frac{1}{2}(\alpha - i\beta) \text{ with } \alpha, \beta \in \mathbb{R}^3 .$$

Then it follows that

$$| \alpha | = | \beta | \neq 0 , \qquad \alpha \cdot \beta = 0$$

and

$$\begin{aligned} X_u(w) &= \alpha \operatorname{Re}(w - w_0)^\nu + \beta \operatorname{Im}(w - w_0)^\nu + o(|w - w_0|^\nu) \\ X_v(w) &= -\alpha \operatorname{Im}(w - w_0)^\nu + \beta \operatorname{Re}(w - w_0)^\nu + o(|w - w_0|^\nu) \end{aligned} \tag{3}$$

as $w \to w_0$, whence

$$X_u(w) \wedge X_v(w) = (\alpha \wedge \beta)|w - w_0|^{2\nu} + o(|w - w_0|^{2\nu}) \text{ as } w \to w_0 .$$

This implies that the surface normal $N(w)$, given by

$$N = | X_u \wedge X_v |^{-1} (X_u \wedge X_v) ,$$

tends to a limit vector N_0 as $w \to w_0$:

$$\lim_{w \to w_0} N(w) = N_0 = | \alpha \wedge \beta |^{-1} (\alpha \wedge \beta) . \tag{4}$$

Consequently, the Gauss map $N(w)$ of a stationary minimal surface $X(w)$ is well defined on all of $B \cup I$ as a continuous mapping into S^2. Therefore *the surface $X(w)$ has a well defined tangent plane at every boundary branch point on I*, and thus at every point $w_0 \in B \cup I$.

Consider now the trace curve $X : I \to \mathbb{R}^3$ of the minimal surface X on the supporting surface S. In order to simplify computations, we assume without loss of generality that I is an interval on the real axis containing w_0 in its interior. Then we can infer from (3) that

$$X_u(w) = \alpha(w - w_0)^\nu + o(| w - w_0 |^\nu) \text{ as } w \to w_0 , \quad w, w_0 \in I ,$$

and writing $w = u, w_0 = u_0$ for $w, w_0 \in I$ we obtain for the unit tangent vector

$$t(u) := |X_u(u)|^{-1} X_u(u)$$

the expansion

$$t(u) = \frac{\alpha(u-u_0)^\nu}{|\alpha||u-u_0|^\nu} + o(1) \quad \text{as } u \to u_0 \,. \tag{5}$$

Therefore *the nonoriented tangent moves continuously through any boundary branch point $u_0 \in I$. The oriented tangent $t(u)$ is continuous if the order ν of u_0 is even, but, for branch points of odd order, the direction of $t(u)$ jumps by 180 degrees when u passes through u_0.*

Finally, by choosing a suitable cartesian coordinate system in $\mathbb{R}^3$, we obtain the expansion

$$\begin{aligned} x(w) + iy(w) &= (x_0 + iy_0) + a(w-w_0)^{\nu+1} + o(|w-w_0|^{\nu+1}) \\ z(w) &= z_0 + o(|w-w_0|^{\nu+1}) \end{aligned} \tag{6}$$

as $w \to w_0$, where $X(w_0) = (x_0, y_0, z_0)$ and $a > 0$; see Section 1.3.

The same reasoning can be used for the investigation of X at a boundary branch point $w_0 \in \text{int}\, C$. We obtain again an expansion of the kind (2) with some $\nu \geq 1$ and some $A \in \mathbb{C}^3$, $A \neq 0$, $A \cdot A = 0$. As $X : C \to \Gamma$ is a monotonic mapping, *the tangent vector*

$$t(\varphi) := |X_\varphi(e^{i\varphi})|^{-1} X_\varphi(e^{i\varphi})$$

of this mapping has to be continuous, and we infer from (2) *that ν is even,* provided that Γ is of class C^2.

The same result can be proved for minimal surfaces $X \in \mathcal{C}(\Gamma)$ which solve Plateau's problem for a closed Jordan curve Γ of class C^2 in $\mathbb{R}^3$. Thus we obtain

Theorem 2. *Let Γ be a closed Jordan curve of class C^2, and suppose that $X \in \mathcal{C}(\Gamma)$ is a minimal surface spanning Γ. Then every boundary branch point $w_0 \in \partial B$ is of even order $\nu = 2\rho$, $\rho \leq 1$, and we have the asymptotic expansion*

$$X_w(w) = A(w-w_0)^{2\rho} + o(|w-w_0|^{2\rho}) \quad \text{as } w \to w_0 \,,$$

where $A \in \mathbb{C}^3$, $A \neq 0$, and $A \cdot A = 0$.

For a detailed discussion we refer to DHKW [1], vol. 2, Chapters 7 and 8, as well to the original papers by Hartman-Wintner [1], Heinz [3], Heinz-Tomi [1], and Hildebrandt-Nitsche [1].

3.5 Geometric Estimates for Minimal Surfaces. The Number of Branch Points

Often one can derive rough estimates on minimal surfaces in the form of *enclosure theorems*. In many cases these results can be viewed as a more or less sophisticated form of the maximum principle for harmonic functions or, more generally, for solutions of elliptic differential equations. The simplest result of this kind is the *convex hull theorem* which is stated as

Theorem 1. *Suppose that $X : \Omega \to \mathbb{R}^3$ is a harmonic mapping of an open, bounded, connected domain Ω of $\mathbb{R}^2$ into $\mathbb{R}^3$ which is continuous on $\overline{\Omega}$. Then $X(\overline{\Omega})$ is contained in the convex hull of its boundary values $X(\partial\Omega)$.*

Proof. For any constant vector A in $\mathbb{R}^3$, the function $h := A \cdot X$ is continuous on $\overline{\Omega}$ and harmonic in Ω. Suppose that $X(\partial\Omega)$ is contained in the halfspace $\mathcal{H} = \{Z \in \mathbb{R}^3 : A \cdot Z \leq d\}, d \in \mathbb{R}$. Then $h(w) \leq d$ for all $w \in \partial\Omega$, and the maximum principle implies $h(w) \leq d$ for all $w \in \overline{\Omega}$. Since any closed convex set is the intersection of its supporting halfspaces, the assertion is proved.

Remark 1. In particular any continuous minimal surface lies in the convex hull of its boundary.

Since a harmonic function assumes an interior maximum if and only if it is constant, the above reasoning shows that if a connected minimal surface $X : \Omega \to \mathbb{R}^3$ touches the boundary $\partial\mathcal{K}$ of a closed convex set $\mathcal{K}$ containing $X(\Omega)$ at some point $X(w_0)$, then either $w_0 \in \partial\Omega$, or X is a planar surface.

To formulate the next result we introduce the cone $\mathcal{K}_0 = \{(x, y, z) \in \mathbb{R}^3 : x^2 + y^2 - z^2 \leq 0\}$ and the two halfcones $\mathcal{K}_0^+ = \mathcal{K} \cap \{z \geq 0\}$ and $\mathcal{K}_0^- = \mathcal{K}_0 \cap \{z \leq 0\}$ intersecting at the vertex 0.

Theorem 2. *Let $\mathcal{K} = \mathcal{K}^+ \cup \mathcal{K}^-$ be a cone such that $\mathcal{K}, \mathcal{K}^+, \mathcal{K}^-$ are congruent to $\mathcal{K}_0, \mathcal{K}_0^+, \mathcal{K}_0^-$ respectively, and suppose that $X : \Omega \to \mathbb{R}^3, \Omega$ bounded, is a minimal surface which is continuous and satisfies $X(\partial\Omega) \subset \mathcal{K}$. Then either Ω is disconnected, or $X(\overline{\Omega})$ lies in exactly one of the two half-cones $\mathcal{K}^+, \mathcal{K}^-$.*

Proof. Let $f(x, y, z) = x^2 + y^2 - z^2$, and set $h := f \circ X$. Then

$$\Delta h = \nabla X \cdot D^2 f(X) \nabla X + Df(X) \cdot \Delta X$$

Since $\Delta X = 0$, we find that $\Delta h = 2(|\nabla x|^2 + |\nabla y|^2 - |\nabla z|^2) = 8(|x_w|^2 + |y_w|^2 - |z_w|^2)$. But from $X_w \cdot X_w = 0$, we have $|z_w|^2 \leq |x_w|^2 + |y_w|^2$, and therefore $\Delta h \geq 0$. We can assume that $\mathcal{K} = \mathcal{K}_0$ whence we have $h(w) \leq 0$ for $w \in \partial\Omega$, and thus the maximum principle implies $h \leq 0$, i.e. $X(\overline{\Omega}) \subset \mathcal{K}$. The asymptotic expansion of $X(w)$ at any $w_0 \in \Omega$ shows that $X(w)$ cannot pass through the vertex of $\mathcal{K}$. Therefore Ω is disconnected, or $X(\Omega)$ is contained in one of the half-cones $\mathcal{K}^+, \mathcal{K}^-$. □

There are many refinements of these results. For example, Osserman and Schiffer [1] have proved that $\mathcal{K}_0^\pm$ in Theorem 2 can be replaced by $\mathcal{C}_0^\pm = \{(x,y,z) : x^2+y^2 \le (\sinh^2\tau)z^2, z \ge 0 \text{ or } z \le 0\}$ where $\tau \sim 1.9968$ is the solution of the equation $\tau \sinh\ \tau = \cosh\ \tau$. The aperture angle of $\mathcal{K}_0$ is 45°, whereas that of $\mathcal{C}_0 = \mathcal{C}_0^+ \cup \mathcal{C}_0^-$ is about 56.47 degrees. The Osserman-Schiffer cone $\mathcal{C}_0$ is optimal in the following sense: for any larger aperture, there exists a part of a catenoid with boundary circles inside the two nappes of the corresponding cone.

The following notation has become customary: A closed set $\mathcal{K}$ in $\mathbb{R}^3$ is said to be *H-convex* if for any $P \in \partial\mathcal{K}$ there is a locally supporting minimal surface $\mathcal{M}$, i.e. $P \in \mathcal{M}$, and $\mathcal{K} \cap B_\varepsilon(P)$ lies on one side of $\mathcal{M} \cap B_\varepsilon(P)$ for $0 < \varepsilon << 1$. If $\partial\mathcal{K}$ is a regular C^2-surface then H-convexity of $\mathcal{K}$ means that the mean curvature of $\partial\mathcal{K}$ with respect to the inward normal is nonnegative.

Then one can prove the following *enclosure theorem* (see e.g. [DHKW], vol. 1, Section 6.2, Osserman-Gulliver-Royden [1], and references to the literature in Section 6.5, pp. 424–426).

Theorem 3. *Let $\mathcal{K}$ be an H-convex set in $\mathbb{R}^3$, Ω a bounded open set in $\mathbb{R}^2$, and $X : \Omega \to \mathbb{R}^3$ a minimal surface of class $C^0(\overline{\Omega}, \mathbb{R}^3)$ satisfying $X(\Omega) \subset \mathcal{K}$ and $X(w_0) \in \operatorname{int}\mathcal{K}$ for some $w_0 \in \Omega$. Then we have $X(\Omega) \subset \operatorname{int}\mathcal{K}$.*

There are numerous refinements of the maximum principle such as the *strong halfspace theorem* (cf. Hoffman-Meeks [1]) or *maximum principles at infinity* (see e.g. Langevin-Rosenberg [1], Kuwert [2], Ripoll-Tomi [1]).

Another kind of geometric a priori estimates for minimal surfaces are the so-called *isoperimetric inequalities*, estimating the area of a minimal surface by the length of its boundary. The prototype of such results is

Theorem 4. *Let $X : B \to \mathbb{R}^3$ be a minimal surface of class $C^0(\overline{B}, \mathbb{R}^3), B = \{w \in \mathbb{C} : |w| < 1\}$, such that $L(\partial X) = \int_C \mid dX \mid < \infty$. Then $D(X) = A(X)$ is finite, and*

$$A(X) \le \frac{1}{4\pi} L^2(\partial X) \tag{1}$$

The equality sign holds if and only if X represents a (simply covered) disk.

A proof of this result as well as refinements can be found in [DHKW], vol. 1, Section 6.3.

By means of an integration by parts one can estimate the length $L(\Sigma)$ of the free trace Σ of a minimal surface satisfying free or partially free boundary conditions. The following result was proved in a sequence of papers by Hildebrandt-Nitsche [4], Küster [1], and Dziuk [8] (see DHKW, vol. 1, Section 6.4). To formulate the result we introduce the following notation: A surface S in $\mathbb{R}^2$ is said to satisfy an *R-sphere condition* if it is a C^2-submanifold of $\mathbb{R}^3$ which is the boundary of an open set U of $\mathbb{R}^3$ that at each of its points P can be touched from both sides by spheres of radius R meeting S only at P.

Theorem 5. *Let S by a smooth support surface satisfying an R-sphere condition, and Γ a rectifiable arc of length $L(\Gamma)$ meeting S only at its endpoints.*

Then the length $L(\Sigma) = \int_I |dX|$ of the free trace Σ (given by $X : I \to \mathbb{R}^3$, $I \subset \partial B$) of a minimal surface $X : B \to \mathbb{R}^3$ stationary in the configuration $< \Gamma, S >$ can be estimated by

$$L(\Sigma) \le L(\Gamma) + (2/R)D(X) \tag{2}$$

where $D(X)$ is the Dirichlet integral of X. This estimate is optimal in the sense that in general the number 2 cannot be replaced by any smaller number.

Finally we want to mention that the Gauss-Bonnet theorem can be used to estimate the number of branch points of a (generalized) minimal surface $X : \Omega \to \mathbb{R}^3$. We just formulate one typical result; for a more comprehensive presentation we refer to [DHKW], vol. 2, Section 7.11. The possibility to estimate the number of branch points of a solution to Plateau's problem was discovered by Nitsche [2], reversing an argument due to Sasaki [1].

Theorem 6. *Let Ω be an m-fold connected, bounded domain in $\mathbb{C}$ whose boundary consists of m closed regular disjoint curves $\gamma_1, \dots \gamma_m$. Secondly, let*

$$X \in C^0(\overline{\Omega}, \mathbb{R}^3) \cap C^2(\Omega, \mathbb{R}^3)$$

be a minimal surface on Ω which maps the γ_j topologically onto closed regular and disjoint Jordan curves $\Gamma_j, 1 \le j \le m$, of class $C^{2,\alpha}, 0 < \alpha < 1$, with the curvature κ. Then $\int_\Omega |K| dA$ and $\int_{\partial\Omega} \kappa_g |dX| = \int_\Gamma \kappa_g ds$ are finite, and we have

$$\int_\Omega K dA + \int_\Gamma \kappa_g ds = 2\pi(2-m) + 2\pi \sum_{w\in\sigma'} \nu(w) + \pi \sum_{w\in\sigma''} \nu(w) \tag{3}$$

where $ds = |dX|$ is the line element of $\Gamma := \bigcup_{j=1}^{m} \Gamma_j$, $\nu(w)$ is the order of a branch point $w \in \Sigma_0$, $\sigma' := \Omega \cap \Sigma_0$, $\sigma'' := \partial\Omega \cap \Sigma_0$, Σ_0 is the set of branch points of X in $\overline{\Omega}$, and κ_g is the geodesic curvature of Γ viewed as a curve on the surface X. In particular, $|\kappa_g| \le \kappa$ and equation (3) implies that

$$2 - m + \sum_{w\in\sigma'} \nu(w) + \frac{1}{2} \sum_{w\in\sigma''} \nu(w) \le \frac{1}{2\pi} \int_\Gamma \kappa ds\,. \tag{4}$$

References *

We particularly refer the reader to the following Lecture notes:

Minimal submanifolds and geodesics. Proceedings of the Japan-United States Seminar on Minimal Submanifolds, including Geodesics, Tokyo 1977, Kagai Publications, Tokyo 1978, Zbl. 423.0012

Seminar on Differential Geometry, edited by S.T. Yau. Ann. Math. Studies **102**, Princeton 1982, 471.00020

Seminar on Minimal Submanifolds, edited by E. Bombieri. Ann. Math. Studies **103**, Princeton 1983, Zbl. 521.00014

Théorie des variétés minimales et applications. Séminaire Palaiseau, oct. 1983– juin 1984. Astérisque **154–155** (1988), Zbl. 635.53001

Geometric Analysis and Computer Graphics. Proceedings of the Conference on Differential Geometry, Calculus of Variations and Computer Graphics, edited by P. Concus, R. Finn, D.A. Hoffman. Math. Sci. Res. Inst. Conference Proc. Series Nr. **17**, Springer, New York 1991, Zbl. 776.68009

The treatise of Dierkes/Hildebrandt/Küster/Wohlrab is abbreviated as DHKW.

Agmon, S., Douglis, A., Nirenberg, L.
1. Estimates near the boundary for solutions of elliptic differential equations satisfying general boundary conditions I. Commun. Pure Appl. Math. **12**, 623–727 (1959), Zbl. 93, 104
2. Estimates near the boundary for solutions of elliptic differential equations satisfying general boundary conditions II. Commun. Pure Appl. Math. **17**, 35–92 (1964), Zbl. 123, 287

Alexander, H., Hoffman, D., Osserman, R.
1. Area estimates for submanifolds of euclidean space. INDAM Sympos. Math. **14**, 445–455 (1974), Zbl. 326.53029

Alexander, H., Osserman, R.
1. Area bounds for various classes of surfaces. Am. J. Math. **97**, 753–769 (1975), Zbl. 326.53030

Almgren, F.J.
1. Some interior regularity theorems for minimal surfaces and an extension of Bernstein's theorem. Ann. Math., II. Ser., **84**, 277–292 (1966), Zbl. 146, 119
2. Plateau's Problem. An Invitation to Varifold Geometry. Benjamin, New York 1966, Zbl. 165, 132
3. Minimal surface forms. Math. Intelligencer **4**, 164–172 (1982), Zbl. 492.53003

Almgren, F.J., Simon, L.
1. Existence of embedded solutions of Plateau's problem. Ann. Sc. Norm. Super. Pisa Cl. Sci., IV. Ser., **6**, 447–495 (1979), Zbl. 417.49051

Almgren, F.J., Taylor, J.E.
1. The geometry of soap films and soap bubbles. Scientific American 82–93 (1976)

---

*For the convenience of the reader, references to reviews in Zentralblatt für Mathematik (Zbl.), compiled by means of the MATH database, and Jahrbuch über die Fortschritte der Mathematik (Jbuch) have, as far as possible, been included in this bibliography by the publisher.

Almgren, F.J., Thurston, W.P.
1. Examples of unknotted curves which bound only surfaces of high genus within their convex hulls. Ann. Math., II. Ser., **105**, 527–538 (1977), Zbl. 353.53001

Alt, H.W.
1. Verzweigungspunkte von H-Flächen. I. Math. Z. **127**, 333–362 (1972), Zbl. 253.58007
2. Verzweigungspunkte von H-Flächen. II. Math. Ann. **201**, 33–55 (1973), Zbl. 257.58005
3. Die Existenz einer Minimalfläche mit freiem Rand vorgeschriebener Länge. Arch. Ration. Mech. Anal. **51**, 304–320 (1973), Zbl. 282.49039

Alt, H.W., Tomi, F.
1. Regularity and finiteness of solutions to the free boundary problem for minimal surfaces. Math. Z. **189**, 227–237 (1985), Zbl. 583.53004

Bakker, G.
1. Kapillarität und Oberflächenspannung. Handbuch der Experimentalphysik, Band 6. Akad. Verlagsges., Leipzig 1928

Barbosa, J.L., do Carmo, M.P.
1. On the size of a stable minimal surface in $\mathbb{R}^3$. Am. J. Math. **98**, 515–528 (1976), Zbl. 332.53006
2. Stability of minimal surfaces and eigenvalues of the Laplacian. Math. Z. **173**, 13–28 (1980), Zbl. 432.53029

Barbosa, J.L., Colares, A.G.
1. Minimal surfaces in $\mathbb{R}^3$. Lect. Notes Math. **1195**. Springer, Berlin Heidelberg New York 1986, Zbl. 609.53001

Beckenbach, E.F.
1. The area and boundary of minimal surfaces. Ann. Math., II. Ser., **33**, 658–664 (1932), Zbl. 5, 262
2. Minimal surfaces in Euclidean n-space. Am. J. Math. **55**, 458–468 (1933), Zbl. 7, 230
3. Bloch's theorem for minimal surfaces. Bull. Am. Math. Soc. **39**, 450–456 (1933), Zbl. 7, 130
4. Remarks on the problem of Plateau. Duke Math. J. **3**, 676–681 (1937), Zbl. 18, 221
5. On a theorem of Fejér and Riesz. J. Lond. Math. Soc. **13**, 82–86 (1938), Zbl. 19, 67
6. Functions having subharmonic logarithms. Duke Math. J. **8**, 393–400 (1941), Zbl. 25, 266
7. The analytic prolongation of a minimal surface. Duke Math. J. **9**, 109–111 (1942)
8. Painlevé's theorem and the analytic prolongation of minimal surfaces. Ann. Math., II. Ser., **46**, 667–673 (1945)

Beckenbach, E.F., Radó, T.
1. Subharmonic functions and minimal surfaces. Trans. Am. Math. Soc. **35**, 648–661 (1933), Zbl. 7, 129
2. Subharmonic functions and surfaces of negative curvature. Trans. Am. Math. Soc. **35**, 662–664 (1933), Zbl. 7, 130

Beeson, M.
1. The behavior of a minimal surface in a corner. Arch. Ration. Mech. Anal. **65**, 379–393 (1977), Zbl. 371.53003
2. On interior branch points of minimal surfaces. Math. Z. **171**, 133–154 (1980), Zbl. 422.53003
3. Some results on finiteness in Plateau's problem, Part I. Math. Z. **175**, 103–123 (1980), Zbl. 437.53006

4. Some results on finiteness in Plateau's problem, Part II. Math. Z. **181**, 1–30 (1980), Zbl. 487.53007
5. The 6π theorem about minimal surfaces. Pac. J. Math. **117**, 17–25 (1985), Zbl. 554.53004

Beeson, M.J., Tromba, A.J.
1. The cusp catastrophe of Thom in the bifurcation of minimal surfaces. Manuscr. Math. **46**, 273–308 (1984), Zbl. 551.58005

Bernstein, S.
1. Sur un théorème de géométrie et ses applications aux équations aux dérivées partielles du type elliptique. Comm. de la Soc. Math. de Kharkov (2-ème sér.) **15**, 38–45 (1915–1917). [German transl.: Math. Z. **26**, 551–558 (1927): Über ein geometrisches Theorem und seine Anwendung auf die partiellen Differentialgleichungen vom elliptischen Typus. Jbuch 53, 670]

Bers, L.
1. Isolated singularities of minimal surfaces. Ann. Math., II. Ser., **53**, 364–386 (1951), Zbl. 43, 159
2. Boundary value problems for minimal surfaces with singularities at infinity. Trans. Am. Math. Soc. **70**, 465–491 (1951), Zbl. 43, 159

Björling, E.G.
1. In integrationem aequationis derivatarum partialum superficiei, cujus in puncto unoquoque principales ambo radii curvedinis aequales sunt signoque contrario. Arch. Math. Phys. (1), **4**, 290–315 (1844)

Böhme, R.
1. Die Zusammenhangskomponenten der Lösungen analytischer Plateauprobleme. Math. Z. **133**, 31–40 (1973), Zbl. 273.49062
2. Stabilität von Minimalflächen gegen Störung der Randkurve. Habilitationsschrift, Göttingen 1974
3. Die Jacobifelder zu Minimalflächen im $\mathbb{R}^3$. Manuscr. Math. **16**, 51–73 (1975), Zbl. 313.58015
4. Über Stabilität und Isoliertheit der Lösungen des klassischen Plateau-Problems. Math. Z. **158**, 211–243 (1978), Zbl. 396.53004
5. New results on the classical problem of Plateau on the existence of many solutions. Sém. Bourbaki, 34e année, 1981/82, Exp. No. 579, Asterisque 92/93, 1–20, Zbl. 535.58014

Böhme, R., Tomi, F.
1. Zur Struktur der Lösungsmenge des Plateauproblems. Math. Z. **133**, 1–29 (1973), Zbl. 273.49061

Böhme, R., Tromba, A.J.
1. The number of solutions to the classical Plateau problem is generically finite. Bull. Am. Math. Soc. **83**, 1043–1044 (1977), Zbl. 369.49020
2. The index theorem for classical minimal surfaces. Ann. Math., II. Ser., **113**, 447–499 (1981), Zbl. 482.58010

Bombieri, E., de Giorgi, E., Giusti, E.
1. Minimal cones and the Bernstein problem. Invent. Math. **7**, 243–268 (1969), Zbl. 183, 259

Bombieri, E., de Giorgi, E., Miranda, M.
1. Una maggiorazione a priori relativa alle ipersurfici minimali non parametriche. Arch. Ration. Mech. Anal. **32**, 255–267 (1969), Zbl. 184, 328

Büch, J.
1. Bifurkation von Minimalflächen und elementare Katastrophen. Manuscr. Math. **55**, 269–306 (1986), 628.58007

Burago, Yu.D., Zalgaller, V.A.
1. Geometric Inequalities. Nauka, Leningrad, 1980, Zbl. 436.52009. [English transl.: Grundlehren math. Wiss., vol. 285. Springer, Berlin Heidelberg New York London Paris Tokyo 1988, Zbl. 633.53002]

Caffarelli, L.A., Riviere, N.M.
1. Smoothness and analyticity of free boundaries in variational inequalities. Ann. Sc. Norm. Super. Pisa Cl. Sci., IV. Ser., **3**, 289–310 (1976), Zbl. 363.35009

Carathéodory, C.
1. Über die gegenseitige Beziehung der Ränder bei der konformen Abbildung des Inneren einer Jordankurve auf einen Kreis. Math. Ann. **73**, 305–320 (1913), Jbuch 44, 757
2. Conformal Representation. Cambridge University Press, 1932, Zbl. 5, 168, 2nd ed. 1952, Zbl. 47, 79
3. Theory of Functions of a Complex Variable. Vols. 1 and 2. Chelsea, New York 1954, Zbl. 55, 303 and Zbl. 56, 671

Carleman, T.
1. Zur Theorie der Minimalflächen. Math. Z. **9**, 154–160 (1921), Jbuch 48, 590

Chen, Y.-W.
1. Branch points, poles and planar points of minimal surfaces in $\mathbb{R}^3$. Ann. Math., II. Ser., **49**, 790–806 (1948), Zbl. 38, 331
2. Existence of minimal surfaces with a simple pole at infinity and condition of transversality on the surface of a cylinder. Trans. Am. Math. Soc. **65**, 331–347 (1949), Zbl. 40, 374
3. Discontinuity and representations of minimal surface solutions. Proc. Conf. on Diff. Eq., Univ. of Maryland, 115–138 (1956), Zbl. 74, 371

Chern, S.S.
1. An elementary proof of the existence of isothermal parameters on a surface. Proc. Am. Math. Soc. **6**, 771–782 (1955), Zbl. 66, 154
2. Minimal surfaces in a Euclidean space of N dimensions. Differential and Combinatorial Topology, a Symposium in Honor of Marston Morse, Princeton Univ. Press, Princeton 1965, 187–198, Zbl. 136, 167
3. Simple proofs of two theorems on minimal surfaces. Enseign. Math. **15**, 53–61 (1969), Zbl. 175, 186

Chern, S.S., Hartman, P., Wintner, A.
1. On isothermic coordinates. Comment. Math. Helvet. **28**, 301–309 (1954), Zbl. 56, 402

Chern, S.S., Osserman, R.
1. Complete minimal surfaces in Euclidean n-space. J. Anal. Math. **19**, 15–34 (1967), Zbl. 172, 228

Choe, J.
1. On the analytic reflection of a minimal surface. Pac. J. Math. **157**, No. 1, 29–36 (1993), Zbl. 786.53039

Concus, P.
1. Numerical solution of the minimal surface equation. Math. Comput. **21**, 340–350 (1967), Zbl. 186, 166
2. Numerical solution of the minimal surface equation by block nonlinear successive overrelaxation. Inf. Process. Proc. IFIP Congr. Edinb. 1968, **1**, 153–158, 1969, Zbl. 198, 501

3. Numerical study of the discrete minimal surface equation in a nonconvex domain. Report LBL-2033, Lawrence Livermore Laboratory, Berkeley 1973

Courant, R.

1. On the problem of Plateau. Proc. Natl. Acad. Sci. USA **22**, 367–372 (1936), Zbl. 15, 28
2. Plateau's problem and Dirichlet's Principle. Ann. Math, II. Ser., **38**, 679–724 (1937), Zbl. 17, 268
3. The existence of a minimal surface of least area bounded by prescribed Jordan arcs and prescribed surfaces. Proc. Natl. Acad. Sci. USA **24**, 97–101 (1938), Zbl. 18, 221
4. Remarks on Plateau's and Douglas' problem. Proc. Natl. Acad. Sci. USA **24**, 519–523 (1938), Zbl. 20, 136
5. Conformal mapping of multiply connected domains. Duke Math. J. **5**, 314–823 (1939), Zbl. 22, 362
6. The existence of minimal surfaces of given topological structure under prescribed boundary conditions. Acta Math. **72**, 51–98 (1940), Zbl. 23, 399
7. Soap film experiments with minimal surfaces. Am. Math. Mon. **47**, 168–174 (1940), Zbl. 24, 417
8. On a generalized form of Plateau's problem. Trans. Am. Math. Soc. **50**, 40–47 (1941), Zbl. 25, 336
9. Critical points and unstable minimal surfaces. Proc. Natl. Acad. Sci. USA **27**, 51–57 (1941)
10. On the first variation of the Dirichlet-Douglas integral and on the method of gradients. Proc. Natl. Acad. Sci. USA **27**, 242–248 (1941)
11. On Plateau's problem with free boundaries. Proc. Natl. Acad. Sci. USA **31**, 242–246 (1945)
12. Dirichlet's principle, conformal mapping, and minimal surfaces. Interscience, New York 1950, Zbl. 40, 346

Courant, R., Davids, N.

1. Minimal surfaces spanning closed manifolds. Proc. Natl. Acad. Sci. USA **26**, 194–199 (1940)

Courant, R., Hilbert, D.

1. Methoden der mathematischen Physik, Vol. 2. Springer, Berlin 1937, Zbl. 17, 397

Courant, R., Hurwitz, A.

1. Funktionentheorie. Springer, Berlin Heidelberg 1922 (first edition), Jbuch 48, 1207, and 1929 (third edition), Jbuch 55, 171

Courant, R., Robbins, H.

1. What is Mathematics? Oxford University Press, London 1941, Zbl. 60, 123

Croke, C.B.

1. Some isoperimetric inequalities and eigenvalue estimates. Ann. Sci. Ec. Norm. Súper., IV. Ser., **13**, 419–435 (1980), Zbl. 465.53032
2. A sharp four dimensional isoperimetric inequality. Comment. Math. Helvet. **59**, 187–192 (1984), Zbl. 552.53017

Darboux, G.

1. Leçons sur la théorie générale des surfaces et les applications géométriques du calcul infinitesimal. 4 vols. Gauthier-Villars, Paris 1887-96, Jbuch 19, 746 and Jbuch 25, 1159

Davids, N.
1. Minimal surfaces spanning closed manifolds and having prescribed topological position. Am. J. Math. **64**, 348–362 (1942)

De Giorgi, E.
1. Frontiere orientate di misura minima. Semin. Mat. Scuola Norm. Sup. Pisa 1–56, 1961
2. Una estensione del teorema di Bernstein. Ann. Sc. Norm. Super. Pisa Cl. Sci. Fis. Mat. Nat., VIII. Ser., **19**, 79–85 (1965), Zbl. 135, 400

De Giorgi, E., Colombini, F., Piccinini, L.C.
1. Frontiere orientate di misura minima e questioni collegate. Scuola Norm. Sup. Pisa, 1972, Zbl. 296.49031

De Giorgi, E., Stampacchia, G.
1. Sulle singolarità eliminabili delle ipersuperficie minimali. Atti Accad. Naz. Lincei, Rend. Cl. Sci. Fis. Mat. Nat., VIII. Ser., **38**, 352–357 (1965), Zbl. 135, 400

Dierkes, U.
1. Maximum principles and nonexistence results for minimal submanifolds. Manuscr. Math. **69**, 203–218 (1990), Zbl. 709.53043

Dierkes, U., Hildebrandt, S., Lewy, H.
1. On the analyticity of minimal surfaces at movable boundaries of prescribed length. J. Reine Angew. Math. **379**, 100–114 (1987), Zbl. 617.53010

Dierkes, U., Hildebrandt, S., Küster, A., Wohlrab, O.
1. (abbreviated as DHKW) Minimal surfaces I, II. Grundlehren math. Wiss., vols. 295 and 296. Springer, Berlin Heidelberg New York, 1992, Zbl. 777.53012, Zbl. 777.53013

do Carmo, M.
1. Stability of minimal submanifolds. Global differential geometry and global analysis. Lect. Notes Math. **838**, 129–139, Springer, Berlin Heidelberg New York 1981, Zbl. 437.53041

Douglas, J.
1. Reduction of the problem of Plateau to an integral equation. Bull. Am. Math. Soc. **33**, 143–144 (1927), Jbuch 53, 385
2. Reduction of the problem of Plateau to the minimization of a certain functional. Bull. Am. Math. Soc. **34**, 405 (1928), Jbuch 54, 441
3. A method of numerical solution of the problem of Plateau. Ann. Math., II. Ser., **29**, 180–188 (1928), Jbuch 54, 509
4. Solution of the problem of Plateau. Bull. Am. Math. Soc. **35**, 292 (1929), Jbuch 55, 405
5. Various forms of the fundamental functional in the problem of Plateau and its relation to the area functional. Bull. Am. Math. Soc. **36**, 49–50 (1930), Jbuch 56, 441
6. Solutions of the problem of Plateau for any rectifiable contour in n-dimensional Euclidean space. Bull. Am. Math. Soc. **36**, 189 (1930), Jbuch 56, 441
7. Solution of the problem of Plateau when the contour is an arbitrary Jordan curve in n-dimensional euclidean space. I, Bull. Am. Math. Soc. **36**, 189–190 (1930); II, Bull. Am. Math. Soc. **36**, 190 (1930), Jbuch 56, 441
8. The problem of Plateau and the theorem of Osgood-Carathéodory on the conformal mapping of Jordan regions. Bull. Am. Math. Soc. **36**, 190–191 (1930), Jbuch 56, 441
9. A general formulation of the problem of Plateau. Bull. Am. Math. Soc. **36**, 50 (1930), Jbuch 56, 441
10. The mapping theorem of Koebe and the problem of Plateau. J. Math. Phys. **10**, 106–130 (1930–31), Jbuch 57, 605

11. Solution of the problem of Plateau. Trans. Am. Math. Soc. **33**, 263–321 (1931), Zbl. 1, 141
12. The problem of Plateau for two contours. J. Math. Phys. **10**, 315–359 (1931), Zbl. 4, 154
13. The least area property of the minimal surface determined by an arbitrary Jordan contour. Proc. Natl. Acad. Sci. USA **17**, 211–216 (1931), Zbl. 1, 340
14. One-sided minimal surfaces with a given boundary. Trans. Am. Math. Soc. **34**, 731–756 (1932), Zbl. 5, 376
15. Seven theorems in the problem of Plateau. Proc. Natl. Acad. Sci. USA **18**, 83–85 (1932), Zbl. 3, 328
16. The problem of Plateau. Bull. Am. Math. Soc. **39**, 227–251 (1933), Zbl. 7, 119
17. An analytic closed space curve which bounds no orientable surface of finite area. Proc. Natl. Acad. Sci. USA **19**, 448–451 (1933), Zbl. 6, 351
18. A Jordan space curve which bounds no finite simply connected area. Proc. Natl. Acad. Sci. USA **19**, 269–271 (1933), Zbl. 6, 176
19. Crescent-shaped minimal surfaces. Proc. Natl. Acad. Sci. USA **19**, 192–199 (1933), Zbl. 6, 128
20. A Jordan space curve no arc of which can form part of a contour which bounds a finite area. Ann. Math., II. Ser., **35**, 100–104 (1934), Zbl. 9, 24
21. Minimal surfaces of general topological structure with any finite number of assigned boundareis. J. Math. Phys. **15**, 105–123 (1936), Zbl. 15, 28
22. Some new results in the problem of Plateau. J. Math. Phys. **15**, 55-64 (1936), Zbl. 13, 417
23. Remarks on Riemann's doctoral dissertation. Proc. Natl. Acad. Sci. USA **24**, 297–302 (1938), Zbl. 19, 218
24. Minimal surfaces of higher topological structure. Proc. Natl. Acad. Sci. USA **24**, 343–353 (1938), Zbl. 19, 268
25. Green's function and the problem of Plateau. Proc. Natl. Acad. Sci. USA **24**, 353–360 (1938), Zbl. 19, 269
26. The most general form of the problem of Plateau. Proc. Natl. Acad. Sci. USA **24**, 360–364 (1938), Zbl. 19, 269
27. Minimal surfaces of higher topological structure. Ann. Math., II. Ser., **40**, 205–298 (1939), Zbl. 20, 374
28. The higher topological form of Plateau's problem. Ann. Sc. Norm. Super. Pisa Cl. Sci., II. Ser., **8**,195–218 (1939), Zbl. 22, 234
29. Green's function and the problem of Plateau. Am. J. Math. **61**, 545–589 (1939), Zbl. 21, 327
30. The most general form of the problem of Plateau. Am. J. Math. **61**, 590–608 (1939), Zbl. 21, 328

Douglas, J., Franklin, P.

1. A step-polygon of a denumerable infinity of sides which bounds no finite area. Proc. Natl. Acad. Sci. USA **19**, 188–191 (1933), Zbl. 6, 350

Dziuk, G.

1. Das Verhalten von Lösungen semilinearer elliptischer Systeme an Ecken eines Gebietes. Math. Z. **159**, 89–100 (1978), Zbl. 375.35025
2. Das Verhalten von Flächen beschränkter mittlerer Krümmung an C^1-Randkurven. Nachr. Akad. Wiss. Gött., II. Math.-Phys. Kl., 21–28 (1979), Zbl. 433.53003
3. On the boundary behavior of partially free minimal surfaces. Manuscr. Math. **35**, 105–123 (1981), Zbl. 477.49024
4. Über quasilineare elliptische Systeme mit isothermen Parametern an Ecken der Randkurve. Analysis **1**, 63–81 (1981), Zbl. 485.35012
5. Über die Stetigkeit teilweise freier Minimalflächen. Manuscr. Math. **36**, 241–251 (1981), Zbl. 468.49023

6. Über die Glattheit des freien Randes bei Minimalflächen. Habilitationsschrift, Aachen, 1982, Zbl. 533.58009
7. C^2-Regularity for partially free minimal surfaces. Math. Z. **189**, 71–79 (1985), Zbl. 564.49024
8. On the length of the free boundary of a minimal surface. Control Cybern. **14**, 161–170 (1985)

Eells, J., Lemaire, L.
1. A report on harmonic maps. Bull. Lond. Math. Soc. **10**, 1–68 (1978), Zbl. 401.58003
2. Selected topics in harmonic maps. CBMS Regional Conf. Ser. **50** (1983), Zbl. 515.58011
3. Another report on harmonic maps. Bull. Lond. Math. Soc. **20**, 385–524 (1988), Zbl. 669.58009

Eschenburg, J.H.
1. Maximum principle for hypersurfaces. Manuscr. Math. **64**, 55–75 (1989), Zbl. 678.53048

Eschenburg, J.H., Tribuzy, R.
1. Branch points of conformal mappings of surfaces. Math. Ann. **279**, 621–633 (1988), Zbl. 653.53046

Federer, H.
1. Geometric Measure Theory. Grundlehren math. Wiss., vol. 153, Springer, Berlin Heidelberg New York 1969, Zbl. 176, 8
2. Some theorems on integral currents. Trans. Am. Math. Soc. **117**, 43–67 (1965), Zbl. 136, 182
3. The singular sets of area minimizing rectifiable currents with codimension one and of area minimizing flat chains modulo two with arbitrary codimension. Bull. Am. Math. Soc. **76**, 767–771 (1970), Zbl. 194, 358

Feinberg, J.M.
1. The isoperimetric inequality for doubly-connected minimal surfaces in $\mathbb{R}^3$. J. Anal. Math. **32**, 249–278 (1977), Zbl. 387.53002
2. Some Wirtinger-like inequalities. Siam J. Math. Anal. **10**, 1258–1271 (1979), Zbl. 426.26008

Finn, R.
1. Isolated singularities of solutions of nonlinear partial differential equations. Trans. Am. Math. Soc. **75**, 385–404 (1953), Zbl. 53, 392
2. A property of minimal surfaces. Proc. Natl. Acad. Sci. USA **39**, 197–201 (1953), Zbl. 51, 125
3. Equilibrium Capillary Surfaces. Springer, New York Berlin Heidelberg Tokyo 1986, Zbl. 583.35002

Fleming, W.H.
1. An example in the problem of least area. Proc. Am. Math. Soc. **7**, 1063–1074 (1957), Zbl. 78, 138
2. On the oriented Plateau problem. Rend. Circ. Mat. Palermo, II. Ser., **11**, 69–90 (1962), Zbl. 107, 313

Fomenko, A.T.
1. Topological Variational Problems (in Russian). Moscow University 1984, Zbl. 679.49001
2. Symmetries of soap films. Comput. Math. Appl. **12B**, 825–834 (1986)

Fomenko, A.T., Tuzhilin, A.A.
1. Elements of the geometry and topology of minimal surfaces in three-dimensional space. Translations of Mathematical Monographs **93**. Am. Math. Soc. 1993, Zbl. 745.53001

Frehse, J.
1. On the regularity of the solution of a second order variational inequality. Boll. Unione Mat. Ital., IV. Ser., **6**, 312–315 (1972), Zbl. 261.49021
2. On variational inequalities with lower dimensional obstacles. Preprint No. **114**, SFB 72, Bonn 1976
3. On Signorini's problem and variational problems with thin obstacles. Ann. Sc. Norm. Super. Pisa Cl. Sci., IV Ser., **4**, 343–363 (1977), Zbl. 353.49020
4. Un problème variationnel bidimensionnel possédant des extremales bornées et discontinues. C.R. Acad. Sci. Paris, Sér. A, **289**, 751–753 (1979), Zbl. 423.35001

Garnier, R.
1. Solution du problème de Riemann pour les systèmes différentiels linéaires due second ordre. Ann. Sci. Ec. Norm. Super., III. Sér, **43**, 177–307 (1926), Jbuch 52, 440
2. Le problème de Plateau. Ann. Sci. Ec. Norm. Super., III. Sér, **45**, 53–144 (1928), Jbuch 54, 748

Gauß, C.F.
1. Allgemeine Auflösung der Aufgabe, die Theile einer gegebenen Fläche auf einer anderen gegebenen Fläche so abzubilden, daß die Abbildung dem Abgebildeten in den kleinsten Theilen ähnlich wird. Astronomische Abhandlungen herausgeg. von H.C. Schumacher, Drittes Heft, Altona (1825)
2. Disquisitiones generales circa superficies curvas. Göttinger Nachr. **6**, 99–146 (1828). German transl.: Allgemeine Flächentheorie. Herausgeg. von A. Wangerin, Ostwald's Klassiker, Engelmann, Leipzig, 1905 (cf. also Dombrowski, P. 150 years after Gaus, "Disquisitiones circa superficies curvas". Astérisque **62**, 1–153 (1979), Zbl. 406.01007

Gergonne, J.D.
1. Questions proposées/résolues. Ann. Math. Pure Appl. **7**, 68, 99–100, 143–147, 156 (1816)

Gerhardt, C.
1. Regularity of solutions of nonlinear variational inequalities. Arch. Ration. Mech. Anal. **52**, 389–393 (1973), Zbl. 277.49003

Gilbarg, D., Trudinger, N.S.
1. Elliptic Partial Differential Equations of Second Order. Grundlehren math. Wiss., vol. 224, Springer, Berlin Heidelberg New York 1977, Zbl. 361.35003. Second edition 1983, Zbl. 562.35001

Giusti, E.
1. Superficie minime cartesiane con ostacoli discontinui. Arch. Ration. Mech. Anal. **40**, 251–267 (1971), Zbl. 214, 112
2. Nonparametric minimal surfaces with discontinuous and thin obstacles. Arch. Ration. Mech. Anal. **49**, 41–56 (1972), Zbl. 257.49015

Goldhorn, K., Hildebrandt, S.
1. Zum Randverhalten der Lösungen gewisser zweidimensionaler Variationsprobleme mit freien Randbedingungen. Math. Z. **118**, 241–253 (1970), Zbl. 202, 205

Grüter, M.
1. Über die Regularität schwacher Lösungen des Systems $\Delta x = 2H(x)x_u \wedge x_v$. Dissertation, Düsseldorf 1979

2. Regularity of weak H-surfaces. J. Reine Angew. Math. **329**, 1–15 (1981), Zbl. 461.53029
3. Eine Bemerkung zur Regularität stationärer Punkte von konform invarianten Variationsintegralen. Manuscr. Math. **55**, 451–453 (1986), Zbl. 594.49027

Grüter, M., Hildebrandt, S., Nitsche, J.C.C.
1. On the boundary behavior of minimal surfaces with a free boundary which are not minima of the area. Manuscr. Math. **35**, 387–410 (1981), Zbl. 483.49037

Grüter, M., Jost, J.
1. On embedded minimal disks in convex bodies. Ann. Inst. Henri Poincaré. Anal. Non Linéaire **3**, 345–390 (1986), Zbl. 617.49017
2. Allard-type regularity results for varifolds with free boundaries. Ann. Sc. Norm. Super. Pisa Cl. Sci., IV. Ser., **13**, 129–169 (1986), Zbl. 615.49018

Gulliver, R.
1. Regularity of minimizing surfaces of prescribed mean curvature. Ann. Math., II. Ser., **97**, 275–305 (1973), Zbl. 246.53053
2. Branched immersions of surfaces and reduction of topological type, I. Math. Z. **145**, 267–288 (1975), Zbl. 308.57015
3. Finiteness of the ramified set for branched immersions of surfaces. Pac. J. Math. **64**, 153–165 (1976), Zbl. 336.55002
4. Branched immersions of surfaces and reduction of topological type, II. Math. Ann. **230**, 25–48 (1977), Zbl. 349.57010
5. A minimal surface with an atypical boundary branch point. Preprint CMA-R38–88, Canberra 1988 (= Proc. Int. Conf., Rio de Janeiro/Bras. 1988, Pitman Monogr. Surv. Pure Appl. Math. **52**, 211–228 (1991; Zbl. 731.53005))

Gulliver, R., Hildebrandt, S.
1. Boundary configurations spanning continua of minimal surfaces. Manuscr. Math. **54**, 323–347 (1986), Zbl. 589.53008

Gulliver, R., Lawson, H.B.
1. The structure of stable minimal surfaces near a singularity. Proc. Symp. Pure Math. **44**, 213–237 (1986), Zbl. 592.53005

Gulliver, R., Lesley, F.D.
1. On boundary branch points of minimizing surfaces. Arch. Ration. Mech. Anal. **52**, 20–25 (1973), Zbl. 263.53009

Gulliver, R., Osserman, R., Royden, H.L.
1. A theory of branched immersions of surfaces. Am. J. Math. **95**, 750–812 (1973), Zbl. 295.53002

Gulliver, R., Scott, P.
1. Least area surfaces can have excess triple points. Topology **26**, 345–359 (1987), Zbl. 636.53011

Gulliver, R.D., Spruck, J.
1. On embedded minimal surfaces. Ann. Math., II. Ser., **103**, 331–347 (1976), Zbl. 347.49033, with a correction in Ann. Math., II. Ser., **109**, 407–412 (1979), Zbl. 398.49024

Gulliver, R., Tomi, F.
1. On false branch points of incompressible branched immersions. Manuscr. Math. **63**, 293–302 (1989), Zbl. 701.53009

Gyemant, A.
1. Kapillarität. In: Handbuch der Physik, Bd. 7, Springer, Berlin 1927, 343–410, Jbuch 53, 772

Haar, A.
1. Über das Plateausche Problem. Math. Ann. **97**, 124–158 (1927), Jbuch 52, 710
2. Zur Variationsrechnung. Abh. Math. Sem, Hamb. Univ. **8**, 1–27 (1931), Jbuch 56, 437

Hall, P.
1. Topological properties of minimal surfaces. Thesis, Warwick 1983
2. Two topological examples in minimal surface theory. J. Differ. Geom. **19**, 475–481 (1984), Zbl. 543.53005
3. On Sasaki's inequality for a branched minimal disc. Preprint, 1985

Hardt, R., Simon, L.
1. Boundary regularity and embedded minimal solutions for the oriented Plateau problem. Ann. Math., II. Ser., **110**, 439–486 (1979), Zbl. 457.49029

Harth, F.P.
1. Minimalflächen mit freiem Rand in Riemannschen Mannigfaltigkeiten. Manuscr. Math. **7**, 35–54 (1972), Zbl. 237.49027
2. Zur Regularität von H-Flächen mit freiem Rand. Math. Z. **150**, 71–74 (1976), Zbl. 352.53004

Hartman, P., Wintner, A.
1. On the local behavior of solutions of nonparabolic partial differential equations. Am. J. Math. **75**, 449–476 (1953), Zbl. 52, 322

Harvey, R., Lawson, B.
1. Extending minimal varieties. Invent. Math. **28**, No. 3, 209–226 (1975), Zbl. 316.49032

Heinz, E.
1. Über die Lösungen der Minimalflächengleichung. Nachr. Akad. Wiss. Gött., II. Math.-Phys. Kl., 51–56 (1952), Zbl. 48, 154
2. On certain nonlinear elliptic differential equations and univalent mappings. J. Anal. Math. **5**, 197–272 (1957), Zbl. 085.08701
3. Über das Randverhalten quasilinearer elliptischer Systeme mit isothermen Parametern. Math. Z. **113**, 99–105 (1970), Zbl. 181.11404
4. Über die analytische Abhängigkeit der Lösungen eines linearen elliptischen Randwertproblems von Parametern. Nachr. Akad. Wiss. Gött., II. Math.-Phys. Kl. 1–20 (1979), Zbl. 436.35010
5. Über eine Verallgemeinerung des Plateauschen Problems. Manuscr. Math. **28**, 81–88 (1979), 415.49027
6. Ein mit der Theorie der Minimalflächen zusammenhängendes Variationsproblem. Nachr. Akad. Wiss. Gött., II. Math.-Phys. Kl. 25–35 (1980), Zbl. 456.49029
7. Minimalflächen mit polygonalem Rand. Math. Z. **183**, 547–564 (1983), Zbl. 507.49027
8. Zum Marx-Shiffmanschen Variationsproblem. J. Reine Angew. Math. **344**, 196–200 (1983), Zbl. 511.49022
9. An estimate for the total number of branch points of quasi-minimal surfaces. Analysis **5**, 383–390 (1985), Zbl. 586.49024

Heinz, E., Hildebrandt, S.
1. Some remarks on minimal surfaces in Riemannian manifolds. Commun. Pure Appl. Math. **23**, 371–377 (1970), Zbl. 188, 421
2. On the number of branch points of surfaces of bounded mean curvature. J. Differ. Geom. **4**, 227-235 (1970), Zbl. 195, 230

Heinz, E., Tomi, F.
1. Zu einem Satz von S. Hildebrandt über das Randverhalten von Minimalflächen. Math. Z. **111**, 372–386 (1969), Zbl. 176, 98

Hildebrandt, S.
1. Über das Randverhalten von Minimalflächen. Math. Ann. **165**, 1–18 (1966), Zbl. 138, 423
2. Über Minimalflächen mit freiem Rand. Math. Z. **95**, 1–19 (1967), Zbl. 158, 399
3. Boundary behavior of minimal surfaces. Arch. Ration. Mech. Anal. **35**, 47–82 (1969), Zbl. 183, 394
4. Ein einfacher Beweis für die Regularität der Lösungen gewisser zweidimensionaler Variationsprobleme unter freien Randbedingungen. Math. Ann. **194**, 316–331 (1971), Zbl. 219.49015
5. Maximum principles for minimal surfaces and for surfaces of continuous mean curvature. Math. Z. **128**, 253–269 (1972), Zbl. 253.53005
6. On the regularity of solutions of two-dimensional variational problems with obstructions. Commun. Pure Appl. Math. **25**, 479–496 (1972), Zbl. 251.49003
7. Interior $C^{1+\alpha}$-regularity of solutions of two-dimensional variational problems with obstacles. Math. Z. **131**, 233–240 (1973), Zbl. 275.49007
8. Liouville's theorem for harmonic mappings and an approach to Bernstein theorems. Seminar on Differential Geometry. Princeton University Press 1982, Ann. Math. Stud. **102**, 107–131 (1982), Zbl. 478.53030
9. Nonlinear elliptic systems and harmonic mappings. Proc. of the 1980 Beijing Symposium on Differential Geometry and Differential Equations, Vol. 1, 481–615, Science Press, Beijing, China 1982, Zbl. 515.58012

Hildebrandt, S., Jäger, W.
1. On the regularity of surfaces with prescribed mean curvature at a free boundary. Math. Z. **118**, 289–308 (1970), Zbl. 204, 115

Hildebrandt, S., Kaul, H.
1. Two-dimensional variational problems with obstructions, and Plateau's problem for H-surfaces in a Riemannian manifold. Commun. Pure Appl. Math. **25**, 187–223 (1972), Zbl. 245.53006

Hildebrandt, S., Kaul, H., Widman, K.-O.
1. An existence theorem for harmonic mappings of Riemannian manifolds. Acta Math. **138**, 1–16 (1977), Zbl. 356.53015

Hildebrandt, S., Nitsche, J.C.C.
1. Minimal surfaces with free boundaries. Acta Math. **143**, 251–272 (1979), Zbl. 444.49035
2. Optimal boundary regularity for minimal surfaces with a free boundary. Manuscr. Math. **33**, 357–364 (1981), Zbl. 465.49029
3. A uniqueness theorem for surfaces of least area with partially free boundaries on obstacles. Arch. Ration. Mech. Anal. **79**, 189–218 (1982), Zbl. 506.49017
4. Geometric properties of minimal surfaces with free boundaries. Math. Z. **184**, 497–509 (1983), Zbl. 517.49029

Hildebrandt, S., Sauvigny, F.
1. Embeddedness and uniqueness of minimal surfaces solving a partially free boundary value problem. J. Reine Angew. Math. **422**, 69–89 (1991), Zbl. 729.53014
2. On one-to-one harmonic mappings and minimal surfaces. Nachr. Akad. Wiss. Goett., II. Math.-Phys. Kl. 1992, No. 3, Zbl. 794.35050
3. Uniqueness of stable minimal surfaces with partially free boundaries. J. Math. Soc. Japan **47**, 423–440 (1995)

Hildebrandt, S., Tromba, A.J.
1. Mathematics and optimal form. Scientific American Library, W.H. Freeman, New York 1985. [French transl.: Mathématiques et formes optimales. Pour la Science, Diff. Belin, Paris 1986. German transl.: Panoptimum, Spektrum der Wissenschaft, Heidelberg 1987, Zbl. 716.53002. Dutch transl.: Architectuur in de Natuur, Wetenschappl. Bibliotheek, Natuur en Techniek, Maastricht/Brussel 1989. Spanish transl.: Matematica y formas optimas. Prensa Cientifica, Barcelona 1990. Japanese transl. 1995]
2. The Parsimonious Universe. Shape and Form in the Natural World. Springer, New York 1996. [German transl.: Kugel, Kreis und Seifenblasen. Optimale Formen in Geometrie und Natur. Birkhäuser, Basel Boston Berlin 1996]

Hoffman, D.A.
1. Embedded minimal surfaces, computer graphics and elliptic functions. Lect. Notes Math. **1156**. Springer, Berlin Heidelberg New York 1985, 204–215, Zbl. 566.53016
2. The computer-aided discovery of new embedded minimal surfaces. Mathematical Intelligencer **9**, 8–21 (1987), Zbl. 616.53007

Hoffman, D.A., Meeks, H.W.
1. The strong halfspace theorem for minimal surfaces. Invent. Math. **101**, No. 2, 373–377 (1990), Zbl. 722.53054

Hoffman, D.A., Osserman, R.
1. The geometry of the generalized Gauss map. Mem. Am. Math. Soc. **236**, 1980, Zbl. 469.53004
2. The area of the generalized Gaussian image and the stability of minimal surfaces in S^n and $\mathbb{R}^n$. Math. Ann. **260**, 437–452 (1982), Zbl. 487.53052
3. The Gauss map of surfaces in $\mathbb{R}^n$. J. Differ. Geom. **18**, 733–754 (1983), Zbl. 535.53004
4. The Gauss map of surfaces in $\mathbb{R}^3$ and $\mathbb{R}^4$. J. Lond. Math. Soc., III. Ser., **50**, 27–56 (1985), Zbl. 547.53001

Hohrein, J.
1. Existence of unstable minimal surfaces of higher genus in manifolds of nonpositive curvature. Heidelberg: Univ. Heidelberg, Naturwiss.-Math. Gesamtfakultät, Zbl. 812.53050

Hopf, E.
1. Bemerkungen zu einem Satz von S. Bernstein aus der Theorie der elliptischen Differentialgleichungen. Math. Z. **29**, 744–745 (1929), Jbuch 55, 289
2. On S. Bernstein's theorem on surfaces $z(x, y)$ of non-positive curvature. Proc. Am. Math. Soc. **1**, 80–85 (1950), Zbl. 39, 169

Imbusch, C.
1. Eine Anwendung des Mountain-Pass-Lemmas auf den Fragenkreis des Plateauschen Problems. Diploma-Thesis, Bonn University, 1997

Jäger, W.
1. Behavior of minimal surfaces with free boundaries. Commun. Pure Appl. Math. **23**, 803–818 (1970), Zbl. 204, 116
2. Ein Maximumprinzip für ein System nichtlinearer Differentialgleichungen. Nachr. Akad. Wiss. Gött., II. Math.-Phys. Kl. 1976, 157–164 (1976), Zbl. 362.35030
3. Das Randverhalten von Flächen beschränkter mittlerer Krümmung bei $C^{1,\alpha}$-Rändern. Nachr. Akad. Wiss. Gött. II. Math.-Phys. Kl. 1977, 45–56 (1977), Zbl. 395.35028

Jarausch, H.
1. Zur numerischen Behandlung von parametrischen Minimalflächen mit finiten Elementen. Dissertation, Bochum (1978)

John, F.
1. Partial Differential Equations. Springer, New York Berlin Heidelberg 1971, Zbl. 209, 400. 4th edition 1982, Zbl. 472.35001

Jost, J.
1. Conformal mappings and the Plateau-Douglas problem in Riemannian manifolds. J. Reine Angew. Math. **359**, 37–54 (1985), Zbl. 568.49025
2. Lectures on harmonic maps (with applications to conformal mappings and minimal surfaces). Lect. Notes Math. **1161**. Springer, Berlin Heidelberg New York 1985, 118–192, Zbl. 577.58010
3. On the regularity of minimal surfaces with free boundaries in a Riemannian manifold. Manuscr. Math. **56**, 279–291 (1986), Zbl. 601.49028
4. Existence results for embedded minimal surfaces of controlled topological type. I. Ann. Sc. Norm. Sup. Pisa (Ser. IV) **13**, 15–50 (1986); II. Ann. Sc. Norm. Super. Pisa Cl. Sci. (Ser. IV) **13**, 401–426 (1986); III. Ann. Sc. Norm. Super. Pisa, Cl. Sci., (IV Ser.) **14**, 165–167 (1987), I: Zbl. 619.49019, II: Zbl. 669.49024, III: 681.49039
5. On the existence of embedded minimal surfaces of higher genus with free boundaries in Riemannian manifolds. In: P. Concus, R. Finn (eds), Variational Methods for Free Surface Interfaces. Proc. Conf. Menlo Park/Calif. 1985. Springer, New York Berlin 1987, 65–75
6. Two-dimensional geometric variational problems. Proc. Int. Congr. Math. 1986, Berkeley, AMS, Providence, 1987, 1094–1100
7. Continuity of minimal surfaces with piecewise smooth boundary. Math. Ann. **276**, 599–614 (1987), Zbl. 606.53006
8. Embedded minimal disks with a free boundary on a polyhedron in $\mathbb{R}^3$. Math. Z. **199**, 311–320 (1988), Zbl. 648.53003
9. Das Existenzproblem für Minimalflächen. Jahresber. Dtsch. Math.-Ver. **91**, No. 1, 1–32 (1988), Zbl. 665.53007
10. Embedded minimal surfaces in manifolds diffeomorphic to the three dimensional ball or sphere. J. Differ. Geom. **30**, 555-577 (1989), Zbl. 682.53056
11. Strings with boundary: A quantization of Plateau's problem. Preprint Bochum, SFB 237, 1989
12. Two-dimensional Geometric Variational Problems. Wiley-Interscience, Chichester New York 1991, Zbl. 729.49001

Jost, J., Struwe, M.
1. Morse-Conley theory for minimal surfaces of varying topological type. Invent. Math. **102**, 465–499 (1990), Zbl. 731.58018

Kaul, H.
1. Ein Einschließungssatz für H-Flächen in Riemannschen Mannigfaltigkeiten. Manuscr. Math. **5**, 103–112 (1971), Zbl. 218.53068
2. Remarks on the isoperimetric inequality for multiply connected H-surfaces. Math. Z. **128**, 271–276 (1972), Zbl. 242.53001
3. Isoperimetrische Ungleichung und Gauß-Bonnet Formel für H-Flächen in Riemannschen Mannigfaltigkeiten. Arch. Ration. Mech. Anal. **45**, 194–221 (1972), Zbl. 234.53054

Kinderlehrer, D.
1. The boundary regularity of minimal surfaces. Ann. Sc. Norm. Sup. Pisa, Sci. Fis. Mat. III Ser., **23**, 711–744 (1969), Zbl. 185, 346

2. Minimal surfaces whose boundaries contain spikes. J. Math. Mech. **19**, 829–853 (1970), Zbl. 193, 393
3. The coincidence set of solutions of certain variational inequalities. Arch. Ration. Mech. Anal. **40**, 231–250 (1971), Zbl. 219.49014
4. The regularity of minimal surfaces defined over slit domains. Pac. J. Math. **37**, 109–117 (1971), Zbl. 214, 105
5. Variational inequalities with lower dimensional obstacles. Isr. J. Math. **10**, 339–348 (1971), Zbl. 263.49008
6. How a minimal surface leaves an obstacle. Acta Math. **130**, 221–242 (1973), Zbl. 268.49050
7. The free boundary determined by the solution of a differential equation. Indiana Univ. Math. J. **25**, 195–208 (1976), Zbl. 336.35031
8. Variational inequalities and free boundary problems. Bull. Am. Math. Soc. **84**, 7–26 (1978), Zbl. 382.35004

Kinderlehrer, D., Nirenberg, L., Spruck, J.
1. Regularity in elliptic free boundary problems, I. J. Anal. Math. **34**, 86–119 (1978); II. Ann. Sc. Norm. Sup. Pisa, Cl. Sci. IV. Ser., **6**, 637–683 (1979), I: Zbl. 402.35045, II: Zbl. 425.35097

Kneser, H.
1. Lösung der Aufgabe 41. Jahresber. Dtsch. Math.-Ver. **35**, 123–124 (1926), Jbuch 52, 498

Koiso, M.
1. On the finite solvability of Plateau's problem for extreme curves. Osaka J. Math. **20**, 177–183 (1983), Zbl. 518.53012
2. On the stability of minimal surfaces in $\mathbb{R}^3$. J. Math. Japan **36**, 523–541 (1984), Zbl. 554.49020
3. On the non-uniqueness for minimal surfaces in $\mathbb{R}^3$. Proc. Diff. Geom., Sendai 1989

Krust, R.
1. Remarques sur le problème extérieur de Plateau. Duke Math. J. **59**, 161–173 (1989), Zbl. 709.49022

Küster, A.
1. An optimal estimate of the free boundary of a minimal surface. J. Reine Angew. Math. **349**, 55–62 (1984), Zbl. 527.53006
2. On the linear isoperimetric inequality. Manuscr. math. **53**, 255–259 (1985), Zbl. 573.53004

Kuwert, E.
1. Embedded solutions for exterior minimal surface problems. Manuscr. Math. **70**, 51–65 (1990), Zbl. 717.49034
2. A bound for minimal graphs with a normal at infinity. Calc. Var. Partial Differ. Equ. **1**, 407–416 (1993), Zbl. 794.49037
3. On solutions of the exterior Dirichlet problem for the minimal surface equation. Ann. Inst. Henri Poincaré Anal. Non linéaire **10**, 445–451 (1993), Zbl. 820.35038
4. Minimizing the energy of maps from a surface into a 2-sphere with prescribed degree and boundary values. Manuscr. Math. **83**, 31–38 (1994), Zbl. 808.58017
5. Area-minimizing immersions of the disk with boundary in a given homotopy class: A general existence theory. Habilitation thesis, Bonn University 1995

Langevin, R., Rosenberg, H.
1. A maximum principle at infinity for minimal surfaces and applications. Duke Math. J. **57**, 819–826 (1988), Zbl. 667.49024

Lawson Jr., H.B.
1. The unknottedness of minimal embeddings. Invent. Math. **11**, 183–187 (1970), Zbl. 205, 520
2. Lectures on Minimal Submanifolds. Publish or Perish press, Berkeley 1971, Zbl. 434.53006 (Publisher Perish Press 1980 Reissue)
3. Some intrinsic characterizations of minimal surfaces. J. Anal. Math. **24**, 151–161 (1971), Zbl. 251.53003
4. The equivariant Plateau problem and interior regularity. Trans. Am. Math. Soc. **173**, 231–249 (1972), Zbl. 279.49043

Lemaire, L.
1. Boundary value problem for harmonic and minimal maps of surfaces into manifolds. Ann. Sc. Norm. Super. Pisa, Cl. Sci. (4), **8**, 91–104 (1982), Zbl. 532.58004

Lesley, F.D.
1. Differentiability of minimal surfaces the boundary. Pac. J. Math. **37**, 123–139 (1971), Zbl. 218.53014

Lévy, P.
1. Surfaces minima et corps convexes en moyenne. C.R. Acad. Sci. Paris Sér. A–B **223**, 881–883 (1946), Zbl. 61, 377
2. Exemples de contours pour lesquels le problème de Plateau a 3 ou $2p + 1$ solutions. C.R. Acad. Sci., Paris **224**, 325–327 (1947), Zbl. 30, 33
3. Le problème de Plateau. Mathematica **23**, 1–45 (1948), Zbl. 37, 348

Lewerenz, F.
1. Eine Bemerkung zu den Marx-Shiffmanschen Minimalvektoren bei Polygonen. Arch. Ration. Mech. Anal. **75**, 199–202 (1981), Zbl. 455.53003

Lewy, H.
1. On minimal surfaces with partially free boundary. Commun. Pure Appl. Math. **4**, 1–13 (1951), Zbl. 43, 160
2. On the boundary behavior of minimal surfaces. Proc. Natl. Acad. Sci. USA **37**, 103–110 (1951), Zbl. 42, 157
3. On a variational problem with inequalities on the boundary. J. Math. Mech. **17**, 861–884 (1968), Zbl. 153, 433
4. On the coincidence set in variational inequalities. J. Differ. Geom. **6**, 497–501 (1972), Zbl. 255.31007

Li, P., Schoen, R., Yau, S.T.
1. On the isoperimetric inequality for minimal surfaces. Ann. Sc. Norm. Super. Pisa Cl. Sci., Ser. IV, Vol. XI.2, 237–244 (1984), Zbl. 551.49020

Lichtenstein, L.
1. Neuere Entwicklung der Potentialtheorie, Konforme Abbildung. Enzykl. Math. Wiss. II C 3 B.G. Teubner, Leibzig 1909–1921, 177–377 (1919), Jbuch 47, 450
2. Beweis des Satzes, daß jedes hinreichend kleine, im wesentlichen stetig gekrümmte, singularitätenfreie Flächenstück auf einen Teil einer Ebene zusammenhängend und in den kleinsten Teilen ähnlich abgebildet wird. Abh. Königl. Preuss. Akad. Wiss. Berlin, Phys.-Math. Kl., Anhang, Abh. VI, 1–49 (1911), Jbuch 42, 710
3. Zur Theorie der konformen Abbildung. Konforme Abbildung nichtanalytischer singularitätenfreier Flächenstücke auf ebene Gebiete. Bull. Acad. Sci. Cracovie, Cl. Sci. Math. Nat. A, 192–217 (1916), Jbuch 46, 547
4. Neuere Entwicklung der Theorie partieller Differentialgleichungen zweiter Ordnung vom elliptischen Typus. Enzykl. Math. Wiss. 2.3.2, B.G. Teubner, Leipzig 1923–1927 (completed 1924), 1277–1334 (1924), Jbuch 50, 324

5. Über einige Hilfssätze der Potentialtheorie. IV. Sitz.-Ber. Sächs. Akad. d. Wiss. Leipzig, Math.-Phys. Kl. **82**, 267–344 (1930), Jbuch 56, 1064

Lin, F.-H.
1. Uniqueness and Nonuniqueness of the Plateau problem. Indiana Univ. Math. J. **36**, No. 4, 843–848 (1987), Zbl. 626.49019
2. Regularity for a class of parametric obstacle problems. Dissertation, Univ. of Minnesota 1985
3. Plateau's problem for H-convex curves. Manuscr. Math. **58**, 497–511 (1987), Zbl. 678.49030

Marx, I.
1. On the classification of unstable minimal surfaces with polygonal boundaries. Commun. Pure Appl. Math. **8**, 235–244 (1955), Zbl. 68, 351

McShane, E.J.
1. Parametrizations of saddle surfaces with applications to the problem of Plateau. Trans. Am. Math. Soc. **35**, 716–733 (1933), Zbl. 7, 119
2. On the analytic nature of surfaces of least area. Ann. Math., II. Ser., **35**, 456–473 (1934), Zbl. 10, 68

Meeks, W.H.
1. Lectures on Plateau's problem. Escola de Geometria Differencial, Universidade Federal do Ceará (Brazil), de 17 a 28 de Julho de 1978
2. Uniqueness theorems for minimal surfaces. Ill. J. Math. **25**, 318–336 (1981), Zbl. 472.53009
3. A survey of the geometric results in the classical theory of minimal surfaces. Bol. Soc. Bras. Mat. **12**, 29–86 (1981), Zbl. 577.53007
4. The topological uniqueness of minimal surfaces in three-dimensional Euclidean space. Topology **20**, 389–410 (1981), Zbl. 458.53003
5. Recent progress on the geometry of surfaces in $\mathbb{R}^3$ and on the use of computer graphics as a research tool. Proc. Int. Congr. Math., Berkeley/Calif. 1986, Vol. 1, 551–560 (1987), Zbl. 669.53006

Meeks, W.H., Rosenberg, H.
1. The maximum principle at infinity for minimal surfaces in flat three-manifolds. Comment. Math. Helv. **69**, 255–270 (1990), Zbl. 713.53008

Meeks, W.H., Simon, L., Yau, S.T.
1. Embedded minimal surfaces, exotic spheres, and manifolds with positive Ricci curvature. Ann. Math. **116**, 621–659 (1982), Zbl. 521.53007

Meeks, W.H., Yau, S.-T.
1. Topology of three-manifolds and the embedding problems in minimal surface theory. Ann. Math. **112**, 441–484 (1980), Zbl. 458.57007
2. The equivariant Dehn's lemma and loop theorem. Comment. Math. Helvet. **56**, 225–239 (1981), Zbl. 469.57005
3. The classical Plateau problem and the topology of three-dimensional manifolds. Topology **21**, 409–442 (1982), Zbl. 489.57002
4. The existence of embedded minimal surfaces and the problem of uniqueness. Math. Z. **179**, 151–168 (1982), Zbl. 479.49026

Micallef, M.J.
1. Stable minimal surfaces in Euclidean space. J. Differ. Geom. **19**, 57–84 (1984), Zbl. 537.32010

Miersemann, E.
1. Zur Regularität verallgemeinerter Lösungen von quasilinearen Differentialgleichungen in Gebieten mit Ecken. Z. Anal. Anwend. (4) **1**, 59–71 (1982), Zbl. 518.35011

Minkowski, H.
1. Kapillarität. Enzykl. Math. Wiss. 5.1.9. Teubner, Leipzig 1903–1921, 558–613 (1907), Jbuch 38, 794

Morgan, F.
1. A smooth curve in $\mathbb{R}^4$ bounding a continuum of area minimizing surfaces. Duke Math. J. **43**, 867–870 (1976), Zbl. 338.53003
2. Almost every curve in $\mathbb{R}^3$ bounds a unique area minimizing surface. Invent. Math. **45**, 253–297 (1978), Zbl. 378.49028
3. A smooth curve in $\mathbb{R}^3$ bounding a continuum of minimal manifolds. Arch. Ration. Mech. Anal. **75**, 193–197 (1980), Zbl. 453.49027
4. On the singular structure of two-dimensional area minimizing surfaces in $\mathbb{R}^n$. Math. Ann. **261**, 101–110 (1982), Zbl. 549.49029
5. On finiteness of the number of stable minimal hypersurfaces with a fixed boundary. Bull. Am. Math. Soc. New. Ser. **13**, 133–136 (1985), Zbl. 572.49021

Morrey, C.B.
1. On the solutions of quasi-linear elliptic partial differential equations. Trans. Am. Math. Soc. **43**, 126–166 (1938), Zbl. 18, 405
2. Multiple integral problems in the calculus of variations and related topics. Univ. California Publ. Math., New Ser. **1**, Nr. 1, 1–130 (1943)
3. The problem of Plateau on a Riemannian manifold. Ann. Math. (2) **49**, 807–851 (1948), Zbl. 33, 396
4. Second order elliptic systems of differential equations. In: Contributions to the theory of partial differential equations, Ann. Math. Stud. **33**, 101–159 (1954), Zbl. 57, 83
5. On the analyticity of the solutions of analytic nonlinear elliptic systems of partial differential equations. Am. J. Math. **80**, I. 198–218, II. 219–234 (1958), I: Zbl. 81, 94, II: Zbl. 81, 94
6. Multiple integral problems in the calculus of variations and related topics. Ann. Sc. Norm. Super. Pisa, Sci. Fis. Mat. III. Ser. **14**, 1–61 (1960), Zbl. 94, 81
7. Multiple Integrals in the Calculus of Variations. Grundlehren math. Wiss., vol. 130. Springer, Berlin Heidelberg New York 1966, Zbl. 142, 387

Morrey, C.B., Nirenberg, L.
1. On the analyticity of the solutions of linear elliptic systems of partial differential equations. Commun. Pure Appl. Math. **10**, 271–290 (1957), Zbl. 82, 94

Morse, M., Tompkins, C.B.
1. The existence of minimal surfaces of general critical types. Ann. of Math. **40**, 443–472 (1939), **42**, 331 (1941), Zbl. 21, 34, Zbl. 24, 328
2. Unstable minimal surfaces of higher topological types. Proc. Natl. Acad. Sci. **26**, 713–716 (1940)
3. The continuity of the area of harmonic surfaces as a function of the boundary representation. Am. J. Math. **63**, 325–338 (1941), Zbl. 26, 124
4. Unstable minimal surfaces of higher topological structure. Duke Math. J. **8**, 350–375 (1941), Zbl. 25, 409
5. Minimal surfaces of non-minimum type by a new mode of approximation. Ann. Math. (2) **42**, 62–72 (1941), Zbl. 24, 417

Nehring, T.
1. Embedded minimal annuli solving an exterior problem. Calc. Var. Partial Differ. Equ. **2**, 373–388 (1994)

Nitsche, J.C.C.
1. Elementary proof of Bernstein's theorem on minimal surfaces. Ann. Math. (2) **66**, 543–544 (1957), Zbl. 79, 377
2. Review of Sasaki [1]. Math. Reviews **25**, Nr. 492 (1963)
3. A supplement to the condition of J. Douglas. Rend. Circ. Mat. Palermo (2) **13**, 192–198 (1964), Zbl. 136, 167
4. On new results in the theory of minimal surfaces. Bull. Am. Math. Soc. **71**, 195–270 (1965), Zbl. 135, 217
5. Ein Einschließungssatz für Minimalflächen. Math. Ann. **165**, 71–75 (1966), Zbl. 138, 424
6. Contours bounding at least three solutions of Plateau's problem. Arch. Ration. Mech. Anal. **30**, 1–11 (1968), Zbl. 164, 142
7. The boundary behavior of minimal surfaces - Kellogg's theorem and branch points on the boundary. Invent. math. **8**, 313–333 (1969), Zbl. 195, 231
8. Concerning the isolated character of solutions of Plateau's problem. Math. Z. **109**, 393–411 (1969), Zbl. 189, 108
9. A variational problem with inequalities as boundary conditions. Bull. Am. Math. Soc. **75**, 450–452 (1969), Zbl. 194, 425
10. Variational problems with inequalities as boundary conditions, or, how to fashion a cheap hat for Giacometti's brother. Arch. Ration. Mech. Anal. **35**, 83–113 (1969), Zbl. 209, 416
11. Concerning my paper on the boundary behavior of minimal surfaces. Invent. Math. **9**, 270 (1970), Zbl. 202, 206
12. An isoperimetric property of surfaces with moveable boundaries. Am. Math. Mon. **77**, 359–362 (1970), Zbl. 205, 109
13. Minimal surfaces with partially free boundary. Least area property and Hölder continuity for boundaries satisfying a chord-arc condition. Arch. Ration. Mech. Anal. **39**, 131–145 (1970), Zbl. 209, 416
14. Minimal surfaces with moveable boundaries. Bull. Am. Math. Soc. **77**, 746–751 (1971), Zbl. 238.49029
15. The regularity of minimal surfaces on the movable parts of their boundaries. Indiana Univ. Math. J. **21**, 505–513 (1971), Zbl. 236.53011
16. A new uniqueness theorem for minimal surfaces. Arch. Ration. Mech. Anal. **52**, 319–329 (1973), Zbl. 285.49001
17. Plateau problems and their modern ramifications. Am. Math. Mon. **81**, 945–968 (1974), Zbl. 298.53004
18. Vorlesungen über Minimalflächen. Grundlehren math. Wiss., vol. 199. Springer, Berlin Heidelberg New York 1975, Zbl. 319.53003. Engl. transl. of Chaps.1–5: Lectures on Minimal Surfaces, Vol. 1: Introduction, Fundamentals, Geometry and Basic Boundary Problems. Cambridge Univ. Press 1989, Zbl. 688.53001
19. Non-uniqueness for Plateau's problem. A bifurcation process. Ann. Acad. Sci. Fenn. Ser. A I **2**, 361–373 (1976), Zbl. 361.49025
20. The regularity of the trace for minimal surfaces. Ann. Sc. Norm. Super. Pisa Cl. Sci., Ser. IV **3**, 139–155 (1976), Zbl. 331.53006
21. Contours bounding at most finitely many solutions of Plateau's problem. Complex Analysis and its Applications, dedicated to I.N. Vekua, Nauka Moscow, Steklov Math. Inst., Moscow 1978, 438–446 (1978), Zbl. 422.49031

22. Uniqueness and non-uniqueness for Plateau's problem – one of the last major questions. Minimal Submanifolds and Geodesics, Proceedings of the Japan-United States Seminar on Minimal Submanifolds, including Geodesics, Tokyo 1977, Kagai Publications, Tokyo 1978, pp. 143–161, Zbl. 436.53008
23. The higher regularity of liquid edqes in aggregates of minimal surfaces. Nachr. Akad. Wiss. Gött., II. Math.-Phys. Kl., Nr. 2, 31–51 (1978), Zbl. 403.53001

Osserman, R.
1. Proof of a conjecture of Nirenberg. Commun. Pure Appl. Math. **12**, 229–232 (1959), Zbl. 86, 362
2. Minimal surfaces in the large. Comment. Math. Helvet. **35**, 65–76 (1961), Zbl. 98, 349
3. On complete minimal surfaces. Arch. Ration. Mech. Anal. **13**, 392–404 (1963), Zbl. 127, 380
4. Global properties of minimal surfaces in E^3 and E^n. Ann. Math., II. Ser., **80**, 340–364 (1964), Zbl. 134, 385
5. A Survey of Minimal Surfaces. Van Nostrand, New York 1969, Zbl. 209, 529
6. Minimal varieties. Bull. Am. Math. Soc. **75**, 1092–1120 (1969), Zbl. 188, 538
7. A proof of the regularity everywhere of the classical solution to Plateau's problem. Ann. Math. (2) **91**, 550–569 (1970), Zbl. 194, 223
8. On the convex hull property of immersed manifolds. J. Differ. Geom. **6**, 267–270 (1971), Zbl. 226.53009
9. Isoperimetric and related inequalities. Differ. Geom. Proc. Symp. Pure Math. Part 1, Stanford 1973 **27**, 207–215 (1975), Zbl. 226.53009
10. Some remarks on the isoperimetric inequality and a problem of Gehring. J. Anal. Math. **30**, 404–410 (1976), Zbl. 333.52006
11. The isoperimetric inequality. Bull. Am. Math. Soc. **84**, 1182–1238 (1978), Zbl. 411.52006
12. Minimal surfaces, Gauss maps, total curvature, eigenvalue estimates, and stability. Proc. Int. Chern. Sympos., Berkeley 1979. Springer, Berlin Heidelberg New York 1980, 199–227, Zbl. 457.53001

Osserman, R., Schiffer, M.
1. Doubly connected minimal surfaces. Arch. Ration. Mech. Anal. **58**, 285–307 (1974/75), Zbl. 352.53005

Plateau, J.A.F.
1. Statique expérimentale et théoretique des liquides soumis aux seule forces moléculaires. Vols. I, II. Gauthier-Villars, Paris 1873, Jbuch 6, 516

Protter, M., Weinberger, H.
1. Maximum Principles in Differential Equations. Prentice-Hall, Englewood, NJ 1967, Zbl. 153, 136

Quien, N., Tomi, F.
1. Nearly planar Jordan curves spanning a given number of minimal immersions of the disc. Arch. Math. **44**, 456–460 (1985), Zbl. 567.53007

Radó, T.
1. Über den analytischen Charakter der Minimalflächen. Math. Z. **24**, 321–327 (1925), Jbuch 61, 546
2. Aufgabe 41. Jahresber. der Deutsch. Math. Ver. **35**, 49 (1926), Jbuch 52, 498
3. Zu einem Satz von S. Bernstein über Minimalflächen im Großen. Math. Z. **26**, 559–565 (1927), Jbuch 53, 670
4. Some remarks on the problem of Plateau. Proc. Natl. Acad. Sci. USA **16**, 242–248 (1930), Jbuch 56, 437
5. The problem of least area and the problem of Plateau. Math. Z. **32**, 763–795 (1930), Jbuch 56, 436

6. On Plateau's problem. Ann. Math. (2) **31**, 457–469 (1930), Jbuch 56, 437
7. On the functional of Mr. Douglas. Ann. Math. (2) **32**, 785–803 (1931), Zbl. 3, 61
8. Contributions to the theory of minimal surfaces. Acta Litt. Sci. Szeged **6**, 1–20 (1932), Zbl. 5, 179
9. On the Problem of Plateau. Ergebnisse der Math. Band 2. Springer, Berlin 1933, Zbl. 7, 118
10. An iterative process in the problem of Plateau. Trans. Am. Math. Soc. **35**, 869–887 (1933), Zbl. 8, 82

Riemann, B.
1. Über die Fläche vom kleinsten Inhalt bei gegebener Begrenzung. Abh. Königl. Ges. d. Wiss. Göttingen, Mathem. Cl. **13**, 3–52 (1867) (K. Hattendorff, edit.)

Ripoll, J., Tomi, F.
1. Maximum principles for minimal surfaces having noncompact boundary and a uniqueness theorem for the helicoid. Preprint 1994

Ritter, I.F.
1. Solution of Schwarz' problem concerning minimal surfaces. Rev. Univ. Nac. de Tucuman, Ser. A Mat. **1**, 49–62 (1940), Zbl. 25, 336

Rosenberg, H.
1. Some recent developments in the theory of properly embedded minimal surfaces in $\mathbb{R}^3$. Séminaire BOURBAKI, Vol. 1991/92, 759 (1992), 64 pp., Zbl. 789.53003

Ruchert, H.
1. A uniqueness result for Enneper's minimal surface. Indiana Univ. Math. J. **30**, 427–431 (1981), Zbl. 492.49022

Sacks, J., Uhlenbeck, K.
1. Existence of minimal immersions of two-spheres. Ann. Math., II. Ser., **113**, 1–24 (1981), Zbl. 462.58014

Sasaki, S.
1. On the total curvature of a closed curve. Jap. J. Math. **29**, 118–125 (1959), Zbl. 133, 142

Sauvigny, F.
1. Ein Eindeutigkeitssatz für Minimalflächen im $\mathbb{R}^p$ mit polygonalem Rand. J. Reine Angew. Math. **358**, 92–96 (1985), Zbl. 553.53003
2. On the Morse index of minimal surfaces in $\mathbb{R}^p$ with polygonal boundaries. Manuscr. Math. **53**,167–197 (1985), Zbl. 607.53004
3. Die zweite Variation von Minimalflächen im $\mathbb{R}^p$ mit polygonalem Rand. Math. Z. **189**, 167–184 (1985), Zbl. 543.53004
4. On the total number of branch points of quasi-minimal surfaces bounded by a polygon. Analysis **8**, 297–304 (1988), Zbl. 657.49017

Schauder, J.
1. Potentialtheoretische Untersuchungen. I. Math. Z. **33**, 602–640 (1931), Zbl. 1, 336
2. Über lineare elliptische Differentialgleichungen zweiter Ordnung. Math. Z. **38**, 257–282 (1934), Zbl. 8, 255

Schneider, R.
1. A note on branch points of minimal surfaces. Proc. Am. Math. Soc. **17**, 1254–1257 (1966), Zbl. 152, 388
2. Ein Eindeutigkeitssatz zum Plateauschen Problem. Preprint 1969

Schoen, R.

1. Uniqueness, symmetry, and embedded minimal surfaces. J. Differ. Geom. **18**, 791–809 (1983), Zbl. 575.53037

Schoen, R., Yau, S.-T.

1. Existence of incompressible minimal surfaces and the topology of three-dimensional manifolds with non-negative scalar curvature. Ann. Math., II. Ser., **110**, 127–142 (1979), Zbl. 431.53051

Schüffler, K.

1. Stabilität mehrfach zusammenhängender Minimalflächen. Manuscr. Math. **30**, 163–197 (1979), Zbl. 426.53005
2. Eine globalanalytische Behandlung des Douglas'schen Problems. Manuscr. Math. **48**, 189–226 (1984), Zbl. 606.58047
3. Function theory and index theory for minimal surfaces of genus 1. Arch. Math. **48**, Part I: 250–266, II: 343–352, III: 446–457 (1987), I: Zbl. 615.58039, II: Zbl. 615.58040, III: Zbl. 615.58041

Schwarz, H.A.

1. Fortgesetzte Untersuchungen über spezielle Minimalflächen. Monatsberichte der Königlichen Akad. Wiss. Berlin, 3–27 (1872). Gesammelte Math. Abhandlungen I, 126–148 (1890), Jbuch 22, 31
2. Gesammelte Mathematische Abhandlungen, Band I und II. Springer, Berlin 1890, Jbuch 22, 31
3. Zur Theorie der Minimalflächen, deren Begrenzung aus geradlinigen Strecken besteht. Sitz.-Ber. Königl. Preuß. Akad. Wiss., Berlin, Phys.-Math. Cl. 1237–1266 (1894), Jbuch 25, 1270

Seidel, W.

1. Über die Ränderzuordnung bei konformen Abbildungen. Math. Ann. **104**, 183–243 (1931), Zbl. 10, 47

Shiffman, M.

1. The Plateau problem for minimal surfaces which are relative minima. Ann. Math. (2) **39**, 309–315 (1938), Zbl. 19, 124
2. The Plateau problem for non-relative minima. Ann. Math., II. Ser., **40**, 834–854 (1939), Zbl. 23, 398
3. The Plateau problem for minimal surfaces of arbitrary topological structure. Am. J. Math. **61**, 853–882 (1939), Zbl. 23, 137
4. Unstable minimal surfaces with any rectifiable boundary. Proc. Natl. Acad. Sci. USA **28**, 103–108 (1942)
5. Unstable minimal surfaces with several boundaries. Ann. Math., II. Ser., **43**, 197–222 (1942)
6. On the isoperimetric inequality for saddle surfaces with singularities. Studies and essays presented to R. Courant. Interscience, New York 1948, 383–394, Zbl. 34, 107
7. On surfaces of stationary area bounded by two circles, or convex curves, in parallel planes. Ann. Math. (2) **63**, 77–90 (1956), Zbl. 70, 168

Simon, L.

1. Survey lectures on minimal submanifolds. Seminar on Minimal Submanifolds. Ann. Math. Stud. **103**, 3–52, Princeton (1983), Zbl. 541.53045

Simons, J.

1. Minimal cones, Plateau's problem, and the Bernstein conjecture. Proc. Natl. Acad. Sci. USA **58**, 410–411 (1967), Zbl. 168, 99

Smyth, B.

1. Stationary minimal surfaces with boundary on a simplex. Invent. Math. **76**, 411–420 (1984), Zbl. 538.53010

Söllner, M.

1. Über die Struktur der Lösungsmenge des globalen Plateau-Problems bei Flächen konstanter mittlerer Krümmung. Dissertation, Bochum 1982, Zbl. 598.58015
2. Plateau's problem for surfaces of constant mean curvature from a global point of view. Manuscr. Math. **43**, 191–217 (1983), Zbl. 525.58016

Steffen, K.

1. Isoperimetric inequalities and the problem of Plateau. Math. Ann. **222**, 97–144 (1976), Zbl. 345.49024

Steffen, K., Wente, H.

1. The non-existence of branch points in solutions to certain classes of Plateau type variational problems. Math. Z. **163**, 211–238 (1978), Zbl. 404.49037

Ströhmer, G.

1. Instabile Minimalflächen in Riemanschen Mannigfaltigkeiten nichtpositiver Schnittkrümmung. J. Reine Angew. Math. **315**, 16–39 (1980), Zbl. 419.53026
2. Instabile Minimalflächen mit halbfreiem Rand. Analysis **2**, 315–335 (1982), Zbl. 524.53008

Struwe, M.

1. On a free boundary problem for minimal surfaces. Invent. Math. **75**, 547–560 (1984), Zbl. 537.35037
2. On a critical point theory for minimal surfaces spanning a wire in $\mathbb{R}^n$. J. Reine Angew. Math. **349**, 1–23 (1984), Zbl. 521.49028
3. A Morse theory for annulus-type minimal surfaces. J. Reine Angew. Math. **368**, 1–27 (1986), Zbl. 605.58018
4. Plateau's problem and the calculus of variations. Mathematical Notes **35**, Princeton University Press, Princeton (1988), Zbl. 694.49028
5. Applications of variational methods to problems in the geometry of surfaces. In: S. Hildebrandt, R. Leis (ed.). Partial differential equations and calculus of variations. Lect. Notes Math. **1357**. Springer, Berlin Heidelberg New York 1988, pp. 359–378, Zbl. 665.49035
6. Minimal surfaces of higher genus and general critical type. Proc. Int. Conf. Microlocal and Nonlinear Analysis, Nankai Inst. 1991. Singapore, World Scientif. 1992
7. Das Plateausche Problem. Jahresber. Dtsch. Math.-Ver. **96**, No. 3, 101–116 (1994), Zbl. 810.53004

Taylor, J.E.

1. Regularity of the singular sets of two-dimensional area-minimizing flat chains modulo 3 in $\mathbb{R}^3$. Invent. Math. **22**, 119–159 (1973), Zbl. 278.49046
2. The structure of singularities in soap-bubble-like and soap-film-like minimal surfaces. Ann. Math., II. Ser., **103**, 489–539 (1976), Zbl. 335.49032
3. Boundary regularity for solutions to various capillarity and free boundary problems. Commun. Partial Differ. Equations **2** (4), 323–357 (1977), Zbl. 357.35010

Tomi, F.

1. Ein einfacher Beweis eines Regularitätssatzes für schwache Lösungen gewisser elliptischer Systeme. Math. Z. **112**, 214–218 (1969), Zbl. 177, 147
2. Minimal surfaces and surfaces of prescribed mean curvature spanned over obstacles. Math. Ann. **190**, 248–264 (1971), Zbl. 198, 148
3. On the local uniqueness of the problem of least area. Arch. Ration. Mech. Anal. **52**, 312–318 (1973), Zbl. 277.49016

4. Bemerkungen zum Regularitätsproblem der Gleichung vorgeschriebener mittlerer Krümmung. Math. Z. **132**, 323–326 (1973), Zbl. 258.35022
5. On the finite solvability of Plateau's problem. Lect. Notes Math. **597**. Springer, Berlin Heidelberg New York, 679–695, 1977, Zbl. 361.53008
6. Plateau's problem for embedded minimal surfaces of the type of the disc. Arch. Math. **31**, 374–381 (1978), Zbl. 399.53002
7. A finiteness result in the free boundary value problem for minimal surfaces. Ann. Inst. Henri Poincaré, Anal. Non Lineaire **3**, 331–343 (1986), Zbl. 603.49028
8. An introduction to Plateau's problem. Adv. Math., Beijing **19** (1), 25–37 (1990), Zbl. 703.53006
9. Plateau's problem for minimal surfaces with a catenoidal end. Arch. Math. **59**, (2), 165–172 (1992), Zbl. 777.53015

Tomi, F., Tromba, A.J.
1. Extreme curves bound an embedded minimal surface of disk type. Math. Z. **158**, 137–145 (1978), Zbl. 361.53052
2. On the structure of the set of curves bounding minimal surfaces of prescribed degeneracy. J. Reine Angew. Math. **316**, 31–43 (1980), Zbl. 421.58002
3. On Plateau's problem for minimal surfaces of higher genus in $\mathbb{R}^3$. Bull. Am. Math. Soc., New. Ser. **13**, 169–171 (1985), Zbl. 589.49026
4. Existence theorems for minimal surfaces of non-zero genus spanning a contour. Mem. Am. Math. Soc. **71** (1988), Zbl. 638.58004
5. The index theorem for higher genus minimal surfaces. Mem. Am. Math. Soc. **560**, 78 p. (1995), Zbl. 845.49018

Tomi, F., Ye, R.
1. The exterior Plateau problem. Math. Z. **205**, No. 2, 233–245 (1990), Zbl. 719.53006

Tonelli, L.
1. Sul problema di Plateau, I & II. Atti. Accad. Naz. Lincei **24**, 333–339, 393–398 (1936) (cf. also: Opere scelte, Vol. III, 328–341), Zbl. 16, 264

Tromba, A.J.
1. On the number of simply connected minimal surfaces spanning a curve. Mem. Am. Math. Soc. No. 194, **12** (1977), Zbl. 389.53003
2. The Morse-Sard-Brown theorem for functionals and the problem of Plateau. Am. J. Math. **99**, 1251–1256 (1977), Zbl. 373.58003
3. On Plateau's problem for minimal surfaces of higher genus in $\mathbb{R}^n$. Preprint No. 580, Bonn, SFB 72, 1983
4. A sufficient condition for a critical point of a functional to be a minimum and its application to Plateau's problem. Math. Ann. **263**, 303–312 (1983), Zbl. 518.58012
5. Degree theory on oriented infinite dimensional varieties and the Morse number of minimal surfaces spanning a curve in $\mathbb{R}^n$. Part II: $n = 3$. Manuscr. Math. **48**, 139–161 (1984), Zbl. 622.58006
6. Degree theory on oriented infinite dimensional varieties and the Morse number of minimal surfaces spanning a curve in $\mathbb{R}^n$. Part I: $n \geq 4$. Trans. Am. Math. Soc. **290**, 385–413 (1985), Zbl. 604.58008
7. On the Morse number of embedded and non-embedded minimal immersions spanning wires on the boundary of special bodies in $\mathbb{R}^3$. Math. Z. **188**, 149–170 (1985), Zbl. 613.58011
8. A proof of the Douglas theorem on the existence of disc-like minimal surfaces spanning Jordan contours in $\mathbb{R}^n$. Astérisque **154–155**, 39–50 (1987), Zbl. 635.53033

9. Open problems in the degree theory for disc minimal surfaces spanning a curve in $\mathbb{R}^3$. In: S. Hildebrandt, R. Leis (ed.): Partial differential equations and calculus of variations Lect. Notes Math. **1357**. Springer, Berlin Heidelberg New York 1988, 379–401, Zbl. 681.58012
10. Intrinsic third derivatives for Pateau's problem and the Morse inequalities for disc minimal surfaces in $\mathbb{R}^3$. Calc. Var. Partial Differ. Equ. **1**, No. 3, 335–353 (1993), Zbl.. 793.53010

Tsuji, M.
1. On a theorem of F. and M. Riesz. Proc. Imp. Acad. Tokyo **18**, 172–175 (1942), Zbl. 60, 359
2. Potential Theory in Modern Function Theory. Maruzen, Tokyo 1959 (2nd ed. Chelsea New York 1975), Zbl. 322.30001

Vekua, I.N.
1. Generalized Analytic Functions. Moskau 1959, Zbl. 92, 297. English transl.: Pergamon Press, Oxford London New York Paris 1962, Zbl. 100, 76

Warschawski, S.E.
1. Über einen Satz von O.D. Kellogg. Nachr. Akad. Wiss. Gött., II. Math.-Phys. Kl., 73–86 (1932), Zbl. 5, 361
2. Über das Randverhalten der zweiten Derivierten Abbildungsfunktion bei konformer Abbildung. Math. Z. **35**, 321–456 (1932), Zbl. 4, 404
3. On the higher derivatives at the boundary in conformal mapping. Trans. Am. Math. Soc. **38**, 310–340 (1935), Zbl. 14, 267
4. On a theorem of L. Lichtenstein. Pac. J. Math. **5**, 835–839 (1955), Zbl. 65, 310
5. On the differentiability at the boundary in conformal mapping. Proc. Am. Math. Soc. **12**, 614–620 (1961), Zbl. 100, 288
6. Boundary derivatives of minimal surfaces. Arch. Ration. Mech. Anal. **38**, 241–256 (1970), Zbl. 209, 418

Weierstraß, K.
1. Mathematische Werke. Vol. 3. Mayer & Müller, Berlin 1903, Jbuch 34, 23
2. Analytische Bestimmung einfach zusammenhängender Minimalflächen, deren Begrenzung aus geradlinigen, ganz im endlichen liegenden Strecken besteht. Math. Werke **3**, 221–238. Mayer & Müller, Berlin 1903, Jbuch 34, 23
3. Untersuchungen über die Flächen, deren Mittlere Krümmung überall gleich Null ist. Math. Werke **3**, 39–52. Mayer & Müller, Berlin 1903, Jbuch 34, 23

Wente, H.
1. The Plateau problem for symmetric surfaces. Arch. Ration. Mech. Anal. **60**, 149–169 (1976), Zbl. 328.53003
2. Large solutions to the volume constrained Plateau problem. Arch. Ration. Mech. Anal. **75**, 59–77 (1980), Zbl. 473.49029

White, B.
1. Existence of least area mappings on N-dimensional domains. Ann. Math., II. Ser., **118**, 179–185 (1983), Zbl. 526.49029
2. Tangent cones to two-dimensional area-minimizing currents are unique. Duke J. Math. **50**, 143–160 (1983), Zbl. 538.49030
3. The least area bounded by multiples of a curve. Proc. Am. Math. Soc. **90**, 230–232 (1984), Zbl. 529.49023
4. Mappings that minimize area in their homotopy classes. J. Differ. Geom. **20**, 433–446 (1984), Zbl. 566.57018
5. Generic regularity of unoriented two-dimensional area minimizing surfaces. Ann. Math., II. Ser., **121**, 595–603 (1985)
6. Homotopy classes in Sobolev spaces and energy minimizing maps. Bull. Am. Math. Soc., New Ser. **13**, 166–168 (1985), Zbl. 595.58011

7. Some recent developments in differential geometry. Math. Intell. **11**, No. 4, 41–47 (1989), Zbl. 701.53004
8. New applications of mapping degrees to minimal surface theory. J. Differ. Geom. **29**, 143–162 (1989), Zbl. 656.58009
9. On the topological type of minimal submanifolds. Topology **31**, No. 2, 445–448 (1992), Zbl. 767.53043
10. Existence of smooth embedded surfaces of prescribed genus that minimize parametric even elliptic functionals on 3-manifolds. J. Diff. Geom. **33**, 413–443 (1991), Zbl. 737.53009
11. The bridge principle for stable minimal surfaces. Calc. Var. **2**, 405–425 (1994)

Wienholtz, D.
1. Der Ausschluß von eigentlichen Verzweigungspunkten bei Minimalflächen vom Typ der Kreisscheibe. Diplomarbeit, München 1994, SFB 256, No. 37, 1996
2. Zum Ausschluß von Randverzweigungspunkten. Thesis, Bonn University, 1996

Wohlrab, O.
1. Zur numerischen Behandlung von parametrischen Minimalflächen mit halbfreien Rändern. Dissertation, Bonn 1985, Zbl. 687.65074
2. Die Berechnung und graphische Darstellung von Randwertproblemen für Minimalflächen. In: H. Jürgens, D. Saupe (editors), Visualisierung in Mathematik und Naturwissenschaften. Springer, Berlin Heidelberg New York 1989

Wolf, K.L.
1. Physik und Chemie der Grenzflächen. Springer, Berlin Göttingen Heidelberg, Vol. 1, 1957, Vol. 2, 1959
2. Tropfen, Blasen und Lamellen. Springer, Berlin Heidelberg New York 1968

Yau, S.T.
1. Survey on partial differential equations in differential geometry. In: Seminar on differential geometry, Ann. Math. Stud. **102**, 1982, 3–71, Zbl. 478.53001
2. Minimal surfaces and their role in differential geometry. Global Riemannian geometry. Proc. Symp., Durham/Engl. 1982, 99–103 (1984), Zbl. 615.53053
3. Nonlinear analysis in geometry. Monogr. 33, Enseign. Math., II. Ser., **33**, 109–158 (1987), Zbl. 631.53002

Ye, R.
1. Randregularität von Minimalflächen. Diplomarbeit, Bonn, 1984
2. A priori estimates for minimal surfaces with free boundary which are not minima of the area. Manuscr. Math. **58**, 95–107 (1987), Zbl. 678.49031
3. Regularity of a minimal surface at its free boundary. Math. Z. **198**, 261–275 (1988), Zbl. 641.58014

Young, L.C.
1. On the isoperimetric ratio for a harmonic surface. Proc. Lond. Math. Soc. (2) **49**, 396–408 (1947), Zbl. 37, 394
2. Some new methods in two-dimensional variational problems with special reference to minimal surfaces. Commun. Pure Appl. Math. **9**, 625–632 (1956), Zbl. 72, 114

IV. The Minimal Surface Equation

Leon Simon

Contents

Introduction

The minimal surface equation (MSE) for functions $u : \Omega \to \mathbb{R}$, Ω a domain of $\mathbb{R}^2$, can be written

$$(1+u_y^2)u_{xx} - 2u_x u_y u_{xy} + (1+u_x^2)u_{yy} = 0\,,$$

or equivalently $u_{xx}+u_{yy}-(1+|Du|^2)^{-1}(u_x^2u_{xx}+2u_xu_yu_{xy}+u_y^2u_{yy}) = 0$, where $u_x = \frac{\partial u(x,y)}{\partial x}$, $u_y = \frac{\partial u(x,y)}{\partial y}$. Generally, for domains $\Omega \subset \mathbb{R}^n$ and functions $u : \Omega \to \mathbb{R}$ depending on the n variables $(x^1, \ldots, x^n) \in \Omega$, $n \geq 2$, the MSE can be written

$$\sum_{i,j=1}^{n} \left(\delta_{ij} - \frac{u_i u_j}{(1+|Du|^2)}\right) u_{ij} = 0\,,$$

where $u_i = D_i u \equiv \frac{\partial u}{\partial x^i}$ and $u_{ij} = D_i D_j u$. Notice that this is a *quasilinear elliptic* equation: that is, it is linear in the second derivatives, and the coefficient matrix $(\delta_{ij} - \frac{u_i u_j}{(1+|Du|^2)})$ is positive definite[1] depending only on the derivatives up to first order. The equation can alternatively be written in "divergence form"

$$\sum_{i=1}^{n} D_i \left(\frac{D_i u}{\sqrt{1+|Du|^2}}\right) = 0\,, \tag{1}$$

which is readily checked using the chain rule and the fact that $\frac{\partial}{\partial p_j}(\frac{p_i}{\sqrt{1+|p|^2}}) = (1+|p|^2)^{-1/2}(\delta_{ij} - \frac{p_i p_j}{1+|p|^2})$.

Because of its geometric significance, the MSE has historically attracted perhaps more interest than any other quasilinear elliptic equation. The geometric significance of the MSE arises from the fact that it is the Euler-Lagrange equation for the area functional

$$\mathcal{A}(u) = \int_\Omega \sqrt{1+|Du|^2}\,,$$

representing the area (n-dimensional Hausdorff measure) of the graph of u. Thus the MSE expresses the fact that the first variation $\delta\mathcal{A}(u)(\zeta) \equiv \frac{d}{dt}\mathcal{A}(u + t\zeta)|_{t=0}$ vanishes for variations $u + t\zeta$ with ζ a smooth function of compact support in Ω. Indeed direct computation shows that, for $\zeta \in C^1(\Omega)$ with compact support,

$$\delta\mathcal{A}(u)(\zeta) = \int_\Omega \sum_{i=1}^{n} \frac{D_i u D_i \zeta}{\sqrt{1+|Du|^2}} \equiv -\int_\Omega \zeta \sum_{i=1}^{n} D_i \left(\frac{D_i u}{\sqrt{1+|Du|^2}}\right),$$

[1] The coefficient matrix $(\delta_{ij} - \frac{u_i u_j}{(1+|Du|^2)})$ has $(n-1)$ eigenvalues equal to 1 and the remaining eigenvalue equal to $(1+|Du|^2)^{-1}$; the MSE is thus a "nonuniformly elliptic" quasilinear equation, because the ratio of the maximum and minimum eigenvalues of the coefficient matrix has unbounded dependence on Du.

where we used integration by parts to give the second identity. Thus we see that $u \in C^2(\Omega)$ is a stationary point of $\mathcal{A}$ if and only if u satisfies the divergence form equation (1), i.e. the MSE.

As a matter of fact, because $\sqrt{1+|p|^2}$ is locally a uniformly convex function of $p = (p_1, \ldots, p_n)$ on $\mathbb{R}^n$, it is easily checked that $\mathcal{A}(u + t(v - u))$ is a strictly convex function of $t \in \mathbb{R}$ (one just directly computes the second derivative with respect to t by differentiation under the integral); then since this function has zero first derivative at $t = 0$ in case u is a solution of the MSE, we conclude that any solution u of the MSE automatically minimizes the area functional $\mathcal{A}(v)$ uniquely among $C^1(\Omega)$ functions v agreeing with u outside a compact subset of Ω. With only slightly more effort it can even be shown (see e.g. Simon (1983b)) that the graph $G = \{(x, u(x)) : x \in \Omega\}$ (thought of as an oriented embedded surface in $\mathbb{R}^{n+1}$) actually minimizes area in $\Omega \times \mathbb{R}$ relative to all oriented locally properly embedded hypersurfaces $M \subset \Omega \times \mathbb{R}$ with $(\overline{M} \backslash M) \cap (\Omega \times \mathbb{R}) = \emptyset$ and $M \backslash K = G \backslash K$ for some compact $K \subset \Omega \times \mathbb{R}$, regardless of the topological type of M. (In fact G minimizes area relative to all integer multiplicity currents T in $\Omega \times \mathbb{R}$ with null boundary in $\Omega \times \mathbb{R}$ and with $G = T$ outside a compact subset of $\Omega \times \mathbb{R}$—for the terminology here, we refer to Simon (1983a); however, if Ω is non-convex, G need not be area minimizing relative to comparisons which are not restricted to stay in $\Omega \times \mathbb{R}$, as shown by an example of Hardt, Lau & Lin (1987).)

Of course solutions of the MSE are alternatively characterized by saying that the graph G (again viewed as an embedded hypersurface in $\mathbb{R}^{n+1}$) has zero mean curvature, because, by a well-known geometric principle (see e.g. Osserman (1986), §1, §3), zero mean curvature is equivalent to stationarity with respect to the area functional.

In view of the above discussion, it is clear that one can hope to gain insight into the qualitative and quantitative properties of solutions by using either geometric or quasilinear PDE techniques. In the case of the 2-dimensional equation there is also the possibility of using complex analysis or quasiconformal techniques, related for example to the fact that the Gauss map of a 2-dimensional minimal surface is *conformal.* (See below for discussion of the Gauss map.)

Here we begin in the 2-dimensional case, describing the main results of the "classical era" in the study of the MSE, which dates back at least to the work of Bernstein in the early part of this century.

We then give a brief survey of the main results proved during the development of the higher dimensional theory (beginning in the 1960's) and conclude with a discussion of more recent developments concerning entire solutions.

§1. Classical (2-Dimensional) Theory

1.1 Local Regularity Results in 2 Dimensions

We recall the Hölder seminorms

$$[f]_{\alpha;\Omega} = \sup_{(x_1,y_1),\,(x_2,y_2)\in\Omega,\ (x_1,y_1)\neq(x_2,y_2)} \frac{|f(x_1,y_1)-f(x_2,y_2)|}{|(x_1,y_1)-(x_2,y_2)|^\alpha},$$

where $\alpha \in (0,1]$, so that $[f]_{\alpha;\Omega} < \infty$ is equivalent to the uniform Hölder continuity of f with Hölder exponent α on Ω; $[f]_{\alpha;\Omega}$ is called the Hölder coefficient, exponent α, of f on Ω.

Recall that, by virtue of the classical work of Morrey (1938) and Nirenberg (1953) (see e.g. Gilbarg & Trudinger (1983) for a detailed discussion of this theory), we know that if u is a C^2 solution of a 2-dimensional equation of the form $au_{xx} + 2bu_{xy} + cu_{yy} = 0$, with coefficient matrix uniformly bounded and uniformly positive definite (that is, $|a|$, $|b|$, $|c| \le \Lambda$, $a\xi^2 + 2b\xi\eta + c\eta^2 \ge \lambda(\xi^2+\eta^2)$ for all $(\xi,\eta) \in \mathbb{R}^2$, where λ, Λ are positive constants), then we have local estimates for the Hölder continuity of the first derivative of u depending only on Λ, λ and $\sup|u|$. More precisely, for each $\theta \in (0,1)$,

$$(1) \qquad R \sup_{B_{\theta R}} |Du| + R^{1+\alpha}[Du]_{\alpha;B_{\theta R}} \le C \sup_{B_R} |u|,$$

where $C > 0$ depends only on θ, λ and Λ, and where $\alpha \in (0,1)$ depends only on λ, Λ. It is to be emphasized that the estimate here requires no continuity hypotheses on the coefficients a, b, c, which in practice might depend on u and its derivatives. No such estimate is valid in case there are more than 2 independent variables (this is one of the main theoretical differences between elliptic equations in 2 and more than 2 variables), although there is a partial result in this direction which is adequate for the study of divergence form quasilinear equations; see the later discussion in §2.1. (There is also such a result for non-divergence form quasilinear elliptic equations with appropriate smoothness restrictions on the coefficients—see Ladyzhenskaya & Ural'tseva (1968) or Gilbarg & Trudinger (1983), Ch. 13.)

The classical Schauder theory (Gilbarg & Trudinger (1983), Ch. 6) tells us furthermore that if $k \in \{0,1,\ldots\}$ and if we assume that the coefficients a, b, c satisfy $\sum_{j=1}^k R^j \max(|D^j a|, |D^j b|, |D^j c|) + R^{k+\alpha} \max([D^k a]_{\alpha;B_R(x_0,y_0)}, [D^k b]_{\alpha;B_R(x_0,y_0)}, [D^k c]_{\alpha;B_R(x_0,y_0)}) \le \Lambda$, then the Hölder coefficients of all the derivatives of u up to order $k+2$ are controlled:

$$(2) \quad \sum_{j=1}^{k+2} R^j \sup_{B_{\theta R}(x_0,y_0)} |D^j u| + R^{k+2+\alpha}[D^{k+2}u]_{\alpha;B_{\theta R}}(x_0,y_0) \le C \sup_{B_R(x_0,y_0)} |u|,$$

where C depends only on θ, α, λ, Λ, and k. (Here $D^j u$ means the collection of all partial derivatives of u of order j.)

Clearly the first inequality above can be applied to solutions of the MSE (with $a(x,y) = 1+u_y^2(x,y)$, $b(x,y) = -u_x(x,y)u_y(x,y)$, $c(x,y) = 1+u_x^2(x,y)$), but the catch is that then the constant Λ depends on $\sup_{B_R} |Du|$, so the inequality only holds with "fixed" C provided we already have a fixed bound on the sup of $|Du|$. Of course if we do have such a fixed bound on $|Du|$ then we can apply the second set of inequalities successively for $k = 0, 1, \ldots,$ in order to deduce that we can bound all the derivatives of all orders of u as in (2), even though the equation is quasilinear rather than linear.

That is, the question of controlling the higher derivatives (as in (2)) for solutions of the MSE, or any quasilinear elliptic equation

$$a(x,y,u,u_x,u_y)u_{xx} + 2b(x,y,u,u_x,u_y)u_{xy} + c(x,y,u,u_x,u_y)u_{yy} = 0\,,$$

with coefficients a, b, c smooth functions of their arguments, reduces to the problem of controlling $\sup|u|$ and $\sup|Du|$. It was therefore a problem of central interest to control the first derivatives of the solutions of the MSE in terms of $\sup|u|$. Such a result was first obtained by R. Finn in 1953 (Finn (1953), (1954)): If u is a C^2 solution of the MSE in a disc $B_R(x_0,y_0) \subset \mathbb{R}^2$, then

$$\sup_{B_{R/2}(x_0,y_0)} |Du(x,y)| \le C_1 \exp\bigl(C_2 R^{-1} \sup_{B_R(x_0,y_0)} |u|\bigr)\,, \tag{3}$$

where C_1, C_2 are fixed constants. Subsequently, using comparison principles, Finn (1963) and Serrin (1963) refined the form of the result, so that we now know for example that if u is a positive solution of the MSE on the disc $x^2+y^2 < 1$, then $|Du(0)| \le C\exp(\frac{\pi u(0)}{2})$, and this is known to be best possible as far as the constant $\frac{\pi}{2}$ in the exponential is concerned.

There are numerous important consequences of the gradient estimate (3) and the consequent derivative estimates (2); indeed such inequalities are always of fundamental importance in the study of quasilinear elliptic and parabolic problems. For example, the estimates immediately imply very strong compactness properties for the family of all solutions to the MSE:

Theorem. *If u_j, $j = 1, 2, \ldots,$ are solutions of the MSE on a domain $\Omega \subset \mathbb{R}^2$, and if $\limsup_{j\to\infty} \sup_K |u_j| < \infty$ for each compact $K \subset \Omega$, then there is a subsequence $u_{j'}$ of u_j converging locally in Ω with respect to the C^k-norm for each k to a $C^\infty(\Omega)$ solution u of the MSE.*

One can easily check that this follows directly from the derivative bounds (2) together with the Arzela-Ascoli Lemma.

Gradient estimates (bounding $|Du|$ in terms of $\sup|u|$) are known for many quasilinear elliptic equations, and there is an extensive literature on them beginning with the work of Bernstein in the early part of this century (see e.g. the general discussion in the book of Gilbarg & Trudinger (1983)). In particular, Finn's 1954 paper actually showed that such a gradient estimate is valid for any "equation of minimal surface type"; an equation $a(x,y,u,u_x,u_y)u_{xx} + 2b(x,y,u,u_x,u_y)u_{xy} + c(x,y,u,u_x,u_y)u_{yy} = 0$

is said to be of minimal surface type if there are positive constants λ, Λ such that $\lambda\sum_{i,j=1}^{2} g^{ij}\xi^i\xi^j \le a\xi_1^2 + 2b\xi_1\xi_2 + c\xi_2^2 \le \Lambda\sum_{i,j=1}^{2} g^{ij}\xi^i\xi^j$, where $g^{ij} = (\delta_{ij} - \frac{u_i u_j}{1+|Du|^2})$ are the coefficients of the MSE. (Finn gives an alternative, but equivalent, characterization of them.) Such equations, and their non-homogeneous analogues, are also discussed in detail in Simon (1977b, c). Finn's 1954 discussion and also the discussion in Simon (1977b, c) is based on the fact that the *Gauss map* of a solution u of such an equation—i.e. the map $\nu = (1+|Du|^2)^{-1/2}(-u_x, -u_y, 1)$, thought of either as a map from the plane into the upper hemisphere of the sphere S^2 or as a map from the graph of u into the upper hemisphere of S^2—is *quasiconformal*.

1.2 Bernstein's Theorem in 2 Dimensions

One of the earliest theorems concerning solutions of the MSE is the famous result of Bernstein (1916) that only the trivial affine solutions $u(x,y) \equiv ax + by + c$ (a, b, c constants) can satisfy the MSE over the entire plane $\mathbb{R}^2$. Bernstein proved this as a special consequence of his remarkable "geometric theorem" which states any bounded C^2 function on $\mathbb{R}^2$ with graph $z = u(x,y)$ having non-negative Gauss curvature (i.e. with u satisfying $u_{xx}u_{yy} - u_{xy}^2 \le 0$) must necessarily have Gauss curvature identically zero. In particular bounded solutions $u \in C^2(\mathbb{R}^2)$ of any equation of the form $au_{xx} + 2bu_{xy} + cu_{yy} = 0$, where $a\xi^2 + 2b\xi\eta + c\eta^2 \ge \xi^2 + \eta^2$, $\sup|a|, |b|, |c| < \infty$, and where the coefficients a, b, c may depend on u and its derivatives, are necessarily constant. This follows because, by rotation of coordinates to diagonalize the Hessian matrix $\begin{pmatrix} u_{xx} & u_{xy} \\ u_{xy} & u_{yy} \end{pmatrix}$, one can show that the solution of any such equation automatically satisfies $u_{xx}u_{yy} - u_{xy}^2 \le -\gamma\,(u_{xx}^2 + 2u_{xy}^2 + u_{yy}^2)$ for suitable positive constant γ (γ determined by any upper bound for $|a|$, $|b|$, $|c|$), and so $u_{xx}u_{yy} - u_{xy}^2 \equiv 0$ implies that the Hessian of u is identically zero, which implies u is affine, which of course implies that u is constant by virtue of the assumed boundedness of u. Bernstein was thus able to prove the stated theorem for $C^2(\mathbb{R}^2)$ solutions u of the MSE, because $\varphi = \arctan u_x$, $\varphi = \arctan u_y$ are bounded solutions of the equation $(1+u_y^2)\varphi_{xx} - 2u_xu_y\varphi_{xy} + (1+u_x^2)\varphi_{yy} = 0$. This last fact is rather unpleasant to check by direct calculation, but one can easily see it by a geometric argument: The equation for φ simply says that φ, thought of as a function on the graph G, is harmonic relative to the intrinsic Laplacian operator on G. (See e.g. Osserman (1986) for discussion of this.) On the other hand the Gauss map of the graph of a solution of the MSE is conformal (see the discussion below), as is the stereographic projection S of $S^2\backslash\{(0,0,-1)\}$ into $\mathbb{R}^2$, and hence also the composition of S and the isometry of S^2 obtained by permuting the 2nd and 3rd coordinate functions (i.e. the map $(y^1,y^2,y^3) \mapsto S(y^1,y^3,y^2)) : S^2\backslash\{(0,-1,0)\} \to \mathbb{R}^2$). Thus, with respect to suitable conformal coordinates for the graph of u, the function $-(1+|Du|^2)^{-1/2}u_x + i(1+|Du|^2)^{-1/2}$ is holomorphic, so that arctan of the ratio of the real and imaginary parts (i.e. $\arctan u_x$) is harmonic in these coordinates, hence harmonic on the graph of u, as required.

Bernstein's original proof of his general geometric theorem actually had a gap in the final stage of the argument (related to connectivity properties of open subsets of $\mathbb{R}^2$); an appropriate connectivity property was formulated and proved by E. Hopf (1950a), who then reworked Bernstein's entire proof in Hopf (1950b). E. Mickle (1950) also published a revision of Bernstein's proof. Actually the geometric theorem (and its proof) applies to any $u \in C^2(\mathbb{R}^2)$ with graph of non-negative Gauss curvature and with *sub-linear* growth; that is, $(x^2+y^2)^{-1/2}u(x,y) \to 0$ uniformly as $\sqrt{x^2+y^2} \to \infty$.

Although the Bernstein/Hopf proof remains the sole known method for proving the general geometric theorem, in more recent times many alternative proofs of the Bernstein theorem for solutions of the MSE were given, some of which are elementary applications of complex variable methods and some of which apply to more general equations. See for example Bers (1951), (1953), Finn (1954), Fleming (1962), Giusti (1984), Nitsche (1957) for various proofs. We here briefly describe a method of proof modified from Simon (1977b, c); although it is not quite the simplest proof, it has the advantage of applying (without significant change) to the whole class of equations of minimal surface type.

The first step in the argument is the observation that there is a good bound on the amount of area of $G = \operatorname{graph} u$ inside the solid 3-dimensional ball B_R of radius R; specifically, for each $R > 0$,

$$|G \cap B_R| \le CR^2 ,$$

where C is a fixed constant not depending on R, and where $|G \cap B_R|$ means area of $G \cap B_R$. For the proof (valid for the graph of a solution of any equation of minimal surface type) we refer to Simon (1977b). Actually for the case of the MSE there is a particularly simple proof of this (and the analogous result in n-dimensions) using a "test function" argument (see e.g. Gilbarg & Trudinger (1983), §16.2).

Having obtained such a bound on the area, the next step involves consideration of the Gauss map $\nu = (\nu^1, \nu^2, \nu^3) \equiv (1+|Du|^2)^{-1/2}(-u_x, -u_y, 1)$ as follows: The Gauss map induces a linear map $\ell : T_X G \to T_X G$, $X = (x, y, u(x,y))$, where $T_X G =$ the tangent space of G at X, defined by

$$\ell(\tau) = D_\tau \nu , \quad \tau \in T_X G ,$$

D_τ just denoting directional differentiation. One easily checks that then ℓ is a *symmetric* map and that the eigenvalues κ_1, κ_2 are geometrically just the principal normal curvatures of the surface G at the point $X = (x, y, u(x,y))$; see Osserman (1986) for discussion. Further, we have the identities

$$\det \ell \, dA = \kappa_1 \kappa_2 \, dA = K \, dA = \nu_3^{-1} d\nu^1 \wedge d\nu^2 \equiv d((1+\nu_3)^{-1}(\nu_1 d\nu_2 - \nu_2 d\nu_1))$$

$$\operatorname{trace} \ell = \kappa_1 + \kappa_2 = (1+|Du|^2)^{-3/2}((1+u_y^2)u_{xx} - 2u_x u_y u_{xy} + (1+u_x^2)u_{yy}) ,$$

where in the first set of identities dA denotes the area form of the graph and where $K = \kappa_1\kappa_2 = \det \ell = (1+|Du|^2)^{-2}(u_{xx}u_{yy} - u_{xy}^2)$ is the Gauss curvature of the graph. This is all valid for an arbitrary graph $z = u(x, y)$ independent of whether or not u satisfies the MSE, but notice that the second identity gives

$$u \text{ satisfies the MSE} \iff \kappa_1 + \kappa_2 = 0 \iff \kappa_1 = -\kappa_2$$

and in particular we see that u satisfies the MSE $\iff$ the Gauss map of graph u is conformal and orientation reversing at all points where $(\kappa_1, \kappa_2) \neq (0, 0)$. Also, if u satisfies the MSE, we can use the first set of identities again to conclude

$$\mathcal{C}^2\, dA = -2d((1+\nu_3)^{-1}(\nu_1 d\nu_2 - \nu_2 d\nu_1)),$$

where $\mathcal{C}^2 \equiv \kappa_1^2 + \kappa_2^2 \equiv \|\ell\|^2 \equiv \sum_{j=1}^3 |d\nu^j|^2$. Now choose a non-negative $\zeta \in C^\infty(\mathbb{R}^3)$ with $\zeta \equiv 0$ for $|X| \geq R$ and notice that (after multiplying through by ζ^2), the above identity implies

$$\zeta^2\mathcal{C}^2\, dA = -2d(\zeta^2(1+\nu_3)^{-1}(\nu_1 d\nu_2 - \nu_2 d\nu_1)) + 4\zeta d\zeta \wedge ((1+\nu_3)^{-1}(\nu_1 d\nu_2 - \nu_2 d\nu_1)).$$

By integration (since Stokes' Theorem says that the first term on the right here integrates to zero) we obtain

$$\begin{aligned}\int_{G\cap B_R} \zeta^2\mathcal{C}^2 &= 4\int_{G\cap B_R} \zeta d\zeta \wedge ((1+\nu_3)^{-1}(\nu_1 d\nu_2 - \nu_2 d\nu_1))) \\ &\leq 4\int_{G\cap B_R} \zeta\, |(1+\nu_3)^{-1}(\nu_1 d\nu_2 - \nu_2 d\nu_1))|\, |d\zeta| \\ &\leq 8\int_{G\cap B_R} \zeta\mathcal{C}\, |d\zeta|.\end{aligned}$$

Using the Cauchy-Schwarz inequality on the right side we thus get

$$\int_{G\cap B_R} \zeta^2\mathcal{C}^2 \leq 64\int_{G\cap B_R} |d\zeta|^2.$$

Finally, in view of the area bound mentioned above (which tells us that $|G \cap B_\rho| \leq C\rho^2$ for all $\rho \in (0, R]$), we can use the "logarithmic cut-off trick" as follows: For $R \geq 2$, we choose $\zeta = \zeta(|X|)$ to be 1 for $|X| < \sqrt{R}$, $\zeta \equiv 0$ for $|X| > R$, and $\zeta = \frac{\log(R/|X|)}{\log\sqrt{R}}$ for $|X| \in (\sqrt{R}, R)$. Plugging this into the above inequality we get

$$\int_{G\cap B_{\sqrt{R}}} \mathcal{C}^2 \leq \frac{C}{(\log R)^2}\int_{G\cap(B_R \setminus B_{\sqrt{R}})} |X|^{-2}.$$

But now

$$\begin{aligned}\int_{G\cap(B_R\setminus B_{\sqrt{R}})}|X|^{-2} &= R^{-2}|G\cap(B_R\setminus B_{\sqrt{R}})| + \int_{G\cap(B_R\setminus B_{\sqrt{R}})}(|X|^{-2}-R^{-2})\\ &= R^{-2}|G\cap(B_R\setminus B_{\sqrt{R}})| + 2\int_{\sqrt{R}}^{R}\rho^{-3}|G\cap(B_\rho\setminus B_{\sqrt{R}})|\,d\rho\\ &\le R^{-2}|G\cap B_R| + 2\int_{\sqrt{R}}^{R}\rho^{-3}|G\cap B_\rho|\,d\rho\\ &\le C + C\log R,\end{aligned}$$

where we used the identity $2\int_{|X|}^{R}\rho^{-3}\,d\rho = |X|^{-2} - R^{-2}$ and Fubini's theorem, together with the area bound $|G\cap B_\rho| \le C\rho^2$. Hence we deduce

$$\int_{G\cap B_{\sqrt{R}}}\mathcal{C}^2 \le \frac{C}{\log R}$$

with C independent of R, and hence by letting $R\uparrow\infty$, we deduce that $\mathcal{C}\equiv 0$, which of course implies that G is a plane as claimed.

We now know that there are many other quasilinear elliptic equations in 2 independent variables which have the "Bernstein property" (that all C^2 solutions defined over all of $\mathbb{R}^2$ are necessarily affine). In particular due to the work of Finn (1954) and Simon (1977c) we know that any equation of minimal surface type has this Bernstein property. A proof of this is obtained by direct modification of the above argument; indeed notice that the above argument continues to be applicable without any essential change if instead of the hypothesis that u satisfies the MSE we require merely that the Gauss map be quasiconformal (so that $\mathcal{C}\le -CK$ for some constant C); since the graph of a solution of any equation of minimal surface type satisfies such an inequality (see the discussion in Simon (1977c)), we thus deduce a Bernstein theorem for any equation of minimal surface type. For other equations in 2 variables having the Bernstein property we refer to Aronsson (1968) and Simon (1995).

In view of the above non-existence of non-trivial entire solutions of the MSE over $\mathbb{R}^2$, it was natural to ask if there are severe restrictions on "exterior" solutions; that is, solutions which are C^2 on $\mathbb{R}^2\setminus\overline{\Omega}$, where Ω is a bounded domain in $\mathbb{R}^2$. Certainly there are non-trivial examples of this type; for example $u(x,y) = \cosh^{-1}(\sqrt{x^2+y^2})$ is such an example over the region $\sqrt{x^2+y^2} > 1$. (The graph of u is just the upper half of the catenoid $(\cosh z)^2 = x^2+y^2$.)

A definitive answer to the above question was given by Bers (1951) who established, using complex analysis methods, the following striking theorem:

Theorem. *Let $\Omega\subset\mathbb{R}^2$ be bounded and $u\in C^2(\mathbb{R}^2\setminus\overline{\Omega})$ satisfy the MSE. Then $Du(x,y)$ has a limit $a\in\mathbb{R}^2$ as $\sqrt{x^2+y^2}\to\infty$.*

Geometrically this may be reformulated to say that graph u has a non-vertical tangent plane at ∞ in the appropriate sense. But notice that for the catenoid example above we have $Du(x,y)\to 0$ as $|x|\to\infty$, whereas u is not asymptotically linear in the precise sense that $|u(x,y) - (ax + by + c)| \to 0$, because $\cosh^{-1}(\sqrt{x^2+y^2})$ grows logarithmically as $\sqrt{x^2+y^2} \to \infty$. Bers actually showed that such logarithmic growth is the worst that can happen (and one can check this from the theorem as stated by a simple linearization argument): that is, for any C^2 solution of the MSE defined over all of $\mathbb{R}^2 \setminus \overline{\Omega}$, where Ω is bounded, we always have constants a, b, c, d such that $|u(x,y) - (ax + by + c) - d\log(\sqrt{x^2+y^2})| \to 0$ as $\sqrt{x^2+y^2} \to \infty$.

The above theorem of Bers was extended to the entire class of equations of minimal surface type in Simon (1977c).

1.3 Curvature Estimates in 2 Dimensions

E. Heinz (1952) proved that 2-dimensional graphs $z = u(x,y)$ defined over a disc $B_R(x_0, y_0)$ have well-controlled curvature away from the boundary cylinder; he in fact showed, assuming that $u(x,y)$ satisfies the MSE in a disc $B_R(x_0,y_0)$, the Gauss curvature K of the graph $z = u(x,y)$ satisfies $|K| \leq CR^{-2}$ at the point $(x_0, y_0, u(x_0, y_0))$, where C is a fixed constant. From the discussion in §1.2 we know that $K = -\frac{1}{2}(\kappa_1^2 + \kappa_2^2) \equiv -\frac{1}{2}\mathcal{C}^2$, so this actually says

$$\mathcal{C}^2 \leq CR^{-2}$$

at the point $(x_0, y_0, u(x_0, y_0))$, with C a fixed constant.

Of course if u is defined on all of $\mathbb{R}^2$, we can let $R \to \infty$ here to deduce that $\kappa_1,\ \kappa_2 \equiv 0$, so graph u is a plane; thus we recover Bernstein's theorem. But quite apart from this application to Bernstein's theorem, the bound of Heinz gives important information about the nature of minimal graphs (Viz., that they cannot be very highly curved at a fixed distance away from the boundary cylinder).

Subsequent to Heinz' proof, various improvements were made to the estimates and proofs by Hopf (1953), Nitsche (1965), Osserman (1960), Finn & Osserman (1964); in particular, the estimate was improved to include a factor $(1 + |Du|^2)^{-1}$ on the right (thus $|K| \leq C(1 + |Du|^2)^{-1}R^{-2}$), so that the curvature estimate improves, and the graph becomes "flatter", as the gradient increases.

Analogous curvature estimates for other variational equations of minimal surface type were established by Jenkins (1961a,b) and Jenkins & Serrin (1963). The estimates were extended in Simon (1977c) to all equations of minimal surface type satisfying the appropriate Hölder continuity restriction on the coefficients; indeed in Simon (1977c) a Hölder continuity estimate for the unit normal of the graph is established for the entire class of equations of minimal surface type. Curvature estimates for solutions of the prescribed mean curvature equations and other non-homogeneous equations of minimal surface type ("equations of mean curvature type") were established by J. Spruck (1974) and L. Simon (1977c).

1.4 The Dirichlet Problem in 2 Dimensions

The Dirichlet problem for the MSE in $\mathbb{R}^2$ is the problem (usually on a bounded domain $\Omega \subset \mathbb{R}^2$) of finding a solution of the MSE in Ω subject to the constraint that the solution should attain prescribed boundary data on $\partial\Omega$. This is sometimes referred to as the non-parametric Plateau problem, since geometrically it is the problem of finding a least area graph over Ω (see the discussion in the introduction concerning area minimizing properties of solutions of the MSE) with prescribed boundary.

Investigations of such problems in the context of general quasilinear elliptic equations go back to Bernstein and earlier. The first observation is that implicit function theory methods, in combination with the $C^{2,\alpha}$ Schauder estimates for linear elliptic equations with Hölder continuous coefficients, lead directly to the solution of this problem in case the domain is suitably smooth (e.g. of class $C^{2,\alpha}$ for some $\alpha \in (0,1)$) and in case the boundary data is suitable smooth *and sufficiently small* (e.g., if the prescribed boundary data φ is of class $C^{2,\alpha}$ and has sufficiently small $C^{2,\alpha}$-norm, depending on Ω). Results of this type go back to Korn (1909), and we refer to the discussion in e.g. Gilbarg & Trudinger (1983) for a modern treatment; if we use the $C^{1,\alpha}$ Schauder theory instead of $C^{2,\alpha}$ Schauder theory then the same existence results apply with $C^{1,\alpha}$ norms in place of $C^{2,\alpha}$.

Due to the work of various mathematicians (including Korn, Bernstein, and Radó) in the early part of the century, it eventually became clear that one could expect a solution of the Dirichlet problem for the MSE on an arbitrary smooth domain Ω with arbitrary continuous boundary data if and only if the domain Ω is convex. For example Radó (1930) establishes the sufficiency of the convexity by showing that the classical solution of the parametric Plateau problem is actually a graph (i.e. it projects simply onto the plane, with smooth inverse map) if the prescribed boundary curve is the graph of a smooth function over the boundary of the convex region Ω. Finn (1965) was the first to show that the convexity of the domain Ω is a necessary condition if we wish to solve the Dirichlet problem on Ω with arbitrary continuous data, and a simple example of Nitsche (1965b) shows that on non-convex domains with a "re-entrant corner" we should not expect the Dirichlet problem to be solvable even for boundary data having small Lipschitz norm. On the other hand work of G. Williams (1984) (in all dimensions) shows that if the domain is smooth then the Dirichlet problem is always solvable with Lipschitz boundary data of sufficiently small Lipschitz norm.

By now there is a very well developed theory of quasilinear PDE, developed by C. B. Morrey (1938), L. Nirenberg (1953) and others in 2-dimensions and by Morrey, Ladyzhenskaya-Ural'tseva, De Giorgi, Nash, Moser, and many others in higher dimensions (see e.g. the books Ladyzhenskaya & Uralt'seva (1968), C. B. Morrey (1966), Gilbarg & Trudinger (1983)), which enable us to put these results for the MSE in a general quasilinear PDE context; see the further discussion in §2.4 below.

For general discussion of the Leray-Schauder approach for the Dirichlet problem of a large class of quasilinear elliptic equations in 2-variables on a convex domain (including the MSE as a very special case) we refer to the book Gilbarg & Trudinger (1983), Chapter 11. Here we must restrict ourselves to just a brief discussion of the method; we have chosen to include such a discussion because it is a classic example of the importance of *a-priori* estimates (as in §1.1 above) in the study of equations like the MSE.

We thus wish to discuss the Dirichlet problem

$$\begin{aligned} a\, u_{xx} + 2b\, u_{xy} + c\, u_{yy} &= 0 \quad \text{in } \Omega \\ u &= \lambda\, \varphi \quad \text{on } \partial\Omega\,, \end{aligned}$$

where φ is a given $C^{2,\alpha}(\overline{\Omega})$ function and a, b, c are the coefficients $1 + u_y^2$, $-u_x u_y$, $1 + u_x^2$ corresponding to the MSE; actually we can assume without additional complication that we are in the general quasilinear setting $a = a(x, y, u, Du)$, $b = b(x, y, u, Du)$, $c = c(x, y, u, Du)$, where $a = a(x, y, z, p)$, $b = b(x, y, z, p)$, $c = c(x, y, z, p)$ are smooth functions of $(x, y, z, p) \in \overline{\Omega} \times \mathbb{R} \times \mathbb{R}^2$ such that $a\xi^2 + 2b\xi\eta + c\eta^2 > 0$ for $(\xi, \eta) \neq (0, 0)$. We begin by defining $\mathcal{B}$ to be the Banach space of $C^{1,\alpha}(\overline{\Omega})$ functions, and we set up a family of operators $T_\lambda : \mathcal{B} \to \mathcal{B}$ by defining $T_\lambda(w)$, for given $(w, \lambda) \in C^{1,\alpha} \times [0, 1]$, to be the solution v of the linear problem

$$\begin{aligned} &a(x, y, w, Dw)v_{xx} + 2b(x, y, w, Dw)v_{xy} + c(x, y, w, Dw)v_{yy} = 0 \quad \text{in } \Omega \\ &v = \lambda\, \varphi \quad \text{on } \partial\Omega\,. \end{aligned}$$

The linear Schauder theory guarantees that there is a unique such v, and that it actually lies in $C^{2,\alpha}(\overline{\Omega}) \subset C^{1,\alpha}(\overline{\Omega})$, so that T_λ is well defined. Of course uniqueness of v follows just from the Hopf maximum principle for elliptic partial differential equations (Gilbarg & Trudinger (1983), Ch. 3). Notice that in this setup, finding a solution of the Dirichlet problem with boundary data φ corresponds precisely to finding a fixed point of the operator T_1, and this is what we now want to do. Since bounded subsets of $C^{2,\alpha}(\overline{\Omega})$ are compact in $C^{1,\alpha}(\overline{\Omega})$, as one checks by calculus, this shows that T_λ is a family of compact operators, and this compactness also makes it rather easy to check that $T_\lambda(w)$ is continuous and compact as an operator from $C^{1,\alpha}(\overline{\Omega}) \times [0, 1]$ into $C^{1,\alpha}(\overline{\Omega})$. Thus we can apply the Leray-Schauder fixed point theorem (Leray & Schauder (1934), Gilbarg & Trudinger (1983), Ch. 11]), which asserts that T_1 has a fixed point provided only that we can prove that there is a fixed constant $M > 0$ such that $\|w\|_{\mathcal{B}} < M$ whenever $(w, \lambda) \in \mathcal{B} \times [0, 1]$ satisfy $T_\lambda(w) = w$. Notice that this simply amounts to proving that there is a fixed constant $M > 0$ such that

$$\|u\|_{C^{1,\alpha}(\overline{\Omega})} < M \tag{1}$$

whenever u is a $C^{2,\alpha}(\overline{\Omega})$ solution of one of the problems

$$a(x,y,u,Du)u_{xx} + 2b(x,y,u,Du)u_{xy} + c(x,y,u,Du)u_{yy} = 0 \quad \text{in } \Omega$$
$$u = \lambda\varphi \quad \text{on } \partial\Omega$$

for $\lambda \in [0,1]$. Here $\|u\|_{C^{1,\alpha}(\overline{\Omega})}$ means (in the notation of §1.1) $\sup_\Omega |u| + \sup_\Omega |Du| + [Du]_{\alpha;\Omega}$. Thus the Leray-Schauder approach reduces the existence question to proving a bound of the form (1); bounds of the form (1) are called *a-priori* bounds because they are fixed bounds on the class of all possible solutions of a specified problem or group of problems, without the necessity of having any advance knowledge about which (if any) of such solutions actually exist or how they were obtained.

The problem of checking (1) turns out, in view of the estimates of §1.1 above, not to be too difficult: First we observe, by virtue of the maximum principle (see e.g. Gilbarg & Trudinger (1983), Chapter 3), that any $C^2(\Omega) \cap C^0(\overline{\Omega})$ solution of the MSE (or indeed of any equation $au_{xx} + 2bu_{xy} + cu_{yy} = 0$, with coefficient matrix uniformly bounded and uniformly positive definite in compact subsets $K \subset \Omega$—that is, for each compact $K \subset \Omega$, $|a|$, $|b|$, $|c| \le \Lambda_K$ and $a\xi^2 + 2b\xi\eta + c\eta^2 \ge \lambda_K(\xi^2 + \eta^2)$ for some constants Λ_K, $\lambda_K > 0$) we have $\sup_{\overline{\Omega}} |u| = \sup_{\partial\Omega} |u|$, so that if u has prescribed boundary data φ then we have

$$\text{(2)} \qquad \sup_{\overline{\Omega}} |u| = \sup_{\partial\Omega} |\varphi| \,.$$

That is, we can uniformly control the supremum of the absolute value of any solution of the Dirichlet problem. Next one observes that the gradient of u also satisfies a maximum principle; indeed this is again valid for solutions of the same general class of equations $au_{xx} + 2bu_{xy} + cu_{yy} = 0$ as above. Because, dividing the equation through by a, and differentiating with respect to the variable y, we find that $\varphi = u_y$ satisfies the equation $\varphi_{xx} + (2a^{-1}b\varphi_x + a^{-1}c\varphi_y)_y = 0$, at least weakly, and this is of the general form $\sum_{i,j=1}^2 D_i(\alpha_{ij}D_j\varphi) = 0$, where we use the alternative notation $D_1 = \frac{\partial}{\partial x}$, $D_2 = \frac{\partial}{\partial y}$ and where the coefficients $\alpha_{11} = 1$, $\alpha_{12} = 0$, $\alpha_{21} = 2a^{-1}b$, $\alpha_{22} = a^{-1}c$ are bounded on compact subsets of Ω and $\sum_{i,j} \alpha_{ij}\xi_i\xi_j \ge a^{-1}\lambda_K(\xi_1^2 + \xi_2^2) \ge \Lambda_K^{-1}\lambda_K(\xi_1^2 + \xi_2^2))$ on compact subsets $K \subset \Omega$. Thus $\varphi = \pm u_y$ must satisfy a maximum principle on Ω (since the solution of any such equation satisfies a maximum principle on Ω). Thus we have $\sup_{\overline{\Omega}} |u_y| = \sup_{\partial\Omega} |u_y|$ for any $C^2(\Omega) \cap C^1(\overline{\Omega})$ solution of the equation. Since the composite of u with any orthogonal transformation satisfies a similar equation, the same argument shows that $\cos\theta\, u_x + \sin\theta\, u_y$ also satisfies a maximum principle for any $\theta \in \mathbb{R}$, and hence $|Du|$ satisfies a maximum principle as claimed. On the other hand if Ω is a uniformly convex domain in $\mathbb{R}^2$ and if φ is a $C^2(\mathbb{R}^2)$ function, then we can construct, corresponding to each point $(x_0, y_0) \in \partial\Omega$, affine functions $\pi_\pm(x,y)$ with the properties that

$$\pi_+(x_0,y_0) = \pi_-(x_0,y_0) = \varphi(x_0,y_0)$$
$$\pi_+(x,y) \ge \varphi(x,y)\,, \qquad \pi_-(x,y) \le \varphi(x,y) \quad \forall\, x,y \in \partial\Omega$$

and such that $\max(|D\pi_+|, |D\pi_-|) \le C$, with C depending only on Ω and Q, where Q is any upper bound for the C^2 norm of φ. Such functions are called upper and lower linear barriers respectively.

Now notice that $u - \pi_\pm$ satisfies exactly the same equation (with the same coefficients a, b, c) as u, and hence the same maximum principle argument that we used for u applies with $u - \pi_\pm$ to show that

$$\pi_-(x,y) \le u(x,y) \le \pi_+(x,y), \quad x, y \in \overline{\Omega},$$

provided that $u \in C^2(\Omega) \cap C^0(\overline{\Omega})$. Then, if $u \in C^2(\Omega) \cap C^1(\overline{\Omega})$, consideration of the normal derivative at the point (x_0, y_0), in light of the above inequalities, shows that $|Du(x_0,y_0)| \le \max(|D\pi_+|, |D\pi_-|, |D\varphi(x_0,y_0)|)$; that is we have a fixed bound, depending on φ and Ω, for $\sup_{\partial\Omega}|Du|$. By virtue of the maximum principle for $|Du|$ discussed above, we then conclude

$$\sup_{\overline{\Omega}} |Du| \le C, \tag{3}$$

where C is a fixed constant determined only by $\varphi, \partial\Omega$; as a matter of fact the argument shows we can choose C so that the inequality applies uniformly to solutions $u \in C^2(\Omega) \cap C^1(\overline{\Omega})$ having boundary data $\lambda\varphi$ in place of φ for any $\lambda \in [0, 1]$. Now something rather nice happens: Since we have a fixed bound on $\sup|u|$ and $\sup|Du|$, depending only on Ω and φ, and not depending on u, then the maximum and minimum eigenvalues of the coefficient matrix are bounded above and below by fixed positive constants independent of the solution u (and independent of λ); hence we can apply the estimates (1) of §1.1, thus giving the required bounds (1) for $\|u\|_{C^{1,\alpha}(\overline{\Omega})}$, as required for the Leray-Schauder procedure. Thus there is a $C^{2,\alpha}(\overline{\Omega})$ solution of the Dirichlet problem as claimed.

Since it is a nice application of the compactness theorem of §1.1, we also discuss now how one can handle boundary data φ which is merely continuous. In this case one first approximates φ uniformly by smooth functions φ_ℓ and applies the above existence to give corresponding solutions u_ℓ of the Dirichlet problem for the MSE with boundary data φ_ℓ. By applying the compactness result of §1.1 above, we see that a subsequence $u_{\ell'}$ converges locally in Ω with respect to the C^k norm for each k to a solution u of the MSE, so all that remains to be checked is that this u is C^0 on $\overline{\Omega}$ and agrees with φ on $\partial\Omega$. But this is easily seen as follows: Let v_1, v_2 be two $C^\infty(\Omega) \cap C^0(\overline{\Omega})$ solutions of the MSE. Then $\sum_{i=1}^2 D_i[A_i(Dv_\ell)] = 0$, $\ell = 1, 2$, where $A_i(p) = \frac{p_i}{\sqrt{1+|p|^2}} \equiv \frac{\partial\sqrt{1+|p|^2}}{\partial p_i}$ for $p = (p_1, p_2) \in \mathbb{R}^2$. But now by calculus $A_i(q) - A_i(p) = \sum_{j=1}^2 a_{ij}(p,q)(q_j - p_j)$, where $a_{ij}(p,q) = \int_0^1 A_{ij}(p + t(q-p))\,dt$, with $A_{ij}(p) = \frac{\partial^2\sqrt{1+|p|^2}}{\partial p_i \partial p_j}$, a positive definite matrix. We thus see that $\varphi = v_1 - v_2$ satisfies the equation $D_i(\tilde{a}_{ij} D_j \varphi) = 0$, where $(\tilde{a}_{ij}) = (a_{ij}(Dv_1, Dv_2))$ is positive definite and continuous in Ω. Thus the maximum principle applies to $\pm\varphi$; that is, the *difference* of any two solutions of the MSE satisfies the maximum principle, and hence we conclude in particular that

$$\sup_{\overline{\Omega}} |u_j - u_\ell| = \sup_{\partial\Omega} |\varphi_j - \varphi_\ell|\,, \quad j,\, \ell = 1, 2, , \ldots,$$

so that $\{u_j\}$ is Cauchy in $C^0(\overline{\Omega})$ because $\{\varphi_j\}$ is Cauchy in $C^0(\partial\Omega)$. Thus u_j converges uniformly on $\overline{\Omega}$.

Notice that this comparison argument also establishes the uniqueness of $C^2(\Omega) \cap C^0(\overline{\Omega})$ solutions of the Dirichlet problem, because it shows that the difference of such solutions satisfies the maximum principle.

§2. Higher Dimensional Theory

2.1 Local Regularity Results

The local regularity theory for the MSE in 2 dimensions described in §1.1 was essentially completed by 1954, and naturally attempts were made to develop an analogous theory in higher dimensions. Before such a development could be completed, there were two main difficulties to be overcome:

(i) There was no known analogue of the Morrey estimate (inequality (1) of §1.1) for linear equations in more than two independent variables, and

(ii) No interior gradient estimate (analogous to inequality (3) of §1.1) for solutions of the MSE in more than 2 independent variables was known.

The difficulty (i) was overcome in the late 1950's independently by De Giorgi (1957) and Nash (1958) who proved a result for divergence form elliptic equations. (Nash also discussed parabolic equations.) They in fact proved that if v is a weak solution on a ball B_R of a divergence form equation $\sum_{i,j=1}^n D_i(a_{ij} D_j v) = 0$, with Dv of class L^2 and with $\sum_{i,j=1}^n a_{ij}\xi_i\xi_j \geq \lambda|\xi|^2$ and $|a_{ij}| \leq \Lambda$ on B_R, where λ, Λ are fixed positive constants, then v is Hölder continuous and there is an estimate of the form

$$R^\alpha [v]_{\alpha;\, B_{R/2}} \leq C \sup_{B_R} |v|\,, \tag{1}$$

where $C > 0$ and $\alpha \in (0,1)$ are fixed constants depending only on n, λ and Λ. This is adequate as a replacement in n-dimensions of the Morrey result 1.1(1), because if u is a solution of the MSE, written in divergence form, then differentiation with respect to x^ℓ ($\ell \in \{1, \ldots, n\}$ arbitrary) shows that $v = D_\ell u$ satisfies the equation

$$\sum_{i,j=1}^n D_i(a_{ij} D_j v) = 0\,,$$

where $a_{ij}(x) = (1+|Du(x)|^2)^{-1/2}(\delta_{ij} - (1+|Du(x)|^2)^{-1}u_i(x)u_j(x))$; i.e. the De Giorgi/Nash result (1) applies to each of the partial derivatives $D_\ell u$ and hence

$$R^\alpha\,[Du]_{\alpha;B_{R/2}} \le C \sup_{B_R} |Du|\,,$$

with C depending on upper and lower bounds for the maximum and minimium eigenvalues of the coefficient matrix a_{ij}. Thus actually (since the coefficient matrix depends continuously on Du) we conclude that

$$R^\alpha\,[Du]_{\alpha;B_{R/2}} \le C\,, \tag{2}$$

with C depending only on an upper bound for $|Du|$ in B_R (and otherwise not depending on R or u). Thus we get an estimate exactly of the form of the 2-dimensional estimate (1) of §1.1, provided only that we already have a bound for $\sup|Du|$. That is, provided the difficulty (ii) above can be resolved positively.

The difficulty (ii) in fact proved to be rather resilient, and it was not until the celebrated paper of Bombieri, De Giorgi & Miranda (1969) that a bound of the form of inequality (3) of §1.1 was established in n dimensions for arbitrary n. Thus:

Theorem. *Let u be a C^2 solution of the MSE on a ball $B_\rho(x_0) \subset \mathbb{R}^n$ ($n \ge 2$ arbitrary). Then*

$$|Du(x_0)| \le C_1 \exp(C_2 R^{-1} \sup_{B_R(x_0)} (u - u(x_0)))\,,$$

where C_1, C_2 are constants depending only on n.

One of the key conceptual ingredients in Bombieri-De Giorgi-Miranda's proof of the above theorem was the realization that at least part of the analysis had to be done "up on the graph" rather than down in $\mathbb{R}^n$; thus there was a rather geometric flavor in some of the analysis. For example, a "De Giorgi/Moser-type" iteration occurs at a crucial stage of the proof, but, unlike the original implementations of such techniques, it is implemented via a Sobolev-type inequality up on the graph, rather than down in the domain. By now such methods are commonplace, but the above paper represented one of the first times that such geometric analysis arguments were used.

The technicalities of the proof were simplified by Trudinger (1972) (indepently simplifications in the same general direction were obtained by Simon (1971) and Michael & Simon (1973)), and the results generalized to many other equations of minimal surface type and mean curvature type by Ladyzhenskaya & Ural'tseva (1970), Simon (1971), (1976b).

A very simple maximum principle proof of the gradient estimate (yielding a bound of less precise, but for most practical purposes, just as useful, form) was given by Korevaar (1986). See also Korevaar & Simon (1989) for a refined version which additionally gives global continuity estimates.

2.2 Entire and Exterior Solutions

In the case of dimension $n \geq 3$, the situation concerning entire and exterior solutions of the MSE proved far more difficult to come to grips with than in the 2-dimensional case. A prime question was whether or not the Bernstein property (that all entire solutions are affine) generalized to dimension $n \geq 3$.

An initial result towards a solution was given by Bombieri, De Giorgi & Miranda (1969):

Theorem. *Suppose* $u \in C^2(\mathbb{R}^n)$ *is a solution of the MSE and suppose that* graph u *is bounded on one side by a cone; that is, there are constants* K_1, $K_2 > 0$ *such that*

$$u(x) \leq K_1 + K_2|x|, \quad x \in \mathbb{R}^n.$$

Then u *is an affine function.*

The argument for this, taken from Bombieri, De Giorgi & Miranda (1969) (and using in part an observation of Moser (1961)), is as follows: By virtue of the gradient estimate of §2.1 above, we have for each $y \in \mathbb{R}^n$ and each $R > 0$

$$\begin{aligned} |Du(y)| &\leq C_1 \exp(C_2 R^{-1} \sup_{B_R(y)} (u(x) - u(y))) \\ &\leq C_1 \exp(C_2 R^{-1} \sup_{B_R(y)} (K_1 + K_2|x| - u(y))) \\ &\leq C_1 \exp(C_2 R^{-1}(K_1 + K_2(R + |y|) - u(y))) \end{aligned}$$

where we used the given growth assumption on u. Since $R^{-1}(K_1 + K_2|y| - u(y)) \to 0$ as $R \to \infty$, this gives

$$|Du(y)| \leq C_1 \exp(C_2 K_2) \quad \forall\, y \in \mathbb{R}^n.$$

That is, Du has bounded gradient on $\mathbb{R}^n$. The proof that u is affine is now completed by simply letting $R \uparrow \infty$ in the inequality (2) of §2.1 in order to deduce that $Du \equiv$ constant.

Of the many proofs of the Bernstein property for MSE in the case $n = 2$ only one, the proof of Fleming (1962), seemed to have any potential for generalization to higher dimensions. Fleming in fact proved a connection between the Bernstein problem and the study of area minimizing surfaces (not necessarily graphs) in Euclidean space. Specifically, he was able to give an argument which shows that (for any dimension $n \geq 2$) a non-affine entire solution of the MSE is asymptotic at infinity to a non-planar area minimizing hypercone (i.e. to an area minimizing cone $\mathbf{C}$ of dimension n in $\mathbb{R}^{n+1}$ with $\mathbf{C}$ not equal to a hyperplane) in an appropriate weak sense—see e.g. Giusti (1984), Fleming (1962), Massari & Miranda (1984), Simon (1983a), (1987) for discussions of the notion of area minimizing hypercone, and the sense in which the graph is asymptotic to such a cone at infinity. One point to keep in mind is that it is not known in general whether or not the cone at infinity is unique. That is, by "blowing down" the graph via a sequence of scalings $\rho_j \to 0$ (this involves consideration of the sequence of scaled graphs $G_j = \{(x,y) \in \mathbb{R}^{n+1} : y = \rho_j u(\rho_j^{-1}x)\}$), one can always take the weak limit of some subsequence in an appropriate sense, and this weak limit is always an area minimizing hypercone, but it is an open question whether or not the cone may vary as we vary the sequence ρ_j (and the subsequence). For a partial result in this direction, we refer to Simon (1987). In any case, Fleming was thus able to deduce that the Bernstein property (entire $\Rightarrow$ affine) holds provided there are no non-planar area minimizing hypercones in $\mathbb{R}^{n+1}$. Since he was able to prove that no such non-trivial hypercones existed in $\mathbb{R}^3$, he could in particular give a new proof of the Bernstein theorem in dimension 2.

Next De Giorgi (1965) made an important improvement in Fleming's method, by showing that in fact, to prove the Bernstein theorem for solutions of the MSE defined on $\mathbb{R}^n$, it sufficed to check that no non-trivial hypercones existed in $\mathbb{R}^n$ (rather than in $\mathbb{R}^{n+1}$), thus in particular proving the Bernstein property in the case $n = 3$. De Giorgi's observation amounted to the fact that a non-affine minimal graph in $\mathbb{R}^{n+1}$ has a tangent cone at ∞ of the cylindrical form $\mathbf{C} \times \mathbb{R}$, with $\mathbf{C}$ a non-planar area minimizing hypercone in $\mathbb{R}^n$, hence his improvement of Fleming's result. For an elegant treatment of this argument see Giusti (1984), Chapter 17 (see also Massari & Miranda (1984)); Giusti's discussion shows that if u is a non-affine entire solution in $\mathbb{R}^n$ then for any sequence $\rho_j \downarrow 0$ there is a subsequence $\rho_{j'}$ such that the boundary of the set $\{x \in \mathbb{R}^n : \rho_{j'} u(\rho_{j'}^{-1}x) \to \infty\}$ is a non-planar area minimizing hypercone $\mathbf{C}$ in $\mathbb{R}^n$, which gives a contradiction if no such area minimizing hypercones $\mathbf{C}$ exist in $\mathbb{R}^n$. Actually it can be shown that the whole "sub-graph" $\{(x,y) \in \mathbb{R}^{n+1} : y < \rho_{j'} u(\rho_{j'}^{-1}x)\}$ approaches the region on one side of the cylindrical cone $\mathbf{C} \times \mathbb{R}$, but this requires more effort, using work of Bombieri & Giusti (1972) as in Miranda (1977a) and Simon (1987), (1989), and in any case it is not needed for the present application to the Bernstein problem.

Ultimately, the Fleming/De Giorgi approach was completely justified when Almgren (1966) (for $n = 4$) and J. Simons (1968) (for $n \leq 7$) were able to prove the appropriate non-existence of area-minimizing hypercones in $\mathbb{R}^n$ (so that the Bernstein theorem was established for $n \leq 7$), and Bombieri, De Giorgi & Giusti (1969) were able to prove that the result is false in dimension $n \geq 8$. Thus the Bernstein property holds for entire solutions of the MSE on $\mathbb{R}^n$ for $n \leq 7$, but there are counterexamples (i.e. non-affine entire solutions) in all dimensions $n \geq 8$. (And hence there are also non-trivial area minimizing hypercones in $\mathbb{R}^n$, $n \geq 8$ by De Giorgi's 1965 result.)

After the first non-affine examples of entire solutions of the MSE were shown to exist (in dimension $n \geq 8$), it was natural to try to obtain more information about the nature (especially with respect to qualitative behaviour) and number of such solutions. The work of Bombieri, De Giorgi & Giusti (1969) showed that there is at least one distinct non-affine solution of the MSE in $\mathbb{R}^n$ for each even $n \geq 8$; of course we here ignore the additional solutions obtained by homotheties and rigid motions of graph u, and by the procedure of "adding redundant variables" (i.e. getting a solution $\tilde{u}$ in $\mathbb{R}^{n+k}$ from a solution u in $\mathbb{R}^n$ by defining $\tilde{u}(x^1, \ldots, x^{n+k}) = u(x^1, \ldots, x^n)$).

Many other non-affine solutions have been shown to exist by Simon (1989). All these examples (and the original examples of Bombieri, De Giorgi & Giusti) have polynomial growth (i.e. there exists constants C, $k > 0$ such that $|u(x)| \leq C(1 + |x|)^k$ for all $x \in \mathbb{R}^n$), and it is interesting to speculate whether or not *all* entire solutions of the MSE have this property. For some partial results in this direction the reader is referred to the discussion in Simon (1989). (See in particular Theorem 5 of Simon (1989).) Another interesting open question is whether or not there exists a solution which is actually a polynomial in the variables $x^1, \ldots, x^n$.

While on the subject of growth of solutions of the MSE, we should mention recent work on *lower* growth bounds for such solutions. Caffarelli, Nirenberg, and Spruck (1990) first proved that if u is a non-affine solution of the MSE, then $|Du(x_k)| \geq C|x_k|^{1/2}$ for some fixed constant $C > 0$ and some sequence $\{x_k\}$ with $|x_k| \to \infty$. Ecker & Huisken (1990) improved this by showing that the same holds with $|x_k|$ in place of $|x_k|^{1/2}$ on the right. (Working independently Nitsche (1989) proved that the same holds with lower bound $|x_k|^\mu$ for any $\mu < 1$.) Finally it is proved in Simon (1996) that the result holds with lower bound $|x_k|^\mu$ for any $\mu < \beta_n$, where $\beta_n = \frac{n-3}{2} - \sqrt{\left(\frac{n-3}{2}\right)^2 - (n-2)}$ (> 1). Notice that this lower bound is essentially best possible for its type, because the non-affine solutions constructed by Bombieri, De Giorgi & Giusti (1969) (and some of the examples in Simon (1989)) have growth bounded above by $C|x|^{\beta_n}$ and have asymptotic growth $C|x|^{\beta_n}$ along some rays.

Concerning *exterior* solutions of the MSE in the case $n \geq 3$, Bers' 1951 theorem (discussed in §1.2 above) has a generalization to $\mathbb{R}^n$, $n \leq 7$. In fact in Simon (1987) a more general result, valid in all dimensions $n \geq 3$, is proved. Specifically:

Theorem. *If $n \geq 2$ and if u is a C^2 solution of the MSE on $\mathbb{R}^n \backslash \Omega$ (Ω bounded open), then either $|Du|$ is bounded and $\lim_{|x|\to\infty} Du(x)$ exists, or else all tangent cones of* graph u *at ∞ have the form of a cylinder $\mathbf{C} \times \mathbb{R}$ $(\subset \mathbb{R}^{n+1})$, with $\mathbf{C}$ an area minimizing hypercone in $\mathbb{R}^n$ with* $\operatorname{sing} \mathbf{C} \neq \emptyset$.

Notice that the required generalization of Bers' Theorem to $n \leq 7$ follows trivially from this general result by virtue of the fact (mentioned in our discussion of the Bernstein Theorem) that there exist no such non-trivial area minimizing hypercones $\mathbf{C}$ in $\mathbb{R}^n$ in case $n \leq 7$.

For equations other than the MSE not so much is known about when a Bernstein-type theorem (entire $\Rightarrow$ affine) holds, or whether or not one can expect reasonable qualitative results about growth properties of non-linear entire solutions (in case they exist). We mention the result of Simon (1977a), where it is shown (using the same dimension reduction idea of De Giorgi (1965) mentioned above, together with the appropriate 2-dimensional curvature estimates from Almgren, Schoen & Simon (1977)) that the Bernstein Theorem generalizes to dimension $n = 3$ in case of variational equations of minimal surface type. A Bernstein result in low dimensions for graphs which are extrema of certain "weighted" area functionals has been proved by Dierkes (1993).

It would be of particular interest to know other classes of equations, other than those mentioned above, for which Bernstein-type theorems hold.

2.3 Curvature Estimates for $n \geq 3$

For solutions of the minimal surface equation over a ball $B_R \subset \mathbb{R}^n$, Schoen, Simon & Yau (1975) and Simon (1976a) established curvature estimates analogous to the 2-dimensional estimates of §1.3 in case $n \leq 7$. Thus if $n \leq 7$ and if u is a C^2 solution of the MSE over a ball $B_R(x_0)$, then $\mathcal{C}^2 \leq C/R^2$, where $\mathcal{C}^2$ denotes the sum of squares of the principal normal curvatures of the graph of u at the point $(x_0, u(x_0))$. Of course for $n \geq 8$ no such curvature estimates can hold, as one sees by considering rescalings of any non-affine entire solution of the MSE.

2.4 The Dirichlet Problem for $n \geq 3$

Given the estimates of §2.1, it is clear that the existence results for the Dirichlet problem for the MSE carry over to the case of dimension ≥ 3 for convex domains.

However in a landmark paper Jenkins & Serrin (1968) pointed out that in more then 2 dimensions the condition of convexity can be relaxed; they established that, for a smooth domain $\Omega \subset \mathbb{R}^n$ with $n \geq 3$, the necessary and sufficient condition for the existence of a $C^0(\overline{\Omega}) \cap C^2(\Omega)$ solution of the Dirichlet problem for arbitrary prescribed continuous boundary data φ is that the boundary $\partial\Omega$ (thought of as an oriented hypersurface in $\mathbb{R}^n$) have non-negative mean curvature, with respect to the inward pointing unit normal, at each point of $\partial\Omega$. Indeed in this instance Jenkins and Serrin were able to show that local "upper and lower barrier functions" exist near each boundary point x_0. Thus if $x_0 \in \partial\Omega$ and if $\varphi \in C^2(\mathbb{R}^n)$, then for each $M > 0$ there are an open U containing x_0 and $C^2(\overline{\Omega} \cap \overline{U})$ functions $\varphi_\pm$ such that $\pm\mathcal{M}(\varphi_\pm) \geq 0$, $\pm\varphi_\pm \geq \pm\varphi$ on $\partial\Omega \cap U$, with equality in some $\partial\Omega$-neighbourhood of x_0, and $\pm\varphi_\pm \geq M$ on $\Omega \cap \partial U$. Here $\mathcal{M}$ is the minimal surface operator defined by $\mathcal{M}(w) = \sum_{i=1}^n D_i \frac{D_i w}{\sqrt{1+|Dw|^2}}$.

Actually, with hindsight it is geometrically almost obvious that such barriers should exist. Indeed the *graphs* of the barrier functions $\varphi_\pm$ can be thought of as normal perturbations of the boundary cylinder $\partial\Omega \times \mathbb{R}$; since this boundary cylinder has non-negative mean curvature (by assumption), it seems geometrically reasonable that the perturbation can be arranged so that the mean curvature remains non-negative. In practice the barriers are most conveniently constructed in the form

$$\varphi_{\pm}(x) = \widetilde{\varphi}_{\pm}(x') \pm \gamma(d) ,$$

where d is the distance function to $\partial\Omega$ and x' is the nearest point projection ($x' = x - d(x)Dd(x)$) of points near $\partial\Omega$ onto $\partial\Omega$, $\widetilde{\varphi}_{\pm}$ are any C^2 functions in a neighbourhood of x_0 such that $\widetilde{\varphi}_{\pm}$ agree with φ near x_0 and $\pm\widetilde{\varphi}_{\pm} \equiv M$ at a fixed (but small) distance from x_0, $\pm\widetilde{\varphi}_{\pm} \geq \pm\varphi$ everywhere, and where γ is a suitably chosen non-negative increasing concave function of d (for example $\gamma(d) = \beta\log(1 + kd)$ with suitable constants β and k will suffice). We refer for example to Gilbarg & Trudinger (1983), §14.1, for a general discussion of such barrier functions.

With such barrier functions in place of the linear functions $\pi_{\pm}$ used in the above discussion of the 2-dimensional Dirichlet problem, it is then clear that the Leray-Schauder method described in §1.4 (and the subsequent approximation argument to handle continuous data) carries over essentially without change to establish existence in case $n \geq 3$. We refer to Gilbarg & Trudinger for the detailed discussion.

Since the paper of Jenkins & Serrin analogous existence results for the Dirichlet problem of various related quasilinear equations were established. For a representative sample of what is presently known about existence and non-existence for the Dirichlet problem for the MSE, and related equations such as the prescribed mean curvature equation, we refer for example to the papers Finn & Giusti (1977), Gerhardt (1974), (1979), Giusti (1972), (1976), Korevaar & Simon (1989), (1995), Lieberman (1983), (1984), Miranda (1974), (1977a), Serrin (1969), Simon (1974), (1976b, c), (1982), Simon & Spruck (1976), Trudinger (1969a, b), (1972), (1973), Ural'tseva (1973), (1975), and Williams (1984), (1986a,b). Some of these papers include also discussion of "partially infinite" or otherwise unbounded Dirichlet data, "unattained" Dirichlet data, and other boundary value problems, including the contact angle problem (when in place of Dirichlet data, one prescribes the angle of contact between the graph and the boundary cylinder). For a discussion of the exterior Dirichlet problem we refer to the discussion and references in Kuwert (1993); for other results concerning the MSE on unbounded domains we refer for example to Collin (1991), Collin & Krust (1991), Earp & Rosenberg (1989), Hwang (1994), and Langevin & Rosenberg (1988).

2.5 Other Local Estimates

There is a considerable literature concerning local regularity (other than the results already mentioned in §§2.1, 2.3 above), including continuity and gradient estimates of various kinds, and estimates up to the boundary in case suitable boundary conditions are assumed to hold. For example many of the results in the references mentioned in §2.4 above depend crucially on such estimates. Some of the estimates make quite a striking contrast to the theory of linear elliptic equations such as Laplace's equation. For example (see Simon (1974) and also Korevaar & Simon (1989)) in an *arbitrary* (possibly very non-smooth) bounded domain in $\mathbb{R}^n$ there is an *a-priori* Hölder continuity estimate for solutions of the MSE which continuously attain Lipschitz boundary data, with the Hölder exponent and Hölder seminorm depending only on an upper bound for the Lipschitz norm of the data and the diameter of the domain. No such estimate holds for solutions of Laplace's equation; if there were such a continuity estimate, then by approximation of Lebesgue's "cusp domain" (described in Courant & Hilbert (1962), Ch. IV, §4.4) by an increasing sequence of smooth domains, we would get (on passing to the limit) a continuous solution of the Dirichlet problem for the Laplacian operator with arbitrary Lipschitz boundary data on the Lebesgue cusp domain, but (as discussed in Courant & Hilbert (1962), Ch. IV, §4.4) this is known to be impossible—there is even smooth data such that no continuous solution exists on such a domain. Also, for solutions of the MSE there are much better results on removability of singularities than one sees for solutions of linear equations. For example, it is known, after contributions from Bers (1951), Finn (1954), Nitsche (1965a), De Giorgi & Stampacchia (1965), Miranda (1977b) and Simon (1977d), that if $\Omega \subset \mathbb{R}^n$ is an arbitrary open set, if K is an arbitrary closed subset of $\mathbb{R}^n$ with zero $(n-1)$-dimensional Hausdorff measure, and if u is a $C^2(\Omega \backslash K)$ solution of the MSE, then u extends to be a smooth solution of the MSE on all of Ω. Again, no such result holds for solutions of Laplace's equation.

§3. Concluding Remarks

This survey is necessarily brief, and the interested reader is strongly recommended to also consult the books Dierkes, Hildebrandt, Küster & Wohlrab (1992), Gilbarg & Trudinger (1983), Giusti (1984), Massari & Miranda (1984), Nitsche (1975), (1989), Osserman (1986), and Simon (1983), and the surveys Nitsche (1965a), Osserman (1984), and Yau (1982) for different perspectives and a wealth of further detailed information concerning related topics not covered here. For example we say nothing here about the minimal surface *system*, although some fascinating results on that are known, albeit much less than the MSE, as for example described in Osserman (1973), (1984), (1986), Lawson & Osserman (1977), Hildebrandt, Jost & Widman (1980), Fischer-Colbrie (1980) and Gregori (1994). Nor does this survey say a great deal about the equation of prescribed mean curvature (i.e. the "non-homogeneous minimal surface equation"), which includes the case of capillary surfaces. For a further discussion of these topics, we refer to the books by Gilbarg & Trudinger (1983) and Finn (1986), and to the further references mentioned there and in §2.4 above. There is also interesting recent work on the parabolic MSE, which is the equation describing evolution of a graph with velocity given by mean curvature. See for example Ecker & Huisken (1989), (1991) and Huisken (1989).

Acknowledgement. The research for this article was partially supported by NSF grant DMS–9504456 at Stanford University.

References *

Almgren, F. (1966): Some interior regularity theorems for minimal surfaces and an extension of Bernstein's theorem. Ann. Math. **84**, 277-292 (1966), Zbl. 146, 119

Almgren, F., Schoen, R., and Simon, L. (1977): Regularity and Singularity Estimates for hypersurfaces minimizing parametric elliptic variational integrals. Acta Math. **139**, 217–265 (1977), Zbl. 386.49030

Aronsson, G. (1968): On the partial differential equation $u_x^2 u_{xx} + 2u_x u_y u_{xy} + u_y^2 u_{yy} = 0$. Ark. Mat. **7**, 395–425 (1968), 162, 422

Bers, L. (1951): Isolated singularities of minimal surfaces. Ann. Math. **53**, 364–386 (1951), Zbl. 43, 159

Bers, L. (1954): Non-linear elliptic equations without non-linear entire solutions. J. Rat. Mech. Anal. **3**, 767–787 (1954), Zbl. 56, 321

Bernstein, S. (1910): Sur la généralisation du problème Dirichlet II. Math. Ann. **69**, 82–136 (1910), Jbuch 41, 427

Bernstein, S. (1916): Über ein geometrisches Theorem und seine Anwendung auf die partiellen Differentialgleichungen vom ellipschen Typus. Math. Zeit. **26** (1927), 551–558 (translation of the original version in Comm. de la Soc. Math. de Kharkov 2-ème sér. **15**, 38–45 (1915–1917))

Bombieri, E., De Giorgi, E., and Miranda, M. (1969): Una maggiorazzione a priori relativa alle ipersuperfici minimali non parametriche. Arch. Rat. Mech. Anal. **32**, 255–267 (1969), Zbl. 184, 328

Bombieri, E., De Giorgi, E., and Giusti, E. (1969): Minimal cones and the Bernstein problem. Invent. Math. **7**, 243–268 (1969), Zbl. 183, 259

Bombieri, E., and Giusti, E. (1972): Harnack's inequality for elliptic differential equations on minimal surfaces. Invent. Math. **15**, 24–46 (1972), Zbl. 227.35021

Collin, P. (1990): Deux exemples de graphes de courbure moyenne constante sur une bande de $\mathbb{R}^2$. C. R. Acad. Sci., Paris, Ser. I Math. **311**, 539–542 (1990), Zbl. 716.53016

Collin, P., and Krust, R. (1991): Le problème de Dirichlet pour l'équation des surfaces minimales sur des domaines non bornés. Bull. Soc. Math. France **119**, 443–462 (1991), Zbl. 754.53013

Caffarelli, L., Nirenberg, L., and Spruck, J. (1990): On a form of Bernstein's theorem. Analyse mathématique et applications 55–56, Gauthier-Villars, Paris 1990, Zbl. 668.35028

Courant, R., and Hilbert, D. (1962): Methods of Mathematical Physics, Vol. II. Interscience Publishers, New York 1962, Zbl. 99, 295

*For the convenience of the reader, references to reviews in Zentralblatt für Mathematik (Zbl.), compiled by means of the MATH database, and Jahrbuch über die Fortschritte der Mathematik (Jbuch) have, as far as possible, been included in this bibliography.

De Giorgi, E. (1957): Sulla differenziabilità e l'analiticità delle estremali degli integrali multipli regolari. Mem. Accad. Sci. Torino Cl. Sci. Fis. Mat. Natur. **3**, 25–43 (1957), Zbl. 84.319

De Giorgi, E. (1961): Frontiere orientate di misura minima. Sem. Mat. Sc. Norm. Sup. Pisa, 1–56 (1961)

De Giorgi, E. (1965): Una estensione del teorema di Bernstein. Ann. Sc. Norm. Sup. Pisa **19**, 79–85 (1965), Zbl. 168, 98

De Giorgi, E., and Stampacchia, G. (1965): Sulle singolarità eliminabili delle ipersuperficie minimali. Atti Accad. Naz. Lincei, Rend., Cl. Sci. Fis. Mat. Nat. **38**, 352–357 (1965), Zbl. 135, 400

Dierkes, U. (1993): A Bernstein result for energy minimizing hypersurfaces. Calc. Var. Partial Differ. Equ. **1**, 37–54 (1993), Zbl. 819.35030

Dierkes, U., Hildebrandt, S., Küster, A., and Wohlrab, O. (1992): Minimal Surfaces, Vols. I, II. Springer-Verlag, Berlin Heidelberg New York 1992, Zbl. 777.53012, Zbl. 777.53013

Earp, R., and Rosenberg, H. (1989): The Dirichlet problem for the minimal surface equation on unbounded planar domains. J. Math. Pures Appl. **68**, 163–183 (1989), Zbl. 696.49069

Ecker, K., and Huisken, G. (1989): Mean curvature evolution of entire graphs. Ann. Math. **130**, 453–471 (1989), Zbl. 696.53036

Ecker, K., and Huisken, G. (1990): A Bernstein result for minimal graphs of controlled growth. J. Differ. Geom. **31**, 397–400 (1990), Zbl. 696.53002

Ecker, K., and Huisken, G. (1991): Interior estimates for hypersurfaces moving by mean curvature. Invent. Math. **105**, 547–569 (1991), Zbl. 725.53009

Federer, H. (1969): Geometric Measure Theory. Springer-Verlag, Berlin Heidelberg New York 1969, Zbl. 176, 8

Finn, R. (1953): A property of minimal surfaces. Proc. Nat. Acad. Sci. USA **39**, 197–201 (1953), Zbl. 51, 125

Finn, R. (1954): On equations of minimal surface type. Ann. Math. **60**, 397–416 (1954), Zbl. 58, 325

Finn, R. (1963): New estimates for equations of minimal surface type. Arch. Rat. Mech. Anal. **14**, 337–375 (1963), Zbl. 133, 46

Finn, R. (1965): Remarks relevant to minimal surfaces and to surfaces of prescribed mean curvature. J. Anal. Math. **14**, 139–160 (1965), Zbl. 163, 346

Finn, R. (1986): Equilibrium Capillary Surfaces. Springer-Verlag, Berlin Heidelberg New York 1986, Zbl. 583.35002

Finn, R., and Giusti, E. (1977): On nonparametric surfaces of constant mean curvature. Ann. Sc. Norm. Sup. Pisa **4**, 13–31 (1977), Zbl. 343.53004

Finn, R., and Osserman, R. (1964): The gauss curvature of nonparametric minimal surfaces. J. Anal. Math. **12**, 351–364 (1964), Zbl. 122, 164

Fischer-Colbrie, D. (1980): Some rigidity theorems for minimal submanifolds of the sphere. Acta. Math. **145**, 29–46 (1980), Zbl. 464.53047

Fleming, W. (1962): On the oriented Plateau problem. Rend. Circ. Mat. Palermo **11**, 69–90 (1962), Zbl. 107, 313

Gerhardt, C. (1974): Existence, regularity, and boundary behaviour of generalized surfaces of prescribed mean curvature. Math. Z. **139**, 173–198 (1974), Zbl. 316.49005

Gerhardt, C. (1979): Boundary value problems for surfaces of prescribed mean curvature J. Math. Pures Appl. **58**, 75–109 (1979), Zbl. 413.35024

Gilbarg, D., and Trudinger, N. (1983): Elliptic Partial Differential Equations of Second Order (2nd ed.), Springer-Verlag, Berlin Heidelberg New York 1983 (1st ed.: Zbl. 361.35003)

Giusti, E. (1972): Boundary behavior of non-parametric minimal surfaces. Indiana Univ. Math. J. **22**, 435–444 (1972–73), Zbl. 262.35020

Giusti, E. (1976): Boundary value problems for non-parametric surfaces of prescribed mean curvature. Ann. Sc. Norm. Sup. Pisa **3**, 501–548 (1976), Zbl. 344.35036

Giusti, E. (1984): Minimal surfaces and functions of bounded variation. Birkhäuser, Boston Basel Stuttgart 1984, Zbl. 545.49018

Gregori, G. (1994): Compactness and gradient bounds for solutions of the mean curvature system in two independent variables. J. Geom. Anal. **4**, 327–360 (1994), Zbl. 940.54907

Hardt, R., Lau, C.-P., and Lin, F.-H. (1987): Nonminimality of minimal graphs. Indiana Univ. Math. J. **36**, 849–855 (1987), Zbl. 637.49008

Heinz, E. (1952): Über die Lösungen der Minimalflächengleichung. Nachr. Akad. Wiss. Göttingen, Math.-Phys. Kl. II, 51–56 (1952), Zbl. 48, 154

Hopf, E. (1950a): A theorem on the accessibility of boundary parts of an open point set. Proc. Am. Math. Soc. **1**, 76–79 (1950), Zbl. 39.189

Hopf, E. (1950b): On S. Bernstein's theorem on surfaces $z(x, y)$ of non-positive curvature. Proc. Am. Math. Soc. **1**, 80–85 (1950), Zbl. 39.169

Hopf, E. (1953): On an inequality for minimal surfaces $z = z(x, y)$. J. Rat. Mech. Anal. **2**, 519–522, 801–802 (1953), Zbl. 51, 126

Hildebrandt, S., Jost, J., and Widman, K. (1980): Harmonic mappings and minimal submanifolds. Invent. Math. **62**, 269–298 (1980), Zbl. 446.58006

Huisken, G. (1989): Nonparametric mean curvature evolution with boundary conditions. J. Differ. Equations **77**, 369–378 (1989), Zbl. 686.34013

Hwang, J.-F. (1994): Growth property for the minimal surface equation in unbounded domains. Proc. Am. Math. Soc. **121**, 1027–1037 (1994), Zbl. 820.35010

Jenkins H. (1961a): On 2-dimensional variational problems in parametric form. Arch. Rat. Mech. Anal. **8**, 181–206 (1961), Zbl. 143, 148

Jenkins, H. (1961b): On quasilinear equations which arise from variational problems. J. Math. Mech. **10**, 705–727 (1961), Zbl. 145, 364

Jenkins, H., and Serrin, J. (1963): Variational problems of minimal surfaces type I. Arch. Rat. Mech. Anal. **12**, 185–212 (1963), Zbl. 122, 396

Jenkins, H., and Serrin, J. (1968): The Dirichlet problem for the minimal surface equation in higher dimensions. J. Reine Angew. Math. **229**, 170–187 (1968), Zbl. 159, 402

Korevaar, N. (1986): An easy proof of the interior gradient bound for solutions to the prescribed mean curvature problem. Proc. Sympos. Pure Math. **45**, Part 2, 81–89 (1986), Zbl. 599.35046

Korevaar, N., and Simon, L. (1989): Continuity estimates for solutions to the prescribed curvature Dirichlet problem. Math. Z. **197**, 457–464 (1989), Zbl. 625.35034

Korevaar, N., and Simon, L. (1995): Equations of mean curvature type with contact angle boundary conditions. Preprint, Stanford 1995

Korn, A. (1909): Über Minimalflächen, deren Randkurven wenig von ebenen Kurven abweichen. Berl. Abhandl. (1909), Jbuch 40, 705

Kuwert, E. (1993): On solutions of the exterior Dirichlet problem for the minimal surface equation. Ann. Inst. H. Poincaré, Anal. Non-lineaire **10**, 445–451 (1993), Zbl. 820.35038

Ladyzhenskaya, O., and Ural'tseva, N. (1968): Linear and quasilinear elliptic equations. Academic Press, New York 1968 (translation of the original version, Moskau (1964), Zbl. 143, 336) (second Russian edition 1973)

Ladyzhenskaya, O., and Ural'tseva, N. (1970): Local estimates for gradients of solutions of nonuniformly elliptic and parabolic equations. Comm. Pure Appl. Math. **23**, 677–703 (1970), Zbl. 193, 72

Langevin, R., and Rosenberg, H. (1988): A maximum principle at infinity for minimal surfaces and applications. Duke Math. J. **57**, 819–826 (1988), Zbl. 667.49024

Lawson, H.B., and Osserman. R. (1977): Non-existence, non-uniqueness, and irregularity of solutions to the minimal surface system. Acta Math. **139**, 1–17 (1977), Zbl. 376.49016

Leray, J., and Schauder, J. (1934): Topologie et équations fonctionelles. Ann. Sci. École Norm. Sup. **51**, 45–78 (1934), Zbl. 9.73

Lieberman, G. (1983): The conormal derivative problem for elliptic equations of variational type. J. Differ. Equations **49**, 218–257 (1983), Zbl. 506.35039

Lieberman, G. (1984): The nonlinear oblique derivative problem for quasilinear elliptic equations. Nonlinear Anal. **8**, 49–65 (1984), Zbl. 541.35032

Michael, J.H., and Simon, L. (1973): Sobolev and mean-value inequalities on generalized submanifolds of $\mathbb{R}^n$. Comm. Pure Appl. Math. **26**, 361–379 (1973), Zbl. 256.53006

Mickle, K.J. (1950): A remark on a theorem of Serge Bernstein. Proc. Am. Math. Soc. **1**, 86–89 (1950), Zbl. 39.169

Miranda, M. (1974): Dirichlet problem with L^1 data for the non-homogeneous minimal surface equation. Indiana Univ. Math. J. **24**, 227–241 (1974/5), Zbl. 293.35029

Miranda, M. (1977a): Superficie minime illimitate. Ann. Sc. Norm. Sup. Pisa, Ser. IV **4**, 313–322 (1977), Zbl. 352.49020

Miranda, M. (1977b): Sulle singolarità eliminabili delle soluzioni dell'equazione delle superficie minime. Ann. Sc. Norm. Sup. Pisa, Ser. IV **4**, 129–132 (1977), Zbl. 344.35037

Massari, U., and Miranda, M. (1984): Minimal surfaces of codimension one. Mathematics Studies **91**, North Holland, Amsterdam New York 1984, Zbl. 565.49030

Morrey, C.B. (1938): On the solutions of quasilinear elliptic partial differential equations. Trans. Am. Math. Soc. **43**, 126–166 (1938), Zbl. 18.405

Morrey, C.B. (1966): Multiple integrals in the calculus of variations. Springer-Verlag, Berlin-Heidelberg-New York 1966, Zbl. 142, 387

Moser, J. (1961): On Harnack's theorem for elliptic differential equations. Comm. Pure Appl. Math. **14**, 577–591 (1961), Zbl. 111, 93

Nash, J. (1958): Continuity of the solutions of parabolic and elliptic equations. Am. J. Math. **80**, 931–954 (1958), Zbl. 96.69

Nirenberg, L. (1953): On nonlinear elliptic partial differential equations and Hölder continuity. Comm. Pure Appl. Math. **6**, 103–156 (1953), Zbl. 50, 98

Nitsche, J.C.C. (1957): Elementary proof of Bernstein's theorem on minimal surfaces. Ann. Math. **66**, 543–544 (1957), Zbl. 79.377

Nitsche, J.C.C. (1965a): On new results in the theory of minimal surfaces. Bull. Am. Math. Soc. **71**, 195–270 (1965), Zbl. 135, 217

Nitsche, J.C.C. (1965b): On the non-solvability of Dirichlet's problem for the minimal surface equation. J. Math. Mech. **14**, 779–788 (1965), Zbl. 133, 144

Nitsche, J.C.C. (1975): Vorlesungen über Minimalflächen. Springer-Verlag, Berlin Heidelberg New York 1975, Zbl. 319.53003

Nitsche, J.C.C. (1989): Lectures on Minimal Surfaces, Vol. 1. Cambridge University Press 1989, Zbl. 688.53001

Osserman, R. (1960): On the Gauss curvature of minimal surfaces. Trans. Am. Math. Soc. **96**, 115–128 (1960), Zbl. 93.343

Osserman, R. (1973): On Bers' theorem on isolated singularities. Indiana Univ. Math. J. **23**, 337–342 (1973), Zbl. 293.53003

Osserman, R. (1984): The minimal surface equation. Seminar on nonlinear partial differential equations. MSRI Publications **2** (S.S. Chern, Ed.), Springer-Verlag (1984), 237–259, Zbl. 557.53033

Osserman, R. (1986): A Survey of Minimal Surfaces. Dover, New York 1986

Radó, T. (1930): The problem of least area and the problem of Plateau. Math. Z. **32**, 763–795 (1930), Zbl. 56, 436

Schoen, R., Simon, L., and Yau, S.-T. (1975): Curvature estimates for minimal hypersurfaces. Acta Math. **134**, 275–288 (1975), Zbl. 323.53039

Serrin, J. (1963): A priori estimates for solutions of the minimal surface equation. Arch. Rat. Mech. Anal. **14**, 376–383 (1963), Zbl. 117, 73

Serrin, J. (1969): The problem of Dirichlet for quasilinear elliptic equations with many independent variables. Philos. Trans. Roy. Soc. London Ser. A **264**, 413–496 (1969), Zbl. 181, 380

Simon, L. (1971): Interior gradient bounds for nonuniformly elliptic equations. PhD thesis, Mathematics Department, University of Adelaide 1971 (Indiana Univ. Math. J. **25**, 821–855 (1976), Zbl. 346.35016)

Simon, L. (1974): Global estimates of Hölder continuity for a class of divergence form elliptic equations. Arch. Rat. Mech. Anal. **56**, 253–272 (1974), Zbl. 295.35027

Simon, L. (1976a): Remarks on curvature estimates for minimal hypersurfaces. Duke Math. J. **43**, 545–553 (1976), Zbl. 348.53003

Simon, L. (1976b): Interior gradient bounds for nonuniformly elliptic equations. Indiana Univ. Math. J. **25**, 821–855 (1976), Zbl. 346.35016

Simon, L. (1976c): Boundary regularity for solutions of the non-parametric least area problem. Ann. Math. **103**, 429–455 (1976), Zbl. 335.49031

Simon, L. (1977a): On some extensions of Bernstein's theorem. Math. Z. **154**, 265–273 (1977), Zbl. 388.49026

Simon, L. (1977b): A Hölder estimate for quasiconformal maps between surfaces in Euclidean space. Acta Math. **139**, 19–51 (1977), Zbl. 402.30022

Simon, L. (1977c): Equations of mean curvature type in 2 independent variables. Pac. J. Math. **69**, 245–268 (1977), Zbl. 354.35040

Simon, L. (1977d): On a theorem of De Giorgi and Stampacchia. Math. Z. **155**, 199–204 (1977), Zbl. 385.49022

Simon, L. (1982): Boundary behaviour of solutions of the non-parametric least area problem. Bull. Aust. Math. Soc. **26**, 17–27 (1982), Zbl. 499.49023

Simon, L. (1983a): Lectures on Geometric Measure Theory. Proceedings of the Centre for Mathematical Analysis, Australian National University, **3**, 1983, Zbl. 546.49019

Simon, L. (1983b): Survey lectures on minimal submanifolds. In: Seminar on minimal submanifolds, Annals of Math. Studies **103** Princeton (1983), 3–52, Zbl. 541.53045

Simon, L. (1989): Entire solutions of the minimal surface equation. J. Differ. Geom. **30**, 643–688 (1989), Zbl. 687.53009

Simon, L. (1995): Asymptotics for exterior solutions of quasilinear elliptic equations. To appear in proceedings of Pacific Rim Geometry Conference, Singapore 1995

Simon, L. (1996): Singular sets and asymptotics in geometrie analysis. Notes of Lipschitz Lectures delivered at the University of Bonn, Summer 1996

Simon, L., and Spruck J. (1976): Existence and regularity of a capillary surface with prescribd contact angle. Arch. Rat. Mech. Anal. **61**, 19–34 (1976), Zbl. 361.35014

Simons, J. (1968): Minimal varieties in Riemannian manifolds. Ann. Math. **88**, 62–105 (1968), Zbl. 181, 497

Spruck, J. (1974): Gauss curvature estimates for surfaces of constant mean curvature. Comm. Pure Appl. Math. **27**, 547–557 (1974), Zbl. 287.53004

Trudinger, N. (1969a): Some existence theorems for quasilinear, non-uniformly elliptic equations in divergence form. J. Math. Mech. **18**, 909–919 (1969), Zbl. 187, 357

Trudinger, N. (1969b): Lipschitz continuous solutions of elliptic equations of the form $\mathcal{A}(Du)D^2u = 0$. Math. Z. **109**, 211–216 (1969), Zbl. 174, 158

Trudinger, N. (1971): The boundary gradient estimate for quasilinear elliptic and parabolic differential equations. Indiana Univ. Math. J. **21**, 657–670 (1971/1972), Zbl. 236.35022

Trudinger, N. (1972): A new proof of the interior gradient bound for the minimal surface equation in n dimensions. Proc. Nat. Acad. Sci. USA **69**, 821–823 (1972), Zbl. 231.53007

Trudinger, N. (1973): Gradient estimates and mean curvature. Math. Z. **131**, 165–175 (1973), Zbl. 253.53003

Ural'tseva, N. (1973): Solvability of the capillary problem. Vestn. Leningr. Univ. No. 19 (Mat. Meh. Astronom. Vyp. **4**), 54–64 (1973), Zbl. 276.35045. English transl.: Vestn. Leningr. Univ., Math. **6**, 363–375 (1979)

Ural'tseva, N. (1975): Sovability of the capillary problem II. Vestn. Leningr. Univ. No. 1 (Mat. Meh. Astronom. Vyp. **1** (1975), Zbl. 303.35026), English transl.: Vestn. Leningr. Univ., Math. **8**, 151–158 (1980)

Williams, G. (1984): The Dirichlet problem for the minimal surface equation with Lipschitz boundary data. J. Reine Angew. Math. **354**, 123–140 (1984), Zbl. 541.35033

Williams, G. (1986a): Solutions of the minimal surface equation – continuous and discontinuous at the boundary. Comm. Partial Differ. Equations **11**, 1439–1457 (1986), Zbl. 605.49030

Williams, G. (1986b): Global regularity for solutions of the minimal surface equation with continuous boundary values. Ann. Inst. H. Poincaré, Anal. Non-linéaire **3**, 411–429 (1986), Zbl. 627.49020

Yau, S.-T. (1982): Survey on partial differential equations in differential geometry. Sem. differential geometry, Ann. Math. Stud. **102** (S.-T. Yau, Ed.), Princeton University Press (1982), 3–71, Zbl. 478.53001

Author Index

Subject Index

Springer and the environment

Printing: Mercedesdruck, Berlin
Binding: Buchbinderei Lüderitz & Bauer, Berlin

GPSR Compliance

The European Union's (EU) General Product Safety Regulation (GPSR) is a set of rules that requires consumer products to be safe and our obligations to ensure this.

If you have any concerns about our products, you can contact us on ProductSafety@springernature.com

In case Publisher is established outside the EU, the EU authorized representative is:

Springer Nature Customer Service Center GmbH
Europaplatz 3
69115 Heidelberg, Germany

Batch number: 10682213

Printed by Printforce, the Netherlands